# WordPress®

## by Lisa Sabin-Wilson

FOREWORD BY Matt Mullenweg

Cofounder of WordPress

## for dummies®

A Wiley Brand

## WordPress® For Dummies®, 9th Edition

Published by: **John Wiley & Sons, Inc.,** 111 River Street, Hoboken, NJ 07030-5774, www.wiley.com

Copyright © 2021 by John Wiley & Sons, Inc., Hoboken, New Jersey

Published simultaneously in Canada

No part of this publication may be reproduced, stored in a retrieval system or transmitted in any form or by any means, electronic, mechanical, photocopying, recording, scanning or otherwise, except as permitted under Sections 107 or 108 of the 1976 United States Copyright Act, without the prior written permission of the Publisher. Requests to the Publisher for permission should be addressed to the Permissions Department, John Wiley & Sons, Inc., 111 River Street, Hoboken, NJ 07030, (201) 748-6011, fax (201) 748-6008, or online at http://www.wiley.com/go/permissions.

**Trademarks:** Wiley, For Dummies, the Dummies Man logo, Dummies.com, Making Everything Easier, and related trade dress are trademarks or registered trademarks of John Wiley & Sons, Inc. and may not be used without written permission. WordPress is a registered trademark of WordPress Foundation. All other trademarks are the property of their respective owners. John Wiley & Sons, Inc. is not associated with any product or vendor mentioned in this book.

LIMIT OF LIABILITY/DISCLAIMER OF WARRANTY: THE PUBLISHER AND THE AUTHOR MAKE NO REPRESENTATIONS OR WARRANTIES WITH RESPECT TO THE ACCURACY OR COMPLETENESS OF THE CONTENTS OF THIS WORK AND SPECIFICALLY DISCLAIM ALL WARRANTIES, INCLUDING WITHOUT LIMITATION WARRANTIES OF FITNESS FOR A PARTICULAR PURPOSE. NO WARRANTY MAY BE CREATED OR EXTENDED BY SALES OR PROMOTIONAL MATERIALS. THE ADVICE AND STRATEGIES CONTAINED HEREIN MAY NOT BE SUITABLE FOR EVERY SITUATION. THIS WORK IS SOLD WITH THE UNDERSTANDING THAT THE PUBLISHER IS NOT ENGAGED IN RENDERING LEGAL, ACCOUNTING, OR OTHER PROFESSIONAL SERVICES. IF PROFESSIONAL ASSISTANCE IS REQUIRED, THE SERVICES OF A COMPETENT PROFESSIONAL PERSON SHOULD BE SOUGHT. NEITHER THE PUBLISHER NOR THE AUTHOR SHALL BE LIABLE FOR DAMAGES ARISING HEREFROM. THE FACT THAT AN ORGANIZATION OR WEBSITE IS REFERRED TO IN THIS WORK AS A CITATION AND/OR A POTENTIAL SOURCE OF FURTHER INFORMATION DOES NOT MEAN THAT THE AUTHOR OR THE PUBLISHER ENDORSES THE INFORMATION THE ORGANIZATION OR WEBSITE MAY PROVIDE OR RECOMMENDATIONS IT MAY MAKE. FURTHER, READERS SHOULD BE AWARE THAT INTERNET WEBSITES LISTED IN THIS WORK MAY HAVE CHANGED OR DISAPPEARED BETWEEN WHEN THIS WORK WAS WRITTEN AND WHEN IT IS READ.

For general information on our other products and services, please contact our Customer Care Department within the U.S. at 877-762-2974, outside the U.S. at 317-572-3993, or fax 317-572-4002. For technical support, please visit https://hub.wiley.com/community/support/dummies.

Wiley publishes in a variety of print and electronic formats and by print-on-demand. Some material included with standard print versions of this book may not be included in e-books or in print-on-demand. If this book refers to media such as a CD or DVD that is not included in the version you purchased, you may download this material at http://booksupport.wiley.com. For more information about Wiley products, visit www.wiley.com.

Library of Congress Control Number: 2020949710

ISBN 978-1-119-69697-1 (pbk); ISBN 978-1-119-69698-8 (ebk); ISBN 978-1-119-69696-4 (ebk)

Manufactured in the United States of America

SKY10022840_112520

# Contents at a Glance

# Table of Contents

# Foreword

There used to be a program from Microsoft called FrontPage, which was the first visual interface for creating websites that I saw. It worked like Microsoft Word and Publisher, so with very little knowledge, I was able to hack together the world's worst website in just a few hours without worrying about what was going on under the hood.

Years later, when I look back at that website, I cringe, but at the time, it was incredibly empowering. The software, though crude, helped me publish something anybody in the entire world could see. It opened a world I had never imagined before.

Now, using software like WordPress, you can have a blog or website light years beyond my first one in both functionality and aesthetics. Just as my first web experience whetted my appetite for more, I hope that your experience entices you to explore the thousands of free plugins, themes, and customizations that are possible with WordPress, many of which are explained in this book.

WordPress is more than just software; it's also a community, a rapidly evolving ecosystem, and a set of philosophies and opinions about how to create the best web experience. When you embrace it, you'll be in good company. WordPress users include old-media organizations such as CNN, *The New York Times*, and *The Wall Street Journal*, along with millions of personal bloggers like me for whom a WordPress blog is a means of expression.

Matt Mullenweg

Cofounder of WordPress

# Introduction

I t was 2003 when I discovered the WordPress blogging software. Way back then (and in Internet years, that's actually quite a lot of time), I used Movable Type as my blogging platform. A friend introduced me to the WordPress software. "Try it," she said. "You'll really like it."

As a creature of habit, I felt reluctant to make the change. But I haven't looked back. I've been with WordPress ever since.

WordPress started as a tool for blogging. Authors, students, parents, business owners, academics, journalists, hobbyists — you name it — use blogs as a matter of course. Over the past decade, WordPress has emerged as the premier content management system (CMS) available on the Internet. WordPress software currently powers 35 percent of the websites you see.

Today, WordPress is much more than a blogging tool. Individuals, organizations, and corporations are using WordPress to build their entire web presence. Word-Press has grown into a valuable solution for everything from selling products on the Internet to running membership sites and blogging. Pretty much anything you think you can do with your website, you can accomplish with WordPress.

To a brand-new user, some aspects of WordPress can be a little bit intimidating. After you start using it, however, you begin to realize how intuitive, friendly, and extensible the software is.

This book presents an insightful look at WordPress. In the book, I cover managing and maintaining your WordPress-powered website through the use of plugins and themes, as well as using the intuitive WordPress Dashboard to manage your content. If you're interested in taking a detailed look at the website-building tool provided by WordPress, you happen to have just the right book in your hands.

# About This Book

This book covers all the important aspects of WordPress that new users need to know to use the software for their own websites. I cover the software package available at `https://wordpress.org` by highlighting important topics, such as these:

>> Installing and setting up the software

>> Navigating the WordPress Dashboard

>> Using the Block Editor to create posts and pages

>> Finding and installing free themes to use on your WordPress website

>> Using basic coding to design your own WordPress theme or modify the one you're using

>> Installing, activating, and managing WordPress plugins

>> Choosing to use the multiple-site WordPress Network option to host a network of websites in your domain

>> Migrating your existing website to WordPress (if you're using a different platform, such as Drupal, Movable Type, or Expression Engine)

With WordPress, you can truly tailor a website to your own tastes and needs. Some sites are packaged with the WordPress software; others are third-party plugins and add-ons created by members of the WordPress user community. You need to invest only a little research, knowledge, and time to put together a site that suits your needs and gives your readers an exciting experience that keeps them coming back for more.

# Foolish Assumptions

I'll never know what assumptions you've made about me at this point, but I can tell you a few things that I already assume about you:

>> You know what a computer is. You can turn it on, and you understand that if you spill coffee on your keyboard, you'll have to run out and get a replacement.

>> You understand how to connect to the Internet and know the basics of using a web browser to surf websites.

>> You have a basic understanding of what websites and blogs are, and you're interested in using WordPress to start your own. Or you already have a website, are already using WordPress, and want to understand the program better so that you can do more cool stuff and stop bugging your geeky best friend whenever you have a question about something.

>> You already have a website on another platform and want to move your website to WordPress.

>> You know what email is. You know what an email address is. You actually have an email address, and you send and receive email on a semiregular basis.

# Icons Used in This Book

Icons emphasize a point to remember, a danger to be aware of, or information that I think you may find helpful. Those points are illustrated as such:

**TIP**

Tips are little bits of information that you may find useful.

**WARNING**

I use this icon to point out dangerous situations.

**TECHNICAL STUFF**

All geeky stuff goes here. I don't use this icon very often, but when I do, you'll know that you're about to encounter technical mumbo-jumbo.

**REMEMBER**

When you see this icon, read the text next to it two or three times to brand it into your brain so that you remember whatever it is that I think you need to remember.

# Beyond the Book

I've put a ton of information between the covers of this book, but at `https://www.dummies.com`, you can find a Cheat Sheet that lists

>> Where to find WordPress support online

>> How to navigate the WordPress Dashboard

>> How to locate a reliable web-hosting provider

When you arrive at `https://www.dummies.com`, type the book's title in the Search field to find the Cheat Sheet.

# Where to Go from Here

This book is a veritable smorgasbord of WordPress information, ideas, concepts, tools, resources, and instructions. Some parts of the book may apply directly to what you want to do with your WordPress blog. Other parts may deal with topics that you're only mildly curious about, so feel free to skim (or skip) those pages.

If you already have WordPress installed on your web server, for example, you can skip Chapter 3. If you aren't interested in digging into the code of a WordPress template and don't want to find out how to apply CSS or HTML to enhance your design, you can skip Chapters 9 through 12. If you have no interest in running more than one website with WordPress, you can skip Chapter 13.

I don't intend for you to read this book from cover to cover (unless you're my mother — then I won't forgive you if you don't). Rather, scan the table of contents and the index to find the information you need.

Long story short: Take what you need, and leave the rest.

# 1

# Introducing WordPress

**IN THIS PART . . .**

Explore all WordPress has to offer.

Discover the basic concepts about publishing a website with WordPress.

Understand the different versions of WordPress and choose the right one for you.

Get ready to use WordPress for your online publishing.

Chapter **1**

# What WordPress Can Do for You

In a world in which technology advances in the blink of an eye, WordPress really does make building websites easy — and free! How else can you get your content out to a potential audience of millions worldwide and spend exactly nothing? There may be no such thing as a free lunch in this world, but you can bet your bottom dollar that there are free websites and blogs. WordPress serves them all up in one nifty package.

The software's free price tag, its ease of use, and the speed at which you can get your website up and running are great reasons to use WordPress to power your personal blog or business website. An even greater reason is the incredibly supportive and passionate WordPress community. In this chapter, I introduce you to the WordPress software so that you can begin to discover how effective it is as a tool for creating your website.

# Discovering the Benefits of WordPress

I work with first-time website owners all the time — folks who are new to the idea of publishing content on the Internet. One of the questions I'm most frequently asked is "How can I run a website? I don't even know how to code or create websites."

Enter WordPress. You no longer need to worry about knowing the code because the WordPress software does the code part for you. When you log in to your website, you have to do only two simple things to publish your thoughts and ideas:

**1.** Write your content.

**2.** Click a button to publish your content.

That's it!

WordPress offers the following competitive advantages as the most popular content management tool on the market:

>> **Diverse options:** Two versions of WordPress are available to suit nearly every type of website owner:

- *WordPress.com:* A hosted turnkey solution; primarily used for blogging

- *WordPress.org:* A self-hosted version to install on the web server of your choice; used for building blogs and websites

I go into detail about each of these versions later in this chapter, in the "Choosing a WordPress Platform" section.

>> **Ease of use:** WordPress setup is quick, and the software is easy to use.

>> **Extensibility:** WordPress is extremely extensible, meaning that you can easily obtain plugins and tools that let you customize it to suit your purposes.

>> **Strong community of users:** WordPress has a large and loyal members-helping-members community via public support forums, blogs, and websites geared to the use of WordPress.

The following sections fill in a few details about these features and point you to places in the book where you can find out more about them.

## Getting set up the fast and easy way

WordPress is one of the only platforms that can brag about a five-minute installation — and stand behind it! Both versions of WordPress take you approximately the same amount of time to set up.

**REMEMBER**

Mind you, five minutes is an *approximate* time for installing the WordPress.org software. This estimate doesn't include the time required to obtain domain registration and web-hosting services or to set up the options in the Dashboard. (You can find information on web-hosting services in Chapter 3.)

When you complete the installation, however, the world of WordPress awaits you. The Dashboard is well organized and easy on the eyes. Everything is clear and logical, making it easy for even a first-time user to see where to go to manage settings and options.

The WordPress software surely has enough meat on it to keep the most experienced developer busy and happy. At the same time, however, it's friendly enough to make a novice user giddy about how easy it is to get started. Each time you use WordPress, you can find out something exciting and new.

## Extending WordPress's capabilities

I've found that the most exciting and fun part of running a WordPress website is exploring the flexibility of the software. Hundreds of plugins and *themes* (designs) are available to let you create a website that functions the way *you* need it to.

**TIP**

If you think of your website as a vacuum cleaner, plugins are the attachments. The attachments don't function alone. When you add them to your vacuum cleaner, however, you add to the functionality of your vacuum, possibly improving its performance.

All WordPress websites are pretty much the same at their core, so by using plugins, you can truly individualize your website by providing additional features and tools that benefit you and your readers. When you come upon a WordPress website that has some really different and cool functions, 98 percent of the time, you can include that function on your own website by using a WordPress plugin. If you don't know what plugin that website is using, try dropping the website owner an email or leave a comment. WordPress website owners usually are eager to share the great tools they discover.

Most plugins are available at no charge. You can find out more about WordPress plugins and where to get them in Chapter 7. Chapter 15 lists my top ten choices for popular WordPress plugins available for download.

In addition to using plugins, you can embellish your WordPress site with templates and themes. WordPress comes with a very nice default theme to get you started. Figure 1-1 shows the default Twenty Twenty theme, created by the team from WordPress, which is displayed by default after you install and set up your site for the first time.

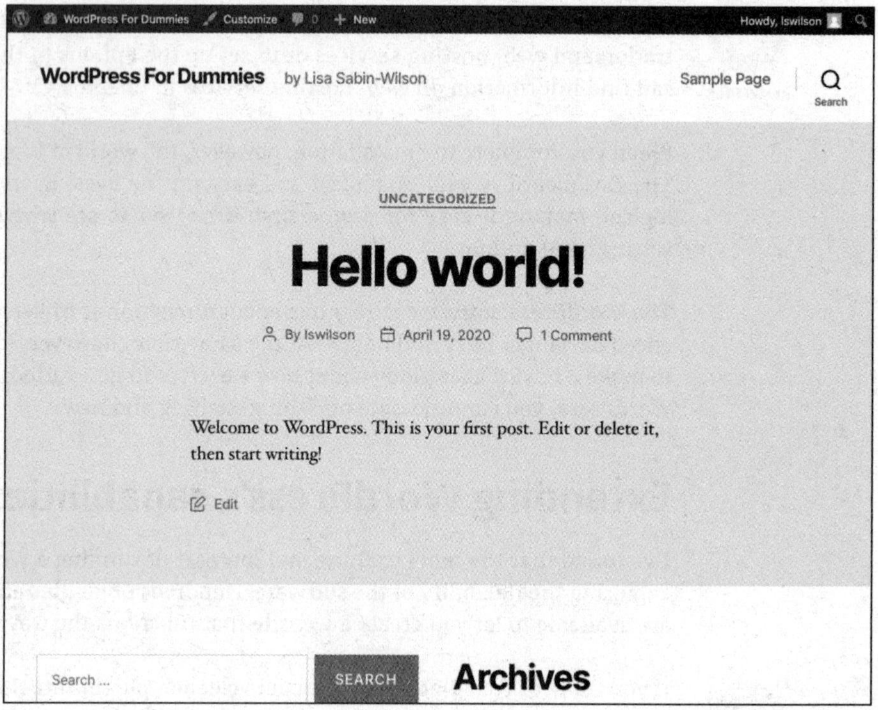

WordPress For Dummies
Customize
0
+ New
Howdy, lswilson
WordPress For Dummies
by Lisa Sabin-Wilson
Sample Page
Search
UNCATEGORIZED
Hello world!
By lswilson
April 19, 2020
1 Comment
Welcome to WordPress. This is your first post. Edit or delete it, then start writing!
Edit
Search ...
SEARCH
Archives


**FIGURE 1-1:** Start a new WordPress website with a theme.

The theme's default style is minimal, with handy settings built into the Customizer that enable you to change the colors and insert an image to use as a header image. (You can find more about tweaking WordPress themes and the Customizer in Chapters 9 through 12.)

REMEMBER

The Twenty Twenty theme (refer to Figure 1-1) includes all the basic elements that you need to start a new WordPress site. You can extend your WordPress site in a hundred ways with plugins and themes released by members of the WordPress community, but this default theme is a nice place to start.

Using some of the thousands of plugins and themes available, you can truly manage many kinds of content on your website. WordPress isn't just for blogging anymore (although it does still excel at it!). Although WordPress became well known as a blogging platform, you can use it to power diverse and dynamic websites that allow you to do things like develop an e-commerce site (selling products online), create a members-only site where your content is curated only for those who have registered and become members of your site, or create a large corporate business site like the one you can see in the Microsoft News Center at `https://news.microsoft.com`.

Using WordPress as a content management system (CMS) frees you from running only a blog on the platform. (See Chapter 12 for more about the technique of designing for WordPress as a CMS.)

## Taking part in the community

Allow me to introduce you to the fiercely loyal folks who make up the user base, better known as the vast WordPress community. This band of merry ladies and gentlemen comes from all around the globe, from California to Cairo, Florida to Florence, and all points in between and beyond.

In March 2005, Matt Mullenweg of WordPress proudly proclaimed that the number of WordPress downloads had reached 900,000 — an amazing landmark in the history of the software. But the real excitement occurred in August 2006, when WordPress logged more than 1 million downloads, and in 2007, when the software had more than 3 million downloads. WordPress downloads have broken the ceiling since then, with more than 30 million downloads by the beginning of 2020, and the number is growing daily. WordPress is easily the most popular CMS available on the web today. By the first half of 2020, it powered approximately 35 percent of all the websites on the Internet in 2020 — roughly two of every six sites you encounter on the World Wide Web.

Don't let the sheer volume of users fool you: WordPress also has bragging rights to the most helpful community on the web. You can find users helping other users in the support forums at `https://wordpress.org/support`. You can also find users contributing to the very helpful WordPress Codex (a collection of how-to documents) at `https://codex.wordpress.org`. Finally, across the Internet, you can find multiple websites about WordPress itself, with users sharing their experiences and war stories in the hope of helping the next person who comes along.

You can subscribe to various mailing lists, too. These lists offer you the opportunity to become involved in various aspects of the WordPress community as well as in the ongoing development of the software.

Joining the WordPress community is easy: Simply start your own website by using one of the two WordPress software options. If you're already publishing on a different platform, such as Blogger or Movable Type, WordPress enables you to easily migrate your current data from that platform to a new WordPress setup. (See Chapter 14 for information about migrating your existing website to WordPress.)

# Choosing a WordPress Platform

One of the realities of running a website today is choosing among the veritable feast of software platforms to find the one that performs the way you need. You want to be sure that the platform you choose has all the options you're looking for. WordPress is unique in that it offers two versions of its software, each designed to meet various needs:

>> The hosted version at WordPress.com: `https://wordpress.com`.

>> The self-installed and self-hosted version available at `https://wordpress.org`. (This book focuses on this version.)

Every WordPress website setup has certain features available, whether you're using the self-hosted software from WordPress.org or the hosted version at WordPress.com. These features include (but aren't limited to)

>> Quick and easy installation and setup

>> Full-featured publishing capability, letting you publish content to the web through an easy-to-use block editor, web-based interface

>> Topical archiving of your posts, using categories

>> Monthly archiving of your posts, with the ability to provide a listing of those archives for easy navigation through your site

>> Comment and trackback tools

>> Automatic spam protection through Akismet

>> Built-in gallery integration for photos and images

>> Media Manager for video and audio files

>> Great community support

>> Unlimited number of static pages, letting you step out of the blog box and into the sphere of running a fully functional website

>> RSS (Really Simple Syndication) capability (see Chapter 2) with RSS 2.0, RSS 1.0, and Atom support

>> Tools for importing content from other blogging systems, such as Blogger, Movable Type, and LiveJournal

Table 1-1 compares the two WordPress versions.

**TABLE 1-1**  **Exploring the Differences between the Two Versions of WordPress**

| Feature | WordPress.org | WordPress.com |
|---|---|---|
| Cost | Free | Free |
| Software download | Yes | No |
| Software installation | Yes | No |
| Web hosting required | Yes | No |
| Custom CSS* control | Yes | $96 per year |
| Template access | Yes | $96 per year |
| Sidebar widgets | Yes | Yes |
| RSS syndication | Yes | Yes |
| Access to core code | Yes | No |
| Ability to install plugins | Yes | $300 per year |
| Theme** installation | Yes | $300 per year |
| Multiauthor support | Yes | Yes |
| Unlimited number of website setups with one account | Yes | Yes |
| Community-based support forums | Yes | Yes |

\* *CSS = Cascading Style Sheets*
\*\* *Limited selection on WordPress.com*

# Choosing the hosted version from WordPress.com

WordPress.com is a free service. If downloading, installing, and using software on a web server sound like Greek to you — and like things you'd rather avoid — the WordPress folks provide a solution for you at WordPress.com.

WordPress.com is a *hosted solution*, which means that it has no software requirement, no downloads, and no installation or server configurations. Everything's done for you on the back end, behind the scenes. You don't even have to worry about how the process happens; it happens quickly, and before you know it, you're making your first post using a WordPress.com solution.

WordPress.com has some limitations, though. You can't install plugins or custom themes, for example, and you can't customize the base code files. Neither are you able to sell advertising or monetize your site at all on WordPress.com unless you pay a $300 annual fee. Also, WordPress.com displays advertisements on your posts and pages to users who aren't logged in to the WordPress.com network (`https://wordpress.com/support/no-ads`). But even with its limitations, WordPress.com is an excellent starting point if you're brand-new to blogging and a little intimidated by the configuration requirements of the self-installed WordPress.org software.

If you don't want or need to create a full website for your business or service and just want to create an online diary of sorts, you would typically use WordPress.com, because it excels at allowing you to get a simple site up and running quickly. As I mention previously, however, if you want to use the thousands of plugins and themes available for WordPress — or if you want to customize your own theme for your website — you're limited to only a few themes on the WordPress.com-hosted service, and you're not able to install your own plugins on the service, either.

The good news is this: If you ever outgrow your WordPress.com-hosted site and want to make a move to the self-hosted WordPress.org software, you can. You can even take all the content from your WordPress.com-hosted site with you and easily import it into your new setup with the WordPress.org software.

## Self-hosting with WordPress.org

The self-installed version from WordPress.org that I cover in this book requires you to download the software from the WordPress website and install it on a web server. Unless you own your own web server, you need to lease one — or lease space on one.

Using a web server typically is referred to as *web hosting,* and unless you know someone who knows someone, hosting generally isn't free. That being said, web hosting doesn't cost a whole lot. You can usually obtain a good web-hosting service for anywhere from $5 to $20 per month, depending on your needs. (Chapter 3 gives you the important details you need to know about obtaining a web host.)

You need to make sure, however, that any web host you choose to work with has the required software installed on the web server. Currently, the minimum software recommendations for WordPress include

>> HTTPS support

>> PHP version 7.3 or later

>> MySQL version 5.6 or later

**TECHNICAL STUFF**

Some web hosting providers haven't yet upgraded to the latest version of PHP. If your web-hosting provider has older PHP or MySQL versions, the WordPress software will work with PHP 5.6.20 or later and MySQL 5.0. These older versions, however, are considered to be *end of life*, which means *discontinued* in the software world. Older versions of PHP and MySQL still work but are no longer supported and, therefore, are susceptible to security vulnerabilities.

After you have WordPress installed on your web server (see the installation instructions in Chapter 3), you can start using it to publish to your heart's content. With the WordPress software, you can install several plugins that extend the functionality of the platform, as I describe in Chapter 7.

You also have full control of the core files and code that WordPress is built on. So if you have a knack for PHP and knowledge of MySQL, you can work within the code to create your own themes and plugins that you think would be good for you and your website. Find information about PHP and MySQL in Chapter 2.

You don't need design ability to make your website look great. Members of the WordPress community have created more than 3,900 WordPress themes, and you can download them for free and install them on your WordPress site. (See Chapter 8.) Additionally, if you're creatively inclined, like to create designs on your own, and know CSS (Cascading Style Sheets), you'll be glad to know that you have full access to the template system within WordPress and can create your own custom themes. (See Part 4.)

**TIP**

The self-hosted WordPress.org software lets you run an unlimited number of websites on one installation of its software platform, on one domain. When you configure the Network options within WordPress to enable a multisite interface, you become administrator of a network of sites. All the options remain the same, but with the Network options configured, you can have additional websites and domains, as well as allow registered users of your website to host their own websites within your network. You can find out more about the WordPress Multisite feature in Chapter 13.

Sites that use the WordPress Network options include the following:

>> **BBC America** (https://www.bbcamerica.com): The BBC America site contains all the shows and movies that the TV network offers. It's a huge WordPress Multisite network, with each show having an individual site.

>> **Boise State University** (https://www.boisestate.edu): Boise State University is Idaho's largest institution of higher education, offering nearly 200 degrees and certificates in 7 colleges. The global navigation and emergency notifications are managed centrally, and updates are pushed out to more than 200 separate WordPress multisite instances within minutes.

>> **Microsoft Windows** (https://blogs.windows.com): Niche-specific blog networks use WordPress to manage the content they publish through various channels on their website about the Windows software — in multiple languages.

# Chapter **2**

# WordPress Basics

A lot happens behind the scenes to make your WordPress blog or website function. The beauty of it is that you don't have to worry about what's happening on the back end to manage and maintain a WordPress site — unless you really want to. In this chapter, I delve a little bit into the technology behind the WordPress platform, including a brief look at PHP and MySQL, two software components required to run WordPress.

This chapter also covers some of the various technologies that help you on your way to running a successful website, such as comments and RSS feeds, as well as information about combatting spam.

## Shining the Spotlight on WordPress

Publishing content is an evolutionary process, and blogs have evolved beyond personal diaries and journals. Undoubtedly, a blog is a fabulous tool for publishing your personal diary of thoughts and ideas, but blogs also serve as excellent tools for business, editorial journalism, news, and entertainment. Sometimes, you find a stand-alone blog that is the sum total of the website; at other times, you find a full website that contains a blog but has other offerings as well (products for sale, memberships, newsletters, forums, and so on). Here are some ways that people use blogs and websites powered by WordPress:

>> **Personal:** This type of blogger creates a blog as a personal journal or diary. You're considered to be a personal blogger if you use your blog mainly to discuss topics that are personal to you or your life, such as your family, your cats, your children, or your interests (such as technology, politics, sports, art, or photography). I maintain a personal blog at `https://lisasabin-wilson.com/blog`.

>> **Business:** This type of site uses the power of blogs to promote a company's business services, products, or both. Blogs are very effective tools for promotion and marketing, and business blogs usually offer helpful information to readers and consumers, such as tips and product reviews. Business blogs also let readers provide feedback and ideas, which can help a company improve its services. My business, WebDevStudios, keeps an active blog within the business website at `https://webdevstudios.com/blog`.

>> **Media/journalism:** More and more popular news outlets, such as Fox News, MSNBC, and CNN, have added blogs to their websites to provide information on current events, politics, and news on regional, national, and international levels. These news organizations often have editorial bloggers as well. *Reader's Digest* is an example of such a publication; its WordPress-powered site is at `https://www.rd.com`.

>> **Citizen journalism:** The emergence of citizen journalism coincided with the swing from old media to new media. In old media, the journalists and news organizations direct the conversation about news topics. With the popularity of blogs and the millions of bloggers who exploded onto the Internet, old media felt a change in the wind. Average citizens, using the power of their voices on blogs, changed the direction of the conversation. Citizen journalists often fact-check traditional media news stories and expose inconsistencies, with the intention of keeping the media or local politicians in check. An example of citizen journalism is Talking Points Memo at `https://talkingpointsmemo.com`.

>> **Professional:** This category is growing every day. Professional bloggers are paid to blog for individual companies or websites. Blog networks, such as Gartner (`https://blogs.gartner.com`), have a full network of staff bloggers. Also, several services match advertisers with bloggers so that the advertisers pay bloggers to make posts about their products. Is it possible to make money as a blogger? Yes, and making money by blogging is common these days. If you're interested in this type of blogging, check out Darren Rowse's ProBlogger (`https://problogger.com`). Rowse is considered to be the grandfather of all professional bloggers because for years, he has provided helpful resources and information about how to make money with blogging.

# Dipping Into WordPress Technologies

The WordPress software is a personal publishing system that uses PHP and MySQL. This platform provides everything you need to create your own website and publish your own content dynamically without having to know how to program those pages yourself. In short, all your content is stored in a MySQL database in your hosting account.

**TECHNICAL STUFF**

*PHP* (which stands for *Hypertext Preprocessor* — and PHP itself originally stood for *personal home page,* as named by its creator, Rasmus Lerdorf) is a server-side scripting language for creating dynamic web pages. When a visitor opens a page built in PHP, the server processes the PHP commands and then sends the results to the visitor's browser. MySQL is an open-source relational database management system (RDBMS) that uses Structured Query Language (SQL), the most popular language for adding, accessing, and processing data in a database. If all that sounds like Greek to you, just think of MySQL as a big filing cabinet in which all the content on your website is stored.

Every time a visitor goes to your website to read your content, he makes a request that's sent to a host server. The PHP programming language receives that request, obtains the requested information from the MySQL database, and then presents the requested information to your visitor through his web browser.

In using the term *content* as it applies to the data that's stored in the MySQL database, I'm referring to your posts, pages, comments, and options that you set up in the WordPress Dashboard. The theme (design) you choose to use for your website — whether it's the default theme, one you create for yourself, or one that you have custom-designed — isn't part of the content, or data, stored in the database assigned to your website. Those files are part of the file system and aren't stored in the database. So create and keep backups of any theme files that you're using. See Chapter 9 for further information on WordPress theme management.

**TIP**

When you look for a hosting service, choose one that provides daily backups of your site so that your content and data won't be lost in case something bad happens. Web-hosting providers that offer daily backups as part of their services can save the day by restoring your site to its original form. You can find more information on choosing a hosting provider in Chapter 3.

## Archiving your publishing history

Packaged within the WordPress software is the capability to maintain chronological and categorized archives of your publishing history — automatically. WordPress uses PHP and MySQL technology to sort and organize everything you publish in an order that you, and your readers, can access by date and category. This archiving process is done automatically with every post or page you publish to your website.

When you create a post on your WordPress website, you can file that post in a category that you specify. This feature makes for a nifty archiving system in which you and your readers can find articles or posts that you've placed within a specific category. A Category archive page on my business website (`https://webdevstudios.com/category/wordpress`; see Figure 2-1) contains an archive of all the posts on the website that were published in the WordPress category.

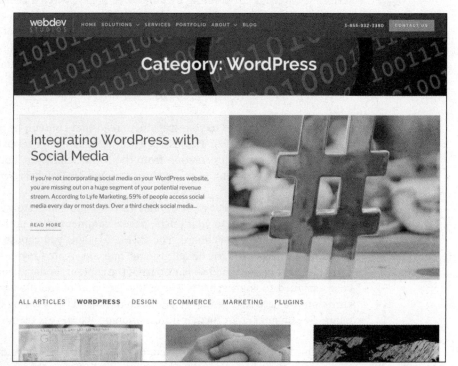

**FIGURE 2-1:**
A category
archive page of
my business blog.

WordPress lets you create as many categories as you want for filing your content and posts by topic. I've seen sites that have just one category and sites that have up to 1,800 categories. WordPress is all about preferences and options for organizing your content. On the other hand, using WordPress categories is your choice. You don't have to use the category feature.

# Interacting with your readers through comments

One of the most exciting and fun aspects of publishing on the web with WordPress is getting feedback from your readers the moment you make a post to your site. Feedback, referred to as *comments,* is akin to having a guestbook on your site. People can leave notes for you that are published to your site, and you can respond and engage your readers in conversation about the topic at hand. See Figure 2-2 and Figure 2-3 for examples. Having this function in your site creates the opportunity to expand the thoughts and ideas that you presented in your post by giving your readers the opportunity to add their two cents' worth.

FIGURE 2-2: Readers use the form to leave their comments.

In the WordPress Dashboard, you have full administrative control over who can and can't leave comments. In addition, if someone leaves a comment with questionable content, you can edit the comment or delete it. You're also free to choose not to allow any comments on your site. Chapter 5 has the information you need about setting up your preferences for comments on your site.

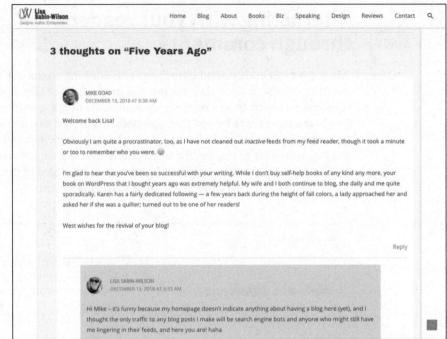

**FIGURE 2-3:** Readers comment on a post on my blog.

**TIP**

Some publishers say that a blog without comments isn't a blog at all because the point of having a blog, in some minds, is to foster communication and interaction between the site authors and the readers. This belief is common in the publishing community because experiencing visitor feedback via comments is part of what made Internet publishing so popular. Allowing comments is a personal choice, however, and you don't have to allow them if you don't want to.

## Feeding your readers

An RSS (Really Simple Syndication) feed is a standard feature that website visitors have come to expect. The What Is RSS? website (www.whatisrss.com) defines RSS as "a format for delivering regularly changing web content. Many news-related sites, weblogs, and other online publishers syndicate their content as an RSS Feed to whoever wants it."

Readers can use feed readers to download your feed — that is, their feed readers automatically discover new content (such as posts and comments) from your blog. Then readers can download that content for their consumption. Table 2-1 lists some of the most popular feed readers on the market today.

**TABLE 2-1**  **Popular RSS Feed Readers**

| Reader | URL | Description |
|---|---|---|
| MailChimp | `https://mailchimp.com` | MailChimp is an email newsletter service. It has an RSS-to-email service that enables you to send your recently published content to your readers via an email subscription service. |
| dlvr.it | `https://dlvrit.com` | Use RSS to autopost to Facebook, Twitter, LinkedIn, Pinterest, and other social media sites. |
| Feedly | `https://feedly.com` | With Feedly, you can keep up with your favorite sites and websites that have syndicated (RSS) content. You have no software to download or install to use this service, but optional applications are available for your use. |

To stay updated with the latest and greatest content you publish to your site, your readers and website visitors can subscribe to your RSS feed. Most platforms allow the RSS feeds to be *autodiscovered* by the various feed readers — that is, the blog reader needs to enter only your site's URL, and the program automatically finds your RSS feed.

WordPress has built-in RSS feeds in several formats. Because the feeds are built into the software platform, you don't need to do anything to provide your readers an RSS feed of your content. Check out Chapter 5 to find out more about using RSS feeds within the WordPress program.

## Tracking back

The best way to understand trackbacks is to think of them as comments, except for one thing: *Trackbacks* are comments left on your site by other sites, not people. Sounds perfectly reasonable, doesn't it? After all, why wouldn't inanimate objects want to participate in your discussion?

Actually, maybe it's not so crazy after all. A trackback happens when you make a post on your site and, within the content of that post, provide a link to a post made by another author on a different site. When you publish that post, your site sends a sort of electronic memo to the site you linked to. That site receives the memo and posts an acknowledgment of receipt in the form of a comment to the post that you linked to on the site. The information contained within the trackback includes a link back to the post on your site that contains the link to the other site — along with the date and time, as well as a short excerpt from your post. Trackbacks are displayed within the comments section of the individual posts.

The memo is sent via a *network ping* (a tool used to test, or verify, whether a link is reachable across the Internet) from your site to the site you link to. This process works as long as both sites support trackback protocol. Almost all major content management systems (CMSes) support the trackback protocol.

Sending a trackback to another site is a nice way of telling the author of that site that you like the information she presented in her post. Most authors appreciate the receipt of trackbacks to their posts from other publishers.

Trackbacks aren't as popular as they were years ago, but they still exist and are tools that some people like to take advantage of. An option in WordPress allows you to turn trackbacks off if you want to; you can find more information in Chapter 5.

# Dealing with comment and trackback spam

The absolute bane of publishing content on the Internet is comment and trackback spam. Ugh. When blogging became the "it" thing on the Internet, spammers saw an opportunity. If you've ever received spam in your email program, you know what I mean. For content publishers, the concept is similar and just as frustrating.

Spammers fill content with open comments with their links but not with any relevant conversation or interaction in the comments. The reason is simple: Websites receive higher rankings in the major search engines if they have multiple links coming in from other sites. Enter software, such as WordPress, with comment and trackback technologies, and these sites become prime breeding ground for millions of spammers.

Because comments and trackbacks are published to your site publicly — and usually with a link to the commenter's website — spammers got their site links posted on millions of sites by creating programs that automatically seek websites with open commenting systems and then hammer those systems with tons of comments that contain links back to their sites.

No one likes spam. Therefore, developers of CMSes such as WordPress spend untold hours in the name of stopping these spammers in their tracks, and for the most part, they've been successful. Occasionally, however, spammers sneak through. Many spammers are offensive, and all of them are frustrating because they don't contribute to the conversations that occur on the websites where they publish their spam comments.

All WordPress systems have one major, excellent thing in common: Akismet, which kills spam dead. Chapter 7 tells you more about Akismet, which is brought to you by Automattic, the maker of WordPress.com.

# Using WordPress as a Content Management System

You see something like the following a lot if you browse websites that publish articles about WordPress: "WordPress is more than a blogging platform; it's a full content management system." A *content management system* (CMS) is a platform that lets you run a full website on your domain. This means that WordPress enables you to create and publish all kinds of content on your site, including pages, blog posts, e-commerce pages for selling products, videos, audio files, and events.

## Exploring the differences between a website and a blog

A website and a blog are two different things. Although a website can contain a blog, a blog doesn't and can't contain a full website. I know that this description sounds confusing, but after you read this section and explore the differences between blogs and websites, you'll have a better understanding.

A *blog* is a chronological display of content, most often posts or articles written by the blog author. Those posts (or articles) are published, usually categorized in topics, and archived by date. Blog posts can have comments activated, which means that readers of a blog post can leave their feedback, and the blog post author can respond, thereby creating an ongoing dialogue between author and reader.

A *website* is a collection of published pages and sections that offer the visitor a variety of experiences or information. Part of the website can be a blog that enhances the overall visitor experience, but it usually includes other sections and features such as the following:

>> **Photo galleries:** This area of your website houses albums and galleries of uploaded photos, allowing your visitors to browse and comment on the photos you display.

>> **E-commerce store:** This feature is a fully integrated shopping cart through which you can upload products for sale, and your visitors can purchase your products via your online store.

>> **Discussion forums:** This area of your website allows visitors to join, create discussion threads, and respond to one another in specific threads of conversation.

- » **Social community:** This section of your website allows visitors to become members, create profiles, become friends with other members, create groups, and aggregate community activity.

- » **Portfolio of work:** If you're a photographer or web designer, for example, you can display your work in a specific section of your site.

- » **Feedback forms:** You can have a page on your website with a contact form that visitors can fill out to contact you via email.

- » **Static pages such as Bio, FAQ (Frequently Asked Questions), or Services:** These pages don't change as often as a blog page does. Blog pages change each time you publish a new post. Static pages contain content that doesn't change very often.

## Viewing examples of blogs and websites

I include a couple of figures in this section to further illustrate the difference between a blog and a website. Figure 2-4 shows what the front page of my business blog at `https://webdevstudios.com/blog` looked like at the time of this writing. Visit that URL in your browser, and you'll notice that the site displays a chronological listing of the most recent blog posts.

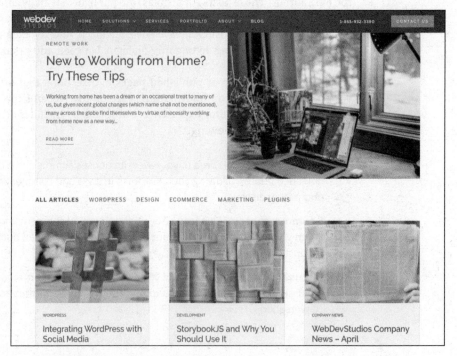

**FIGURE 2-4:** My business blog uses WordPress as a blogging tool.

My business website at https://webdevstudios.com also uses WordPress to power the entire site, not just the blog. This full site includes a static front page of information that acts as a portal to the rest of the site, on which you can find a blog; a portfolio of work; a contact form; and various landing pages, including service pages that outline information about the different services we offer (https://webdevstudios.com/services). Check out Figure 2-5 for a look at this page on our website; it's quite different from the blog section of the site.

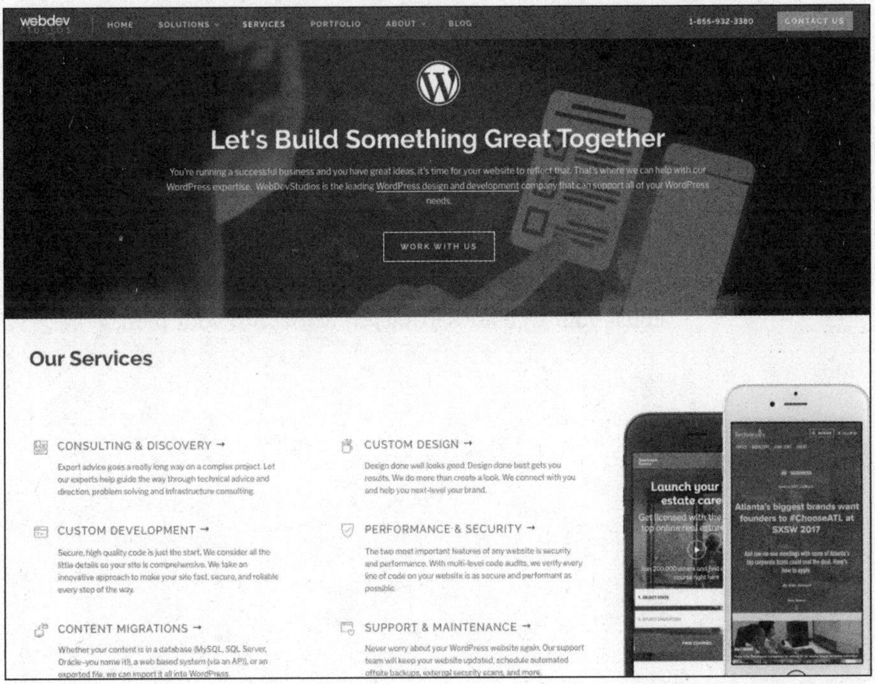

**FIGURE 2-5:** My business website uses WordPress as a CMS.

Using WordPress as a CMS means that you're using it to create not just a blog, but an entire website full of sections and features that offer a different experience for your visitors.

# Moving On to the Business of Publishing

Before getting started with publishing, you need to take a long look at your big plans for your website. A word of advice: Organize your plan of attack before you start. Have a good idea of what types of information you want to publish, how you want to present and organize that information, and what types of services and interaction you want to provide your audience.

Ask these questions out loud: "What am I going to write about?" and "Am I going to have a blog on my website at all?" Go ahead — ask. Do you have answers? Maybe you do, and maybe not, but either way, it's all right. There's no clear set of ground rules you must follow. Having an idea of what you expect to write about in your blog makes planning your attack a little easier. You may want to write about your personal life. Maybe you plan to share only some of your photography and provide very little commentary to go along with it. Or maybe you're a business owner and want to blog about your services and the news within your industry.

Having an idea of your subject matter helps you determine how you want to deliver that information. My design blog, for example, is where I write about web design projects, client case studies, and news related to design and blogging. You won't find pictures of my cats there, but you can find those pictures on my personal blog. I keep the two blogs separate, in much the same way that most people like to keep a distinct line of separation between their personal and professional lives, no matter what industry they work in.

When you have your topic and plan of delivery in mind, you can move forward and adjust your website settings to work with your plan.

# 2

# Setting Up WordPress

**IN THIS PART . . .**

Discover everything you need to know about the WordPress software version found on `https://wordpress.org`.

Set up your website base camp by registering a domain and getting set up with a web-hosting provider.

Install WordPress on your own hosted server and begin configuring the settings and options to personalize your site the way you want it.

Establish your own publishing routine by discovering and using the tools that WordPress provides.

Chapter **3**

# Setting Up Your WordPress Base Camp

Before you can start using WordPress, you have to set up your base camp. Doing so involves more than simply downloading and installing the WordPress software. You also need to establish your *domain* (your website address) and your *web-hosting service* (the place that houses your website). Although you initially download your WordPress software onto your hard drive, your web host is where you install it.

Obtaining a web server and installing software on it are much more involved projects than simply obtaining an account with the hosted version of WordPress that's available at WordPress.com (covered in Chapter 1). You need to consider many factors in this undertaking, as well as cope with a learning curve, because setting up your website through a hosting service involves using some technologies that you may not feel comfortable with at first.

This chapter takes you through the basics of those technologies, and by the last page of this chapter, you'll have WordPress successfully installed on a web server with your own domain name.

# Establishing Your Domain

You've read all the hype. You've heard all the rumors. You've seen the flashy websites on the web powered by WordPress. But where do you start?

The first steps in installing and setting up a WordPress website are deciding on a domain name and then purchasing the registration of that name through a domain registrar. A *domain name* is the *unique* web address that you type in a web browser's address bar to visit a website. Some examples of domain names are WordPress.org and Google.com.

**TIP**

I emphasize the word *unique* because no two domain names can be the same. If someone else has registered the domain name you want, you can't have it. Sometimes, it takes a bit of time to find a domain that isn't already in use and is available for you to use.

## Understanding domain name extensions

When registering a domain name, be aware of the *extension* that you want. The `.com`, `.net`, `.org`, `.info`, `.me`, `.us`, or `.biz` extension that you see tacked onto the end of any domain name is the *top-level domain extension*. When you register your domain name, you're asked to choose the extension you want for your domain (as long as it's available, that is).

## DOMAIN NAMES: DO YOU OWN OR RENT?

When you "buy" a domain name, you don't really own it. Rather, you're purchasing the right to use that domain name for the period of time specified in your order. You can register a domain name for one year or up to ten years. Be aware, however, that if you don't renew the domain name when your registration period ends, you lose it — and most often, you lose it right away to someone who preys on abandoned or expired domain names. Some people keep a close watch on expiring domain names, and as soon as the buying window opens, they snap up the names and start using them for their own websites in the hope of taking full advantage of the popularity that the previous owners worked so hard to attain for those domains.

A word to the wise here: Just because you registered your domain as a .com doesn't mean that someone else doesn't, or can't, own the very same domain name with a .net. So if you register MyDogHasFleas.com, and it becomes a hugely popular site among readers with dogs that have fleas, someone else can come along and register MyDogHasFleas.net and run a similar site to yours in the hope of riding the coattails of your website's popularity and readership.

If you want to avert this problem, you can register your domain name with all available extensions. My personal website, for example, has the domain name lisasabin-wilson.com, but I also own lisasabin-wilson.net just in case someone else out there has the same combination of names.

## Considering the cost of a domain name

Registering a domain costs you anywhere from $5 to $500 per year, depending on what service you use for a registrar and what options (such as storage space, bandwidth, privacy options, search-engine submission services, and so on) you apply to your domain name during the registration process.

When you pay the domain registration fee today, you need to pay another registration fee when the renewal date comes up again in a year, or two, or five — however many years you chose to register your domain name for. (See the nearby sidebar "Domain names: Do you own or rent?") Most registrars give you the option of signing up for a service called Auto Renew to automatically renew your domain name and bill the charges to the credit card you set up for that account. The registrar sends you a reminder a few months in advance, telling you that it's time to renew. If you don't have Auto Renew set up, you need to log in to your registrar account before it expires and renew your domain name manually.

## Registering your domain name

Domain registrars are certified and approved by the Internet Corporation for Assigned Names and Numbers (ICANN; https://www.icann.org). Although hundreds of domain registrars exist today, those in the following list are popular because of their longevity in the industry, competitive pricing, and variety of services offered in addition to domain name registration (such as web hosting and website traffic builders):

>> **GoDaddy:** https://www.godaddy.com

>> **Register.com:** https://www.register.com

>> **Network Solutions:** https://www.networksolutions.com

>> **WordPress:** https://wordpress.com/domains

No matter where you choose to register your domain name, here are the steps you can take to accomplish this task:

1. **Decide on a domain name.**

   Doing a little planning and forethought here is necessary. Many people think of a domain name as a *brand* — a way of identifying their websites or blogs. Think of potential names for your site; then you can proceed with your plan.

2. **Verify the domain name's availability.**

   In your web browser, enter the URL of the domain registrar of your choice. Look for the section on the registrar's website that lets you enter the domain name (typically, a short text field) to see whether it's available. If the domain name isn't available as a .com, try .net or .info.

3. **Purchase the domain name.**

   Follow the domain registrar's steps to purchase the name, using your credit card. After you complete the checkout process, you receive an email confirming your purchase, so use a valid email address during the registration process.

The next step is obtaining a hosting account, which I cover in the next section.

**REMEMBER**

Some domain registrars have hosting services that you can sign up for, but you don't have to use those services. Often, you can find hosting services for a lower cost than most domain registrars offer. It just takes a little research.

# Finding a Home for Your Website

When you've registered your domain, you need to find a place for it to live: a web host. Web hosting is the second piece of the puzzle that you need before you begin working with the WordPress software.

A *web host* is a business, group, or person that provides web server space and bandwidth for file transfer to website owners who don't have it. Usually, web-hosting services charge a monthly or annual fee — unless you're fortunate enough to know someone who's willing to give you server space and bandwidth for free. The cost varies from host to host, but you can obtain web-hosting services starting at $10 to $30 per month or more, depending on your needs.

Some web hosts consider WordPress to be a *third-party application*. What this means to you is that the host typically won't provide technical support on the use of WordPress (or any other software application) because support isn't included in your hosting package. To find out whether your chosen host supports WordPress, always ask first. As a WordPress user, you can find WordPress support in the official forums at `https://wordpress.org/support`.

**TIP**

Several web-hosting providers have WordPress-related services available for additional fees. These services can include technical support, plugin installation and configuration, and theme design services.

Web-hosting providers generally provide at least these services with your account:

>> Hard drive space

>> Bandwidth (transfer)

>> Domain email with web mail access

>> Secure File Transfer Protocol (SFTP) access

>> Comprehensive website statistics

>> MySQL database(s)

>> PHP

Because you intend to run WordPress on your web server, you need to look for a host that provides the minimum requirements needed to run the software on your hosting account, which are

>> PHP version 7.3 (or later)

>> MySQL version 5.6 (or later) *or* MariaDB version 10.0 (or later)

**TIP**

The easiest way to find out whether a host meets the minimum requirements for running the WordPress software is to check the FAQ (frequently asked questions) section of the host's website, if it has one. If not, find the contact information for the hosting company and fire off an email, or use the live chat on their website if they have it, requesting information on exactly what it supports.

## Getting help with hosting WordPress

The popularity of WordPress has given birth to services on the web that emphasize the use of the software. These services include WordPress designers, WordPress consultants, and — yes — web hosts that specialize in using WordPress.

Many of these hosts offer a full array of WordPress features, such as automatic WordPress installation included with your account, a library of WordPress themes, and a staff of support technicians who are very experienced in using WordPress.

Here are some of those providers:

>> **Bluehost:** https://bluehost.com

>> **GoDaddy:** https://godaddy.com

>> **Pagely:** https://pagely.com

>> **WP Engine:** https://wpengine.com

**WARNING**

A few web-hosting providers offer free domain name registration when you sign up for hosting services. Research this topic and read the terms of service, because that free domain name may come with conditions. Many of my clients have gone this route only to find out a few months later that the web-hosting provider has full control of the domain name, and they aren't allowed to move that domain off the host's servers for a set period (usually, a year or two) or for eternity. It's always best to have the control in *your* hands, not someone else's, so try to stick with an independent domain registrar, such as Network Solutions.

## Dealing with disk space and bandwidth

Web-hosting services provide two very important things with your account:

>> **Disk space:** The amount of space you can access on the web servers' hard drive, generally measured in megabytes (MB) or gigabytes (GB).

>> **Bandwidth transfer:** The amount of transfer your site can do per month. Typically, traffic is measured in gigabytes.

Think of your web host as a garage that you pay to park your car in. The garage gives you the place to store your car (disk space). It even gives you the driveway so that you, and others, can get to and from your car (bandwidth). It won't, however, fix the rockin' stereo system (WordPress or any other third-party software application) that you've installed unless you're willing to pay a few extra bucks for that service.

### Managing disk space

*Disk space* is nothing more complicated than the hard drive on your computer. Each hard drive has the capacity, or space, for a certain number of files. An 80GB (gigabyte) hard drive can hold 80GB worth of data — no more. Your hosting account provides you a limited amount of disk space, and the same concept applies. If your web host provides you 80GB of disk space, that's the limit on the file size that you're allowed to have. If you want more disk space, you need to upgrade your space limitations. Most web hosts have a mechanism in place for you to upgrade your allotment.

Starting out with a small WordPress blog doesn't take much disk space at all. A good starting point for disk space is 10GB to 20GB. If you need additional space, contact your hosting provider for an upgrade.

REMEMBER

The larger your website, the more space you need. Also, the more visitors and traffic your website has, the more bandwidth you require (see the next section).

## Choosing the size of your bandwidth pipe

*Bandwidth* refers to the amount of data that is carried from point A to point B within a specific period (usually, a second or two). I live out in the country — pretty much the middle of nowhere. I get my water from a private well that lies buried in the backyard somewhere. Between my house and the well are pipes that bring the water into my house. The pipes provide a free flow of water to our home so that everyone can enjoy long, hot showers while I labor over dishes and laundry, all at the same time. Lucky me!

The very same concept applies to the bandwidth available with your hosting account. Every web-hosting provider offers a variety of bandwidth limits on the accounts it offers. When I want to view your website in my browser window, the bandwidth is essentially the "pipe" that lets your data flow from your "well" to my computer and appear on my monitor. The bandwidth limit is kind of like the pipe connected to my well: It can hold only a certain amount of water before it reaches maximum capacity and won't bring water from the well any longer. Your bandwidth pipe size is determined by how much bandwidth your web host allows for your account; the larger the number, the bigger the pipe. A 50MB bandwidth limit makes for a smaller pipe than does a 100MB limit.

Web hosts are pretty generous with the amount of bandwidth they provide in their packages. Like disk space, bandwidth is measured in gigabytes (GB). Bandwidth provision of 50GB to 100GB is generally a respectable amount to run a website with a blog.

REMEMBER

Websites that run large files — such as video, audio, or photo files — generally benefit from more disk space (compared with sites that don't have large files). Keep this point in mind when you're signing up for your hosting account: If your site requires big files, you need more space. Planning will save you a few headaches down the road.

WARNING

Be wary of hosting providers that offer things like unlimited bandwidth, domains, and disk space. Those offers are great selling points, but what the providers don't tell you outright (you may have to look into the fine print of the agreement) is that although they may not put those kinds of limits on you, they limit your site's CPU use.

CPU (which stands for *central processing unit*) is the part of a computer (web server, in this case) that handles all the data processing requests sent to your web servers whenever anyone visits your site. Although you may have unlimited bandwidth to handle a large amount of traffic, if a high spike in traffic increases your site's CPU use, your host will throttle your site because it limits CPU use.

What do I mean by *throttle*? I mean that the host shuts down your site — turns it off. The shutdown isn't permanent; it lasts maybe a few minutes to an hour. The host does this to kill any connections to your web server that are causing the spike in CPU use. Your host eventually turns your site back on, but the inconvenience happens regularly with many clients across various hosting environments.

When you look into different web hosting providers, ask about their policies on CPU use and what they do to manage a spike in processing. It's better to know up front than to find out after your site's been throttled.

# Understanding Secure File Transfer Concepts

This section introduces you to the basic elements of SFTP (Secure File Transfer Protocol), which is a method of transferring files in a secure environment. SFTP provides an additional layer of security beyond what you get with regular FTP, as it uses SSH (Secure Shell) and encrypts sensitive information, data, and passwords so that they aren't clearly transferred within the hosting network. Encrypting the data ensures that anyone monitoring the network can't read the data freely — and, therefore, can't obtain information that should be secured, such as passwords and usernames.

SFTP offers two ways of moving files from one place to another:

» **Uploading:** Transferring files from your local computer to your web server

» **Downloading:** Transferring files from your web server to your local computer

You can do several other things with SFTP, including the following, which I discuss later in this chapter:

» **View files:** After you log in via SFTP, you can see all the files that are located on your web server.

» **View date modified:** You can see the date when a file was last modified, which can be helpful when you're trying to troubleshoot problems.

>> **View file size:** You can see the size of each file on your web server, which is helpful if you need to manage the disk space on your account.

>> **Edit files:** Almost all SFTP clients allow you to open and edit files through the client interface, which is a convenient way to get the job done.

>> **Change permissions:** Commonly referred to as CHMOD (an acronym for Change Mode), the permission settings on a file control what type of read/write/execute permissions the files on your web server have.

SFTP is a convenient utility that gives you access to the files located on your web server, which makes managing your WordPress website a bit easier.

TIP

I highly recommend using SFTP instead of FTP (File Transfer Protocol), because it's a secure connection to your web host. If your web hosting provider doesn't provide SFTP connections for you, strongly consider switching to a hosting provider that does. Almost all hosting providers these days provide SFTP as the standard protocol for transferring files.

## Setting up SFTP on your hosting account

Many web hosts today offer SFTP as part of their hosting packages, so just confirm that your hosting provider makes SFTP available to you for your account. cPanel is by far the most popular hosting account management software used by hosts on the web, eclipsing other popular tools such as Plesk and NetAdmin. You find the settings within the hosting account management interface, where you can set up the SFTP account for your website.

TIP

In this chapter, I use cPanel as the example. If your hosting provider gives you a different interface to work with, the concepts are the same, but you need to ask your hosting provider for the specific details you need to adapt these directions to your environment.

Mostly, SFTP for your hosting account is set up automatically. Figure 3-1 shows the User Manager page in cPanel, where you set up user accounts for SFTP access.

Follow these steps to get to this page and set up your SFTP account:

1. **Log in to cPanel for your hosting account.**

   Typically, you browse to http://*yourdomain*.com/cpanel to bring up the login screen for your cPanel. Enter your hosting account username and password in the login fields and then click OK.

2. **Browse to the User Manager page.**

   Click the SFTP Accounts link or icon in your cPanel to open the User Manager page (refer to Figure 3-1).

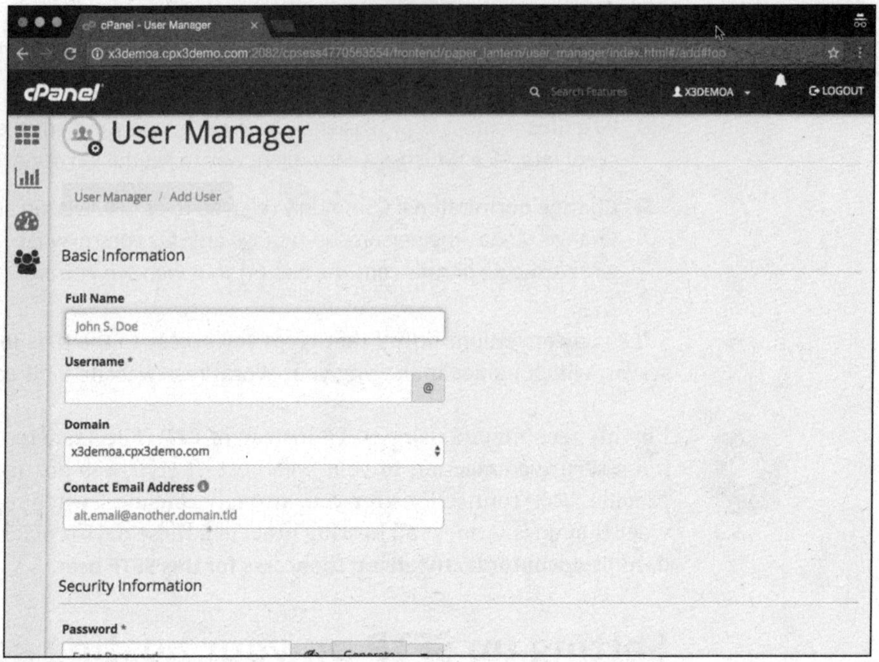

**FIGURE 3-1:**
The User
Manager page
within cPanel.

**3.** **View the existing SFTP account.**

If your hosting provider automatically sets you up with an SFTP account, you see it in the User Manager section. Ninety-nine percent of the time, the default SFTP account uses the same username and password combination as your hosting account or the login information you used to log in to your cPanel in Step 1.

If the SFTP Accounts page doesn't display a default SFTP user in the User Management section, you can easily create one in the Add SFTP Account section by following these steps:

**1.** **Fill in the provided fields.**

The fields of the User Manager page ask for your name, desired username, domain, and email address.

**2.** **Type your desired password in the Password field.**

You can choose to type in your own password or click the Password Generator button to have the server generate a secure password for you. Retype the password in the Password (Again) field to validate it.

**3.** **Check the Strength indicator.**

The server tells you whether your password is Very Weak, Weak, Good, Strong, or Very Strong (not shown in Figure 3-1, but you find it farther down on the same screen). You want to have a very strong password for your SFTP account that's difficult for hackers and malicious Internet users to guess and crack.

**4.** **In the Services section, click the Disabled icon within the FTP section.**

This action changes the icon label to Enabled and enables FTP for the user you're creating.

**5.** **Indicate the space limitations in the Quota field.**

Because you're the site owner, leave the radio-button selection set to Unrestricted. (In the future, if you add a new user, you can limit the amount of space, in megabytes [MB], by selecting the radio button to the left of the text field and typing the numeric amount in the text box, such as 50MB.)

**6.** **(Optional) Type the directory access for this SFTP user.**

cPanel fills in this information for you: something like `public_html/user`. (The user, in this case, is the same username you entered in Step 1.) Leaving this field as is gives the new SFTP user access only to a folder with his or her username. For the purposes of installing WordPress, you want this account to have access to the `public_html` folder, so remove the `/user` portion of the directory access so that the field contains only `public_html`.

**7.** **Click the Create button.**

You see a new screen with a message that the account was created success-fully. Additionally, you see the settings for this new user account.

**8.** **Copy and paste the settings into a blank text editor window (such as Notepad for PC or TextEdit for Mac users).**

The settings for the user account are the details you need to connect to your web server via SFTP. Save the following settings:

- *Username: username@yourdomain.com*
- *Password: yourpassword*
- *Host name: yourdomain.com*
- *SFTP Server Port:* 22
- *Quota:* Unlimited MB

Username, Password, and SFTP Server Port are specific to your domain and the information you entered in the preceding steps.

**TIP**

Ninety-nine point nine percent of the time, the SFTP Server Port is 22. Be sure to double-check your SFTP settings to make sure that this is the case, because some hosting providers use different port numbers for SFTP.

**REMEMBER**

At any time, you can revisit the User Accounts page to delete the user accounts you've created, change the quota, change the password, and find the connection details specific to that account.

## Connecting to the web server via SFTP

SFTP programs are referred to as SFTP *clients* or SFTP *client software*. Whatever you decide to call it, an SFTP client is software that you use to connect to your web server to view, open, edit, and transfer files to and from your web server.

Using SFTP to transfer files requires an *SFTP* client. Many SFTP clients are available for download. Here are some good (and free) ones:

>> **SmartFTP (PC):** https://www.smartftp.com/en-us/download

>> **FileZilla (PC or Mac):** https://filezilla-project.org

>> **Transmit (Mac):** https://panic.com/transmit

For the purposes of this chapter, I use the FileZilla SFTP client (https://filezilla-project.org) because it's very easy to use, and the cost is free ninety-nine (that's open-source geek-speak for free!).

Figure 3-2 shows a FileZilla client that's not connected to a server. By default, the left side of the window displays a directory of files and folders on the local computer.

The right side of the window displays content when the FileZilla client is connected to a web server — specifically, directories of the web server's folders and files.

**REMEMBER**

If you use a different SFTP client software from FileZilla, you need to adapt your steps and practice for the specific SFTP client software you're using.

Connecting to a web server is an easy process. The SFTP settings you saved from Step 8 in "Setting up SFTP on your hosting account" earlier in this chapter are the same settings you see in your cPanel User Manager page if your SFTP was set up automatically for you:

>> **Username:** *username@yourdomain.com*

>> **Password:** *yourpassword*

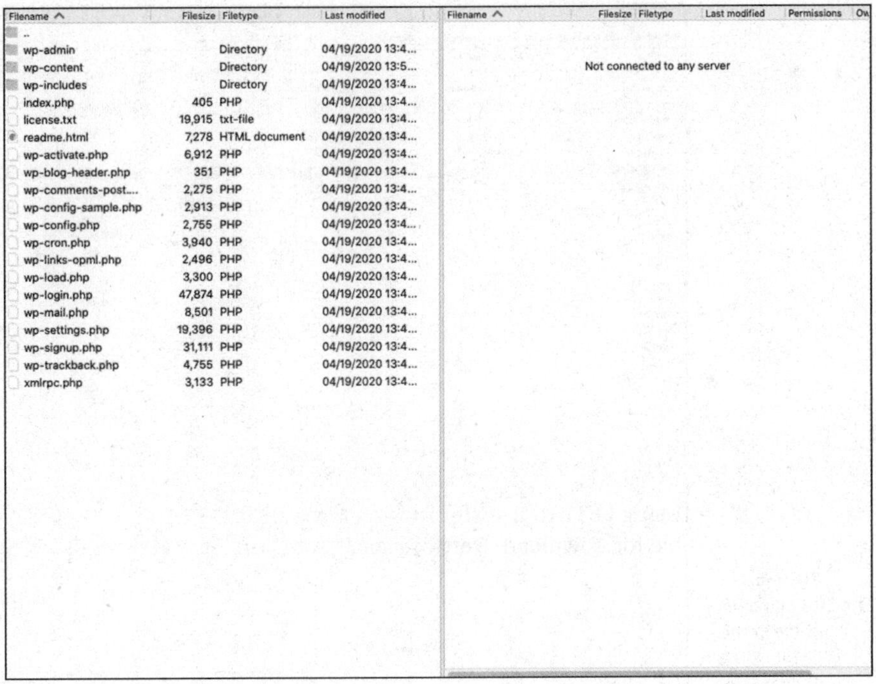

| Filename ∧ | Filesize | Filetype | Last modified | | Filename ∧ | Filesize | Filetype | Last modified | Permissions | Ow |
|---|---|---|---|---|---|---|---|---|---|---|
| .. | | | | | | | | | | |
| wp-admin | | Directory | 04/19/2020 13:4... | | | | | | | |
| wp-content | | Directory | 04/19/2020 13:5... | | | | Not connected to any server | | | |
| wp-includes | | Directory | 04/19/2020 13:4... | | | | | | | |
| index.php | 405 | PHP | 04/19/2020 13:4... | | | | | | | |
| license.txt | 19,915 | txt-file | 04/19/2020 13:4... | | | | | | | |
| readme.html | 7,278 | HTML document | 04/19/2020 13:4... | | | | | | | |
| wp-activate.php | 6,912 | PHP | 04/19/2020 13:4... | | | | | | | |
| wp-blog-header.php | 351 | PHP | 04/19/2020 13:4... | | | | | | | |
| wp-comments-post.... | 2,275 | PHP | 04/19/2020 13:4... | | | | | | | |
| wp-config-sample.php | 2,913 | PHP | 04/19/2020 13:4... | | | | | | | |
| wp-config.php | 2,755 | PHP | 04/19/2020 13:4... | | | | | | | |
| wp-cron.php | 3,940 | PHP | 04/19/2020 13:4... | | | | | | | |
| wp-links-opml.php | 2,496 | PHP | 04/19/2020 13:4... | | | | | | | |
| wp-load.php | 3,300 | PHP | 04/19/2020 13:4... | | | | | | | |
| wp-login.php | 47,874 | PHP | 04/19/2020 13:4... | | | | | | | |
| wp-mail.php | 8,501 | PHP | 04/19/2020 13:4... | | | | | | | |
| wp-settings.php | 19,396 | PHP | 04/19/2020 13:4... | | | | | | | |
| wp-signup.php | 31,111 | PHP | 04/19/2020 13:4... | | | | | | | |
| wp-trackback.php | 4,755 | PHP | 04/19/2020 13:4... | | | | | | | |
| xmlrpc.php | 3,133 | PHP | 04/19/2020 13:4... | | | | | | | |

**FIGURE 3-2:**
Mozilla FileZilla
SFTP client
software.

>> **Server:** *yourdomain.com*

>> **SFTP Server Port:** 22

>> **Quota:** Unlimited MB

This process is where you need that information. To connect to your web server via the FileZilla SFTP client, follow these few steps:

1. **Open the SFTP client software on your local computer.**

   Locate the program on your computer, and click (or double-click) the program icon to launch the program.

2. **Choose File ⇨ Site Manager to open the Site Manager utility.**

   The Site Manager utility appears, as shown in Figure 3-3.

3. **Click the New Site button.**

4. **Type a name for your site that helps you identify the site.**

   This site name can be anything you want it to be, because it isn't part of the connection data you add in the next steps. (I used *WPFD* — short for *WordPress For Dummies*.)

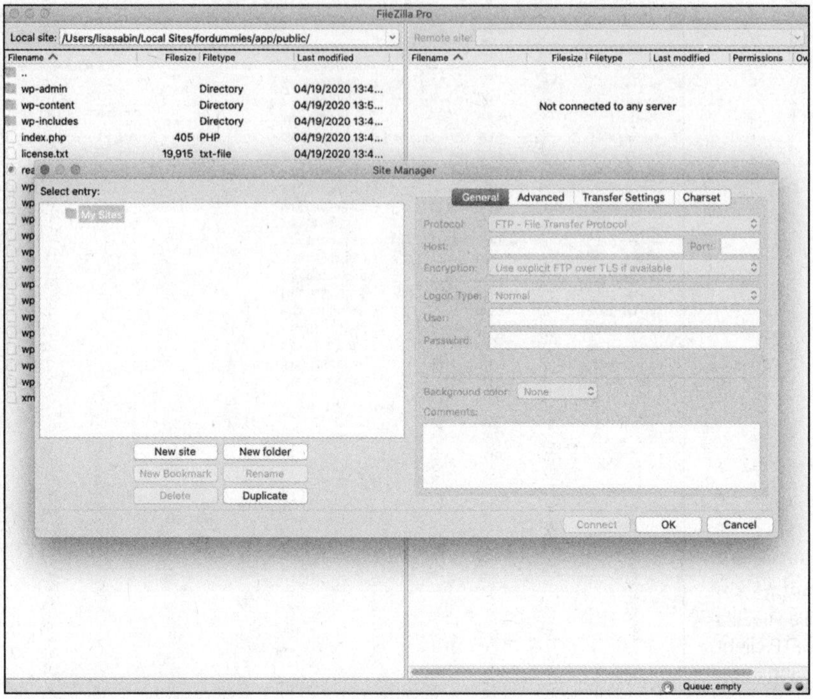

**FIGURE 3-3:**
The Site Manager utility in the FileZilla SFTP client software.

5. **Enter the SFTP server in the Host field.**

   Host is the same as the SFTP server information provided to you when you set up the SFTP account on your web server. In the example, the SFTP server is `wpaio.sftp.wpengine.com`, so that's entered in the Host field, as shown in Figure 3-4.

6. **Enter the SFTP port in the Port field.**

   Typically, SFTP uses port 22, and this setting generally never changes. My host, on the other hand, uses port 2222 for SFTP. In case your host is like mine and uses a port other than 22, double-check your port number and enter it in the Port field, as shown in Figure 3-4.

7. **Select the server type.**

   FileZilla asks you to select a server type (as do most SFTP clients). Choose SFTP – SSH File Transfer Protocol from the Protocol drop-down menu, as shown in Figure 3-4.

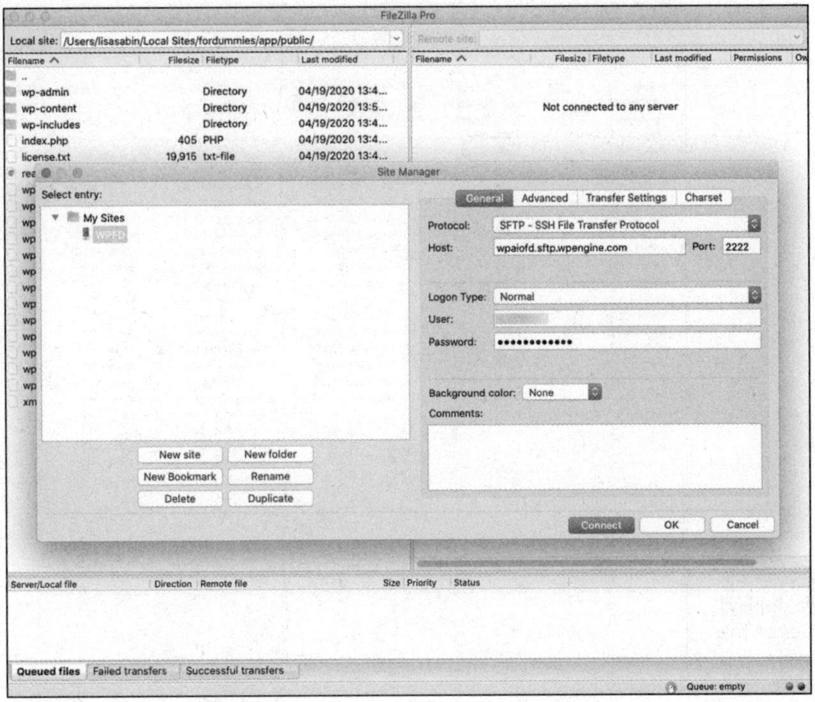

FIGURE 3-4:
The FileZilla Site
Manager utility
with SFTP account
information
filled in.

**8.** **Select the logon type.**

FileZilla gives you several logon types to choose among (as do most SFTP clients). Choose Normal from the Logon Type drop-down menu.

**9.** **Enter your username in the Username field.**

This entry is the username given to you in the SFTP settings.

**10.** **Type your password in the Password field.**

This entry is the password given to you in the SFTP settings.

**11.** **Click the Connect button.**

This step connects your computer to your web server. The directory of folders and files from your local computer displays on the left side of the FileZilla SFTP client window, and the directory of folders and files on your web server displays on the right side, as shown in Figure 3-5.

Now you can take advantage of all the tools and features SFTP has to offer you!

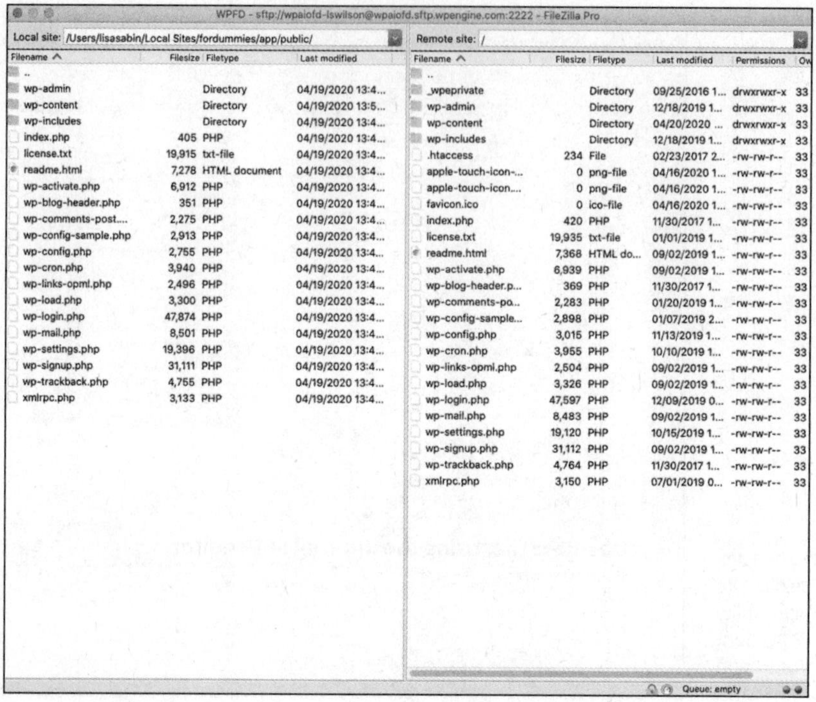

**FIGURE 3-5:**
FileZilla displays local files on the left and server files on the right.

# Transferring files from point A to point B

Now that your local computer is connected to your web server, transferring files between the two couldn't be easier. Within the SFTP client software, you can browse the directories and folders on your local computer on the left side and browse the directories and folders on your web server on the right side.

SFTP clients make it easy to transfer files from your computer to your hosting account by using a drag-and-drop method.

Two methods of transferring files are

» **Uploading:** Generally, transferring files from your local computer to your web server. To upload a file from your computer to your web server, click the file you want to transfer from your local computer, and drag and drop it on the right side (the web server side).

» **Downloading:** Transferring files from your web server to your local computer. To download a file from your web server to your local computer, click the file you want to transfer from your web server, and drag and drop it on the left side (the local computer side).

Downloading files from your web server is an efficient, easy, and smart way of backing up files to your local computer. It's always a good idea to keep your files safe, especially things like theme files and plugins, which are covered in Part 3 of this book.

## Editing files via SFTP

At times, you need to edit certain files that live on your web server. You can use the methods described in the preceding section to download a file, open it, edit it, save it, and then upload it to your web server. Another way is to use the edit feature built into most SFTP client software by following these steps:

1. **Connect the SFTP client to your web server.**

2. **Locate the file you want to edit.**

3. **Open the file, using the internal SFTP editor.**

   Right-click the file, and choose View/Edit from the shortcut menu. (Remember that I'm using FileZilla; your SFTP client may use different labels, such as Open or Edit.) FileZilla, like most SFTP clients, uses a program (such as Notepad for a PC or TextEdit for Mac) designated for text editing that already exists on your computer. In some rare cases, your SFTP client software may have its own internal text editor.

4. **Edit the file to your liking.**

5. **Save the changes you made to the file.**

   Click the Save icon or choose File ⇨ Save.

6. **Upload the file to your web server.**

   After you save the file, FileZilla alerts you that the file has changed and asks whether you want to upload the file to the server.

7. **Click the Yes button.**

   The newly edited file replaces the old one.

That's all there is to it. Use the SFTP edit feature to edit, save, and upload files as needed.

When you edit files with the SFTP edit feature, you're editing files in a "live" environment, meaning that when you save the changes and upload the file, the changes take effect immediately and affect your live website. For this reason, I strongly recommend downloading a copy of the original file to your local computer before making changes. That way, if you happen to make a typo in the saved file and your website goes haywire, you have a copy of the original file to upload to restore the file to its original state.

**TECHNICAL STUFF**

Programmers and developers are people who generally are more technologically advanced than your average user. These folks typically don't use SFTP for editing or transferring files. Instead, they use a version-control system called Git. Git manages the files on your web server through a versioning system that has a complex set of deployment rules for transferring updated files to/from your server. Most beginners don't use such a system for this purpose, but Git *is* a system that beginners can use. If you're interested in using Git, you can find a good resource to start with at Elegant Themes (`https://www.elegantthemes.com/blog/resources/git-and-github-a-beginners-guide-for-complete-newbies`).

## Changing file permissions

Every file and folder on your web server has a set of assigned attributes, called *permissions*, that tells the web server three things about the folder or file. On a very simplistic level, these permissions include:

>> **Read:** This setting determines whether the file/folder is readable by the web server.

>> **Write:** This setting determines whether the file/folder is writable by the web server.

>> **Execute:** This setting determines whether the file/folder is executable by the web server.

Each set of permissions has a numeric code assigned to it, identifying what type of permissions are assigned to that file or folder. A lot of permissions are available, so here are the most common ones that you run into when running a WordPress website:

>> **644:** Files with permissions set to 644 are readable by everyone and writable only by the file/folder owner.

>> **755:** Files with permissions set to 755 are readable and executable by everyone, but they're writable only by the file/folder owner.

>> **777:** Files with permissions set to 777 are readable, writable, and executable by anyone. For security reasons, you shouldn't use this set of permissions on your web server unless absolutely necessary.

Typically, folders and files within your web server are assigned permissions of 644 or 755. Usually, you see PHP files, or files that end with the .php extension, with permissions set to 644 if the web server is configured to use PHP Safe Mode.

**TIP**

This section gives you a basic look at file permissions because usually, you don't need to mess with file permissions on your web server. In case you do need to dig further, you can find a great reference on file permissions at https://www.elated.com/articles/understanding-permissions.

You may find yourself in a situation in which you're asked to edit and change the file permissions on a particular file on your web server. For WordPress sites, this situation usually happens when you're dealing with plugins or theme files that require files or folders to be writable by the web server. This practice is referred to as CHMOD. When someone says, "You need to CHMOD that file to 755," you'll know what he is talking about.

Here are some easy steps for using your SFTP program to CHMOD a file or edit its permissions on your web server:

1. **Connect the SFTP client to your web server.**

2. **Locate the file you want to CHMOD.**

3. **Open the file attributes for the file.**

   Right-click the file on your web server, and choose File Permissions from the shortcut menu. (Your SFTP client, if not FileZilla, may use different terminology.)

   The Change File Attributes dialog box appears, as shown in Figure 3-6.

4. **Type the correct file permissions number in the Numeric Value field.**

   This number is assigned to the permissions you want to give the file. Most often, the plugin or theme developer tells you which permissions number to assign to the file or folder — typically, 644 or 755. (The permissions in Figure 3-6 are assigned the value of 644.)

5. **Click OK to save the file.**

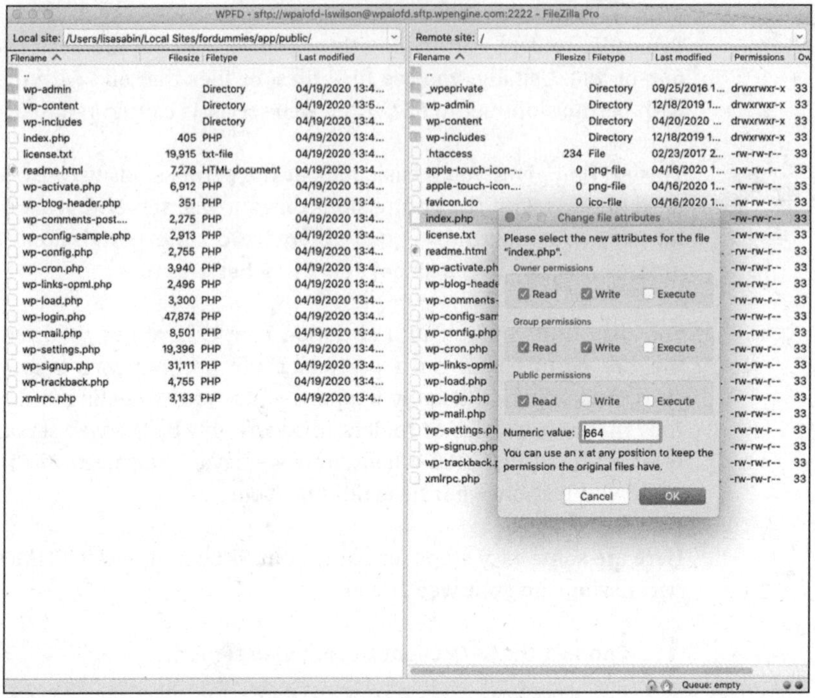

**FIGURE 3-6:**
The Change File Attributes dialog box.

# Installing WordPress

By the time you're finally ready to install WordPress, you should have done the following things:

» Purchased the domain name registration for your account

» Obtained a hosting service on a web server for your blog

» Established your hosting account's username, password, and Secure File Transfer Protocol (SFTP) address (see Chapter 2)

» Acquired an SFTP client for transferring files to your hosting account (see Chapter 2)

If you missed doing any of these items, you can go back to the beginning of this chapter to read the portions you need.

Some hosting providers have their own WordPress installers that can make installing WordPress easier by providing things like a step-by-step wizard or an easy interface. Check with your hosting provider to see whether it provides a WordPress installer for you to use.

TIP

# Exploring Preinstalled WordPress

The WordPress software has become such a popular publishing tool that almost all hosting providers available today provide WordPress for you in a couple of ways:

» Already installed on your hosting account when you sign up

» A user dashboard with a utility for installing WordPress from within your account management system

**TIP**

If your hosting provider doesn't give you access to an installation utility, skip to "Installing WordPress manually" later in this chapter for the steps to install WordPress via SFTP.

One of the most popular web hosts for managed WordPress hosting is a service called WP Engine, which you can find at `https://wpengine.com`. The service provides a handy, easy-to-use installation utility that's built right into your account dashboard at WP Engine to allow you to get up and running with WordPress right away.

You may not be using WP Engine, so your host may have a slightly different utility, but the basic concept is the same. Be sure to apply the same concepts to whatever kind of utility your hosting provider gives you. To access the account dashboard of WP Engine, follow these steps:

1. **Log in to the WP Engine user dashboard.**

   (a) *Browse to* `https://my.wpengine.com` *to bring up the login screen.*

   (b) *Enter the email address you used to sign up, enter your password, and then click Log In. The page refreshes and displays the dashboard for your account.*

2. **Click the Add Site button (see Figure 3-7).**

   The Add Site page displays in your browser window.

3. **Type the name of your new WordPress installation in the Site Name field.**

   This name is the temporary domain name of your new website. As shown in Figure 3-8, I will use *wpfd as the site name*, which stands for *WordPress For Dummies*. This step creates the domain name `wpfd.wpengine.com`.

4. **Choose Ungrouped from the Site Group drop-down list.**

   This step creates a new site in your account that isn't grouped with any other site.

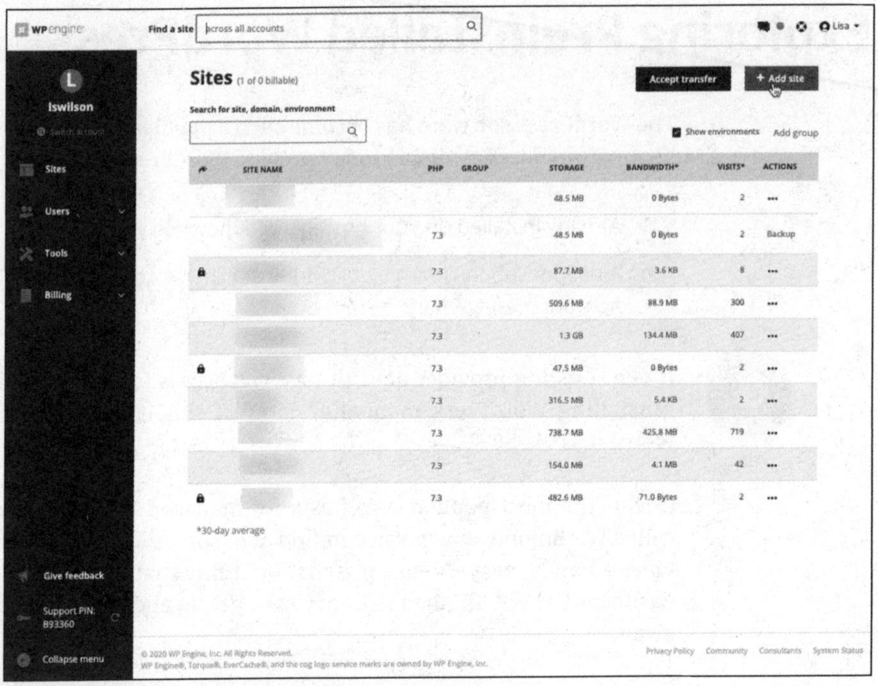

FIGURE 3-7:
The Add Site
button within a
WP Engine
account
dashboard.

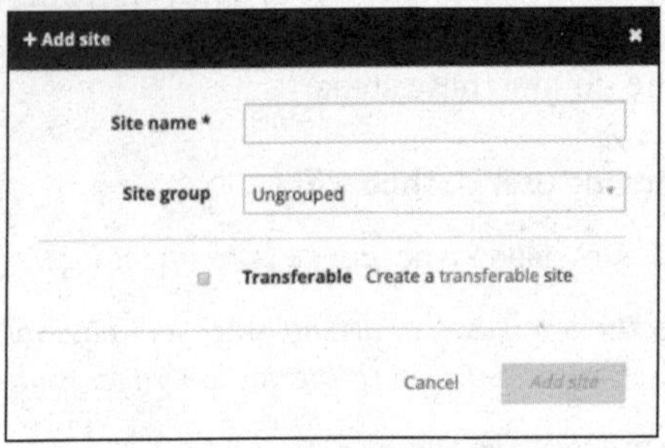

FIGURE 3-8:
The Add Site
module at WP
Engine.

**5.** **Leave the Create a Transferable Site check box unselected.**

There may come a day where you want to create a WordPress installation that can be transferred between two WP Engine accounts, but today isn't that day. You can read about the process at `https://wpengine.com/support/transfer-wp-engine-environment/`.

TECHNICAL
STUFF

**6.** **Click the Add Site button.**

This step creates the WordPress installation in your account and takes you to the Overview page, where a message states that your WordPress installation is being created. When the installation is ready to use, you receive an email from WP Engine.

In my experience, WP Engine always has the most up-to-date version of WordPress available for installation. Be sure to check that your hosting provider is supplying the latest version of WordPress with its installation utility.

## Installing WordPress manually

Doing a manual install of WordPress is where the rubber meets the road — that is, where you're putting WordPress's famous five-minute installation to the test. Set your watch and see whether you can meet that five-minute goal.

The famous five-minute installation includes only the time it takes to install the software. It doesn't include the time to

>> Register a domain name

>> Obtain and set up your web-hosting service

>> Download, install, configure, and learn how to use the SFTP software

Without further ado, go get the latest version of the WordPress software at `https://wordpress.org/download`.

WordPress gives you two compression formats for the software: `zip` and `tar.gz`. I recommend getting the `.zip` file because it's the most common format for compressed files.

Download the WordPress software to your computer, and decompress (unpack or unzip) it to a folder on your computer's hard drive. These steps are the first in the installation process for WordPress. Having the program on your own computer isn't enough, however; you also need to *upload* (transfer) it to your web server account (the one you obtained in "Finding a Home for Your Website" earlier in this chapter).

Before installing WordPress on your web server, make sure that you have a MySQL database set up and ready to accept the WordPress installation. The next section tells you what you need to know about MySQL.

## Setting up the MySQL database

The WordPress software is a personal publishing system that uses a PHP-and-MySQL platform, which provides everything you need to create your own website and publish your own content dynamically without knowing how to program those pages yourself. In short, all your content (options, posts, comments, and other pertinent data) is stored in a MySQL database in your hosting account.

Every time visitors go to your blog to read your content, they make a request that's sent to your server. The PHP programming language receives that request, obtains the requested information from the MySQL database, and then presents the requested information to your visitors through their web browsers.

Every web host is different in how it gives you access to set up and manage your MySQL database(s) for your account. In this section, I use cPanel, a popular hosting interface. If your host provides a different interface, the same basic steps apply, but the setup in the interface that your web host provides may be different.

To set up the MySQL database for your WordPress website with cPanel, follow these steps:

1. **Log in to the administration interface with the username and password assigned to you by your web host.**

   I'm using the cPanel administration interface, but your host may provide NetAdmin or Plesk, for example.

2. **Locate the MySQL Database Administration section.**

   In cPanel, click the MySQL Databases icon.

3. **Choose a name for your database, and enter it in the Name text box.**

   Note the database name, because you'll need it during installation of WordPress later.

4. **Click the Create Database button.**

   You get a message confirming that the database has been created.

5. **Click the Back button on your browser toolbar.**

   The MySQL Databases page displays in your browser window.

6. **Locate MySQL Users on the MySQL Databases page.**

   Scroll approximately to the middle of the page to locate this section.

7. **Choose a username and password for your database, enter them in the Username and Password text boxes, and then click the Create User button.**

   A confirmation message appears, stating that the username was created with the password you specified.

   For security reasons, make sure that your password isn't something that sneaky hackers can easily guess. Give your database a name that you'll remember later. This practice is especially helpful if you run more than one MySQL database in your account. If you name a database *WordPress* or *wpblog,*

for example, you can be reasonably certain a year from now, when you want to access your database to make some configuration changes, that you know exactly which credentials to use.

Make sure that you note the database name, username, and password that you set up during this process. You need them before officially installing WordPress on your web server. Jot these details on a piece of paper, or copy and paste them into a text editor window; either way, make sure that you have them handy.

**8.** **Click the Back button on your browser toolbar.**

The MySQL Databases page displays in your browser window.

**9.** **In the Add Users to Database section of the MySQL Databases page, choose the user you just set up from the User drop-down list and then choose the new database from the Database drop-down list.**

The MySQL Account Maintenance, Manage User Privileges page appears in cPanel.

**10.** **Assign user privileges by selecting the All Privileges check box.**

Because you're the *administrator* (or owner) of this database, you need to make sure that you assign all privileges to the new user you just created.

**11.** **Click the Make Changes button.**

The resulting page displays a message confirming that you've added your selected user to the selected database.

**12.** **Click the Back button on your browser toolbar.**

You return to the MySQL Databases page.

The MySQL database for your WordPress website is complete, and you're ready to proceed to the final step of installing the software on your web server.

## Uploading the WordPress files

To upload the WordPress files to your host, return to the folder on your computer where you unpacked the WordPress software that you downloaded earlier. You'll find all the files you need in a folder called /wordpress (see Figure 3-9).

Using your SFTP client, connect to your web server, and upload all these files to the root directory of your hosting account.

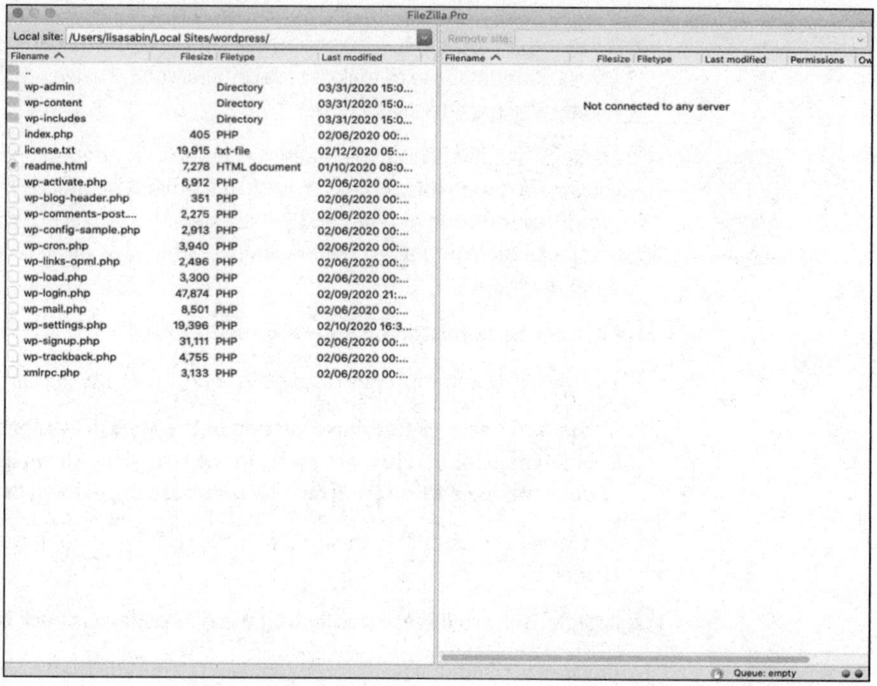

FileZilla Pro

Local site: /Users/lisasabin/Local Sites/wordpress/

Remote site:

| Filename ∧ | Filesize | Filetype | Last modified |
|---|---|---|---|
| .. | | | |
| wp-admin | | Directory | 03/31/2020 15:0... |
| wp-content | | Directory | 03/31/2020 15:0... |
| wp-includes | | Directory | 03/31/2020 15:0... |
| index.php | 405 | PHP | 02/06/2020 00:... |
| license.txt | 19,915 | txt-file | 02/12/2020 05:... |
| readme.html | 7,278 | HTML document | 01/10/2020 08:0... |
| wp-activate.php | 6,912 | PHP | 02/06/2020 00:... |
| wp-blog-header.php | 351 | PHP | 02/06/2020 00:... |
| wp-comments-post.... | 2,275 | PHP | 02/06/2020 00:... |
| wp-config-sample.php | 2,913 | PHP | 02/06/2020 00:... |
| wp-cron.php | 3,940 | PHP | 02/06/2020 00:... |
| wp-links-opml.php | 2,496 | PHP | 02/06/2020 00:... |
| wp-load.php | 3,300 | PHP | 02/06/2020 00:... |
| wp-login.php | 47,874 | PHP | 02/09/2020 21:... |
| wp-mail.php | 8,501 | PHP | 02/06/2020 00:... |
| wp-settings.php | 19,396 | PHP | 02/10/2020 16:3... |
| wp-signup.php | 31,111 | PHP | 02/06/2020 00:... |
| wp-trackback.php | 4,755 | PHP | 02/06/2020 00:... |
| xmlrpc.php | 3,133 | PHP | 02/06/2020 00:... |

| Filename ∧ | Filesize | Filetype | Last modified | Permissions | Ow |
|---|---|---|---|---|---|

Not connected to any server

Queue: empty

**FIGURE 3-9:**
WordPress software files to be uploaded to your web server.

**TIP**

If you don't know what your root directory is, contact your hosting provider and ask "What is my root directory for my account?" Every hosting provider's setup is different. On my web server, my root directory is the `public_html` folder; some of my clients have a root directory in a folder called `httpdocs`. The answer depends on what type of setup your hosting provider has. When in doubt, ask!

Here are a few things to keep in mind when you're uploading your files:

» **Upload the *contents* of the `/wordpress` folder to your web server — not the folder itself.** Most SFTP client software lets you select all the files and drag and drop them to your web server. Other programs have you highlight the files and click a Transfer button.

» **Choose the correct transfer mode.** File transfers via SFTP have two forms: ASCII and binary. Most SFTP clients are configured to autodetect the transfer mode. Understanding the difference as it pertains to this WordPress installation is important so that later you can troubleshoot any problems you have:

  • *Binary transfer mode* is how images (such as `.jpg`, `.gif`, `.bmp`, and `.png` files) are transferred via SFTP.

  • *ASCII transfer mode* is for everything else (text files, `.php` files, JavaScript, and so on).

For the most part, it's a safe bet to make sure that the transfer mode of your SFTP client is set to autodetect. But if you experience issues with how those files load on your site, retransfer the files, using the appropriate transfer mode.

>> **You can choose a different folder from the root.** You aren't required to transfer the files to the root directory of your web server. You can make the choice to run WordPress on a subdomain, or in a different folder on your account. If you want your blog address to be http://*yourdomain*.com/blog, transfer the WordPress files into a folder named /blog.

>> **Choose the right file permissions.** *File permissions* tell the web server how these files can be handled on your server — whether they're files that can be written to. As a general rule, .php files need to have a permission (CHMOD) of 644, whereas file folders need a permission of 755. Almost all SFTP clients let you check and change the permissions on the files if you need to. Typically, you can find the option to change file permissions in the menu options of your SFTP client.

**TECHNICAL STUFF**

Some hosting providers run their PHP software in a more secure format called *safe mode.* If this is the case with your host, you need to set the .php files to 644. If you're unsure, ask your hosting provider what permissions you need to set for .php files.

## Last step: Running the installation script

The final step in the installation procedure for WordPress is connecting the Word-Press software you uploaded to the MySQL database. Follow these steps:

**1.** **Type the URL of your website in the address bar of your web browser** (http://*yourdomain*.com/wp-admin/install.php).

If you chose to install WordPress in a different folder from the root directory of your account, make sure that you indicate this fact in the URL for the install script. If you transferred the WordPress software files to a folder called /blog, for example, point your browser to the following URL to run the installation: http://*yourdomain*.com/blog/wp-admin/install.php.

Assuming that you did everything correctly, you should see the first step in the installation process, as shown in Figure 3-10.

**TIP**

If you have troubles at any point in the installation, refer to Table 3-1 at the end of these steps for some troubleshooting tips.

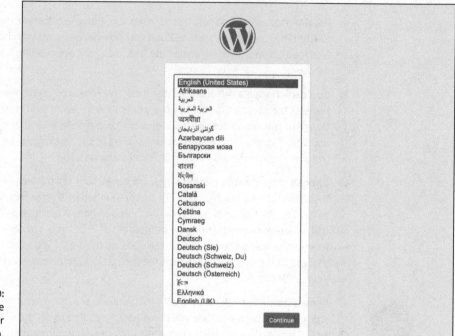

FIGURE 3-10:
Choose the
language for your
installation.

2. **Select your preferred language from the list provided on the setup page (refer to Figure 3-10).**

   At this writing, WordPress is available in 87 languages. For these steps, I'm using English (United States).

3. **Click the Continue button.**

   You see a new page with a welcome message from WordPress and instructions that you need to gather the MySQL information you saved earlier in this chapter.

4. **Click the Let's Go button.**

   A new page loads, displaying fields you need to fill out.

5. **Dig out the database name, username, and password that you saved (refer to "Setting up the MySQL database" earlier in this chapter), and use that information to fill in the following fields, as shown in Figure 3-11:**

   - *Database Name:* Type the database name you used when you created the MySQL database before this installation. Because hosts differ in configurations, enter either the database name or the database name with your hosting account username appended.

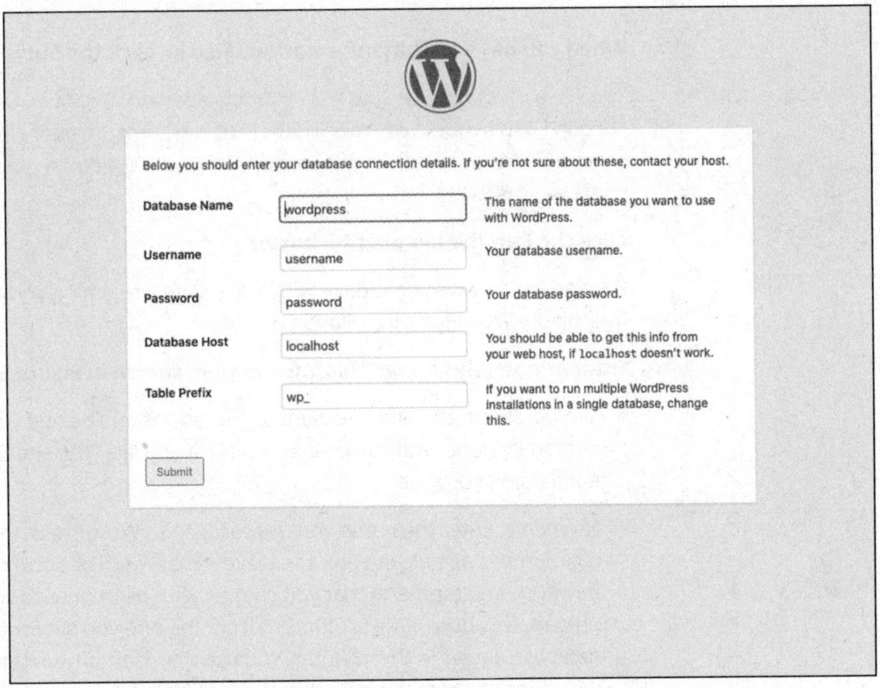

**FIGURE 3-11:**
Entering the database name, username, and password.

If you named your database *wordpress,* for example, enter **wordpress** in this text box. If your host requires you to append the database name with your hosting account username, enter ***username*_wordpress**, substituting your hosting username for *username.* My username is *lisasabin,* so I'd enter **lisasabin_wordpress**.

- *Username:* Type the username you used when you created the MySQL database before this installation. Depending on what your host requires, you may need to append this username to your hosting account username.

- *Password:* Type the password you used when you set up the MySQL database. You don't need to append the password to your hosting account username here.

- *Database Host:* Ninety-nine percent of the time, you'll leave this field set to *localhost.* Some hosts, depending on their configurations, have different hosts set for the MySQL database server. If *localhost* doesn't work, contact your hosting provider to find out the MySQL database host.

- *Table Prefix:* Leave this field set to *wp_.*

**6.** **When you have all that information filled in, click the Submit button.**

You see a message that says `All right, sparky! You've made it through this part of the installation. WordPress can now communicate with your database. If you're ready, time now to run the install!`

**7.** **Click the Run the Installation button.**

You see another welcome page with a message welcoming you to the famous five-minute WordPress installation process.

**8.** **Enter or possibly change this information, shown in Figure 3-12:**

- *Site Title:* Enter the title you want to give your site. The title you enter isn't written in stone; you can change it later, if you like. The site title also appears on your site.

- *Username:* Enter the name you use to log in to WordPress. By default, the username is *admin,* and you can leave it that way. For security reasons, however, I recommend that you change your username to something unique. This username is different from the one you set for the MySQL database earlier in this chapter. You use this username when you log in to WordPress to access the Dashboard (see Chapter 4), so be sure to make it something you'll remember.

- *Password:* Type your desired password in the text box. If you don't enter a password, one is generated automatically for you. For security reasons, it's a good thing to set a different password here from the one you set for your MySQL database; just don't get the passwords confused.

**TIP**

For security reasons (and so that other people can't make a lucky guess), passwords should be at least seven characters long and use as many characters in as many combinations as possible. Use a mixture of upper-case and lowercase letters, numbers, and symbols (such as ! " ? $ % ^ &).

- *Your Email:* Enter the email address at which you want to be notified of administrative information about your blog. You can change this address later, too.

- *Search Engine Visibility:* By default, this check box (shown in Figure 3-12) is selected, which lets the search engines index the content of your blog and include your blog in search results. To keep your blog out of the search engines, deselect this check box. (See Chapter 13 for information on search engine optimization.)

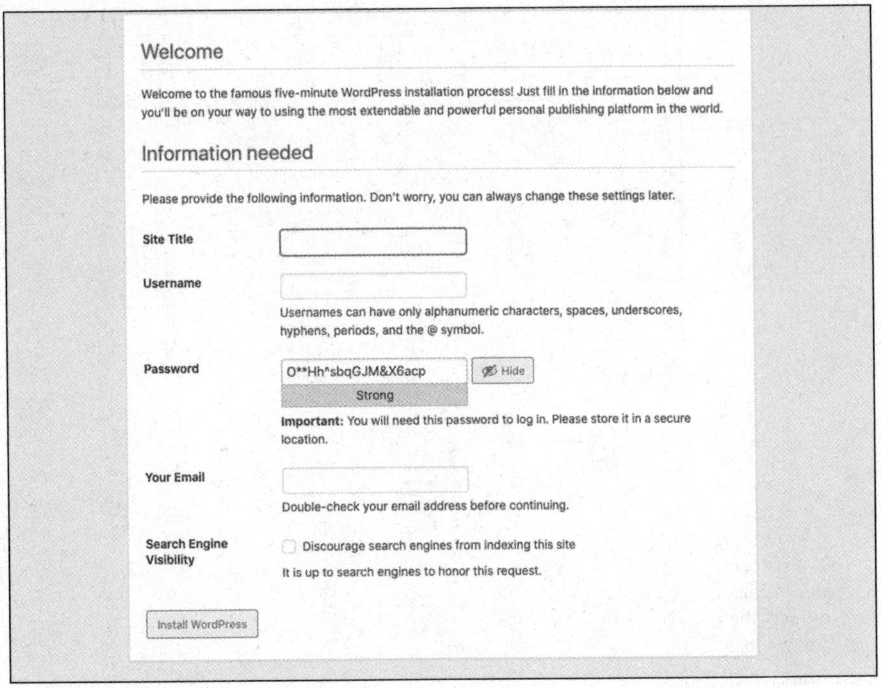

**Welcome**

Welcome to the famous five-minute WordPress installation process! Just fill in the information below and you'll be on your way to using the most extendable and powerful personal publishing platform in the world.

**Information needed**

Please provide the following information. Don't worry, you can always change these settings later.

| | |
|---|---|
| **Site Title** | |
| **Username** | |
| | Usernames can have only alphanumeric characters, spaces, underscores, hyphens, periods, and the @ symbol. |
| **Password** | O**Hh^sbqGJM&X6acp  👁 Hide |
| | Strong |
| | **Important:** You will need this password to log in. Please store it in a secure location. |
| **Your Email** | |
| | Double-check your email address before continuing. |
| **Search Engine Visibility** | ☐ Discourage search engines from indexing this site |
| | It is up to search engines to honor this request. |

Install WordPress

**FIGURE 3-12:**
Finishing the WordPress installation.

### 9. Click the Install WordPress button.

The WordPress installation machine works its magic and creates all the tables within the database that contain the default data for your site. WordPress displays the login information you need to access the WordPress Dashboard. Make note of this username and password before you leave this page; scribble them on a piece of paper or copy them into a text editor, such as Notepad (PC) or TextEdit (Mac).

**REMEMBER**

After you click the Install WordPress button, you receive an email with the login information and login URL. This information is handy if you're called away during this part of the installation process. So go ahead and let the dog out, answer the phone, brew a cup of coffee, or take a 15-minute power nap. If you somehow get distracted away from this page, the email sent to you contains the information you need to log in to your WordPress website.

### 10. Click the Log In button to log in to WordPress.

**TIP**

If you happen to lose this page before clicking the Log In button, you can always find your way to the login page by entering your domain followed by the call to the login file (such as http://*yourdomain*.com/wp-login.php).

You know that you're finished with the installation process when you see the login page, shown in Figure 3-13.

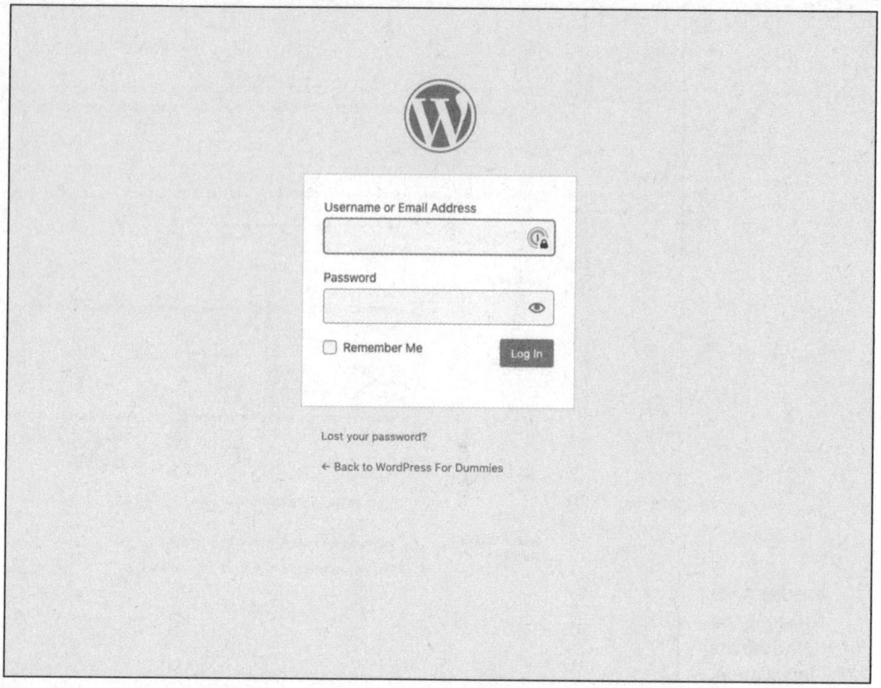

**FIGURE 3-13:**
You've successfully installed WordPress when you see the login page.

**TABLE 3-1**     **Common WordPress Installation Problems**

| Error Message | Common Cause | Solution |
|---|---|---|
| Error Connecting to the Database | The database name, user-name, password, or host was entered incorrectly. | Revisit your MySQL database to obtain the database name, username, and password, and re-enter that information. |
| Headers Already Sent | A syntax error occurred in the wp-config.php file. | Open the wp-config.php file in a text editor. The first line should contain only this line: <?php. The last line should contain only this line: ?>. Make sure that those lines contain nothing else — not even white space. Save the file changes. |
| 500: Internal Server Error | Permissions on .php files are set incorrectly. | Try setting the permissions (CHMOD) on the .php files to 666. If that doesn't work, set them to 644. Each web server has different settings for how it lets PHP execute on its servers. |
| 404: Page Not Found | The URL for the login page is incorrect. | Make sure that the URL you're using to get to the login page is the same as the location of your WordPress installation (such as http://yourdomain.com/wp-login.php). |
| 403: Forbidden Access | An index.html or index.htm file exists in the Word-Press installation directory. | WordPress is a PHP application, so the default home page is index.php. Look in the WordPress installation folder on your web server. If the folder contains an index.html or index.htm file, delete it. |

If you experience any problems during the installation process, Table 3-1 covers some of the common problems users encounter.

So do tell — how much time does your watch show for the installation? Was it five minutes? Drop me a line on Twitter (@LisaSabinWilson), and let me know whether WordPress stood up to its famous five-minute-installation reputation. I'm a curious sort.

The good news is — you're done! Were you expecting a marching band? WordPress isn't that fancy . . . yet. Give it time, though; if anyone can produce one, the folks at WordPress can.

Let me be the first to congratulate you on your newly installed WordPress site! When you're ready, log in, and familiarize yourself with the Dashboard, which I describe in Chapter 4.

Chapter **4**

# Understanding the WordPress.org Dashboard

With WordPress.org successfully installed, you can explore your new website software. This chapter guides you through the preliminary setup of your new WordPress site using the Dashboard.

When you create a website with WordPress, you spend a lot of time in the Dashboard, which is where you make all the exciting behind-the-scenes stuff happen. In this panel, you find all the settings and options that enable you to set up your site just the way you want it. (If you still need to install and configure WordPress, check out Chapter 3.)

Feeling comfortable with the Dashboard sets you up for a successful entrance into the WordPress world. Expect to tweak your WordPress settings several times throughout the life of your website. In this chapter, as I go through the various sections, settings, options, and configurations available to you, understand that nothing is set in stone. You can set options today and change them at any time.

# Logging In to the Dashboard

I find that the direct approach (also known as jumping in) works best when I want to get familiar with a new software tool. To that end, just follow these steps to log in to WordPress and take a look at the guts of the Dashboard:

1. **Open your web browser, and type the WordPress login-page address (or URL) in the address box.**

   The login-page address looks something like this:

   `http://www.yourdomain.com/wp-login.php`

   **TIP**

   If you installed WordPress in its own folder, include that folder name in the login URL. If you installed WordPress in a folder ingeniously named wordpress, the login URL becomes

   `http://www.yourdomain.com/wordpress/wp-login.php`

2. **Type your username or email address in the Username or Email Address text box and your password in the Password text box.**

   **REMEMBER**

   In case you forget your password, WordPress has you covered. Click the Lost Your Password? link (located near the bottom of the page), enter your username or email address, and then click the Get New Password button. WordPress resets your password and emails the new password to you.

   After you request a password, you receive an email from your WordPress installation. The email contains a link that you need to click to reset your password.

3. **Select the Remember Me check box if you want WordPress to place a cookie in your browser.**

   The cookie tells WordPress to remember your login credentials the next time you show up. The cookie set by WordPress is harmless and stores your WordPress login on your computer. Because of the cookie, WordPress remembers you the next time you visit. Also, because this option tells the browser to remember your login, I don't advise checking this option on public computers. Avoid selecting Remember Me when you're using your work computer or a computer at an Internet café.

   *Note:* Before you set this option, make sure that your browser is configured to allow cookies. (If you aren't sure how to do this, check the help documentation of the Internet browser you're using.)

4. **Click the Log In button.**

   After you log in to WordPress, you see the Dashboard page.

# Navigating the Dashboard

You can consider the Dashboard to be a control panel of sorts because it offers several links and areas that provide information about your website, starting with the actual Dashboard page, shown in Figure 4-1.

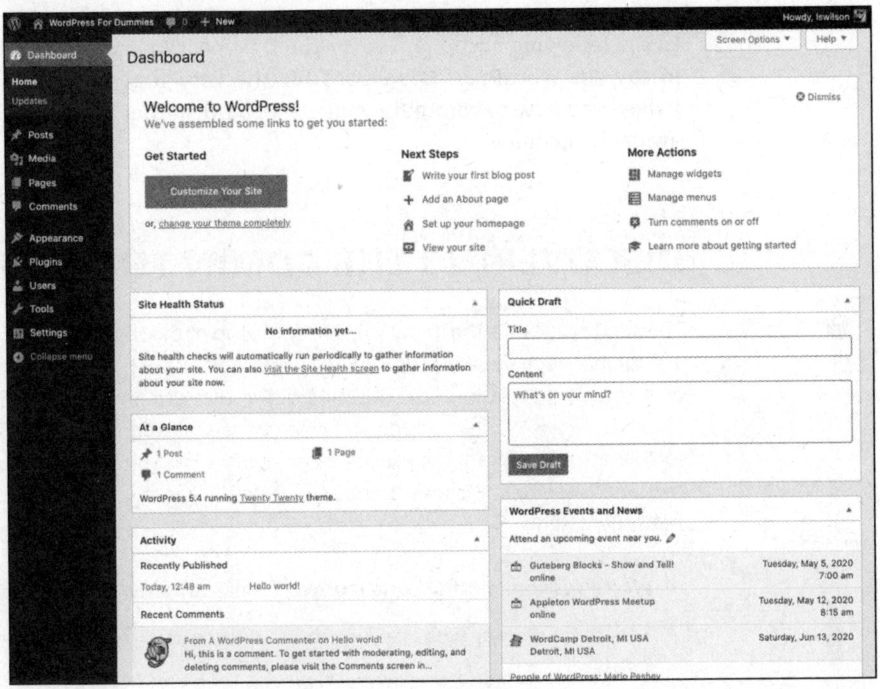

**FIGURE 4-1:**
Log in to the WordPress Dashboard.

You can change how the WordPress Dashboard looks by changing the order of the modules that appear in it (such as At a Glance and Activity). You can expand (open) and collapse (close) a module by clicking the small gray arrow to the right of its title. This feature is really nice because it allows you to use the Dashboard for just those modules that you use regularly.

The concept is simple: Keep the modules you use all the time open, and close the ones that you use only occasionally; you can open those modules only when you really need them. You save space and can customize your Dashboard to suit your own needs. WordPress remembers the way you set your Dashboard, so if you close certain modules today, they remain closed every time you visit the Dashboard until you open them again.

When you view your Dashboard for the first time, all the modules appear in the expanded (open) position by default (refer to Figure 4-1).

**TIP**

The navigation menu of the WordPress Dashboard appears on the left side of your browser window. When you need to get back to the main Dashboard page, click the Dashboard link at the top of the navigation menu on any of the screens within your WordPress Dashboard.

In the following sections, I cover the Dashboard page as it appears when you log in to your WordPress Dashboard for the very first time. Later in this chapter, I show you how to configure your Dashboard so that it best suits how you use the available modules.

## DISCOVERING THE ADMIN TOOLBAR

The admin toolbar is the menu you see at the top of the Dashboard (refer to Figure 4-1). The toolbar appears at the top of every page on your site by default, and it appears at the top of every page of the Dashboard if you set it to do so in your profile settings. The nice thing is that the only person who can see the admin toolbar is you because the toolbar displays only for the user who is logged in. The admin toolbar contains short-cuts that take you to the most frequently viewed areas of your WordPress Dashboard, from left to right:

- **WordPress links:** This shortcut provides links to various WordPress.org sites.

- **The name of your website:** This shortcut takes you to the front page of your website.

- **Comments page:** The next link is a comment balloon icon; click it to visit the Comments page of your Dashboard.

- **New:** Hover your mouse over this shortcut, and you find links titled Post, Media, Page, and User. Click these links to go to the Add New Post, Upload New Media, Add New Page, and Add New User pages, respectively.

- **Your photo and name display:** Hover your mouse pointer over this shortcut to open a drop-down menu with links to two areas of your Dashboard: Edit Your Profile and Log Out.

Again, the admin toolbar is visible at the top of your site only to you, no matter what page you're on, as long as you're logged in to your WordPress site.

# Welcome to WordPress!

This module, shown in Figure 4-2, appears at the top of your Dashboard screen the first time you log in to your new WordPress installation. It can stay there, if you want it to. Also notice a small link on the right side of that module labeled Dismiss. That link allows you to remove this module if you'd rather not have it there.

**FIGURE 4-2:**
The Welcome to
WordPress!
module provides
helpful links to
get you started.

The makers of the WordPress software have done extensive user testing to discover what items users want to do immediately when they log in to a new WordPress site. The result of that user testing is a group of links presented in the Welcome to WordPress! module:

>> **Get Started:** This section contains a button that, when clicked, opens the Customizer, where you can customize the active theme. Additionally, this section provides a link that takes you to the Themes page, where you can change your theme. Chapter 8 contains tons of information about choosing a theme, as well as customizing it to look the way you want it to.

>> **Next Steps:** This section provides links to various areas within the WordPress Dashboard to get you started publishing content, including writing your first post and adding an About page. Additionally, the View Your Site link in this section opens your site, allowing you to view what it looks like to your visitors.

>> **More Actions:** This section contains a few links that help you manage your site, including a link to manage widgets or menus and turn comments on or off. This section also contains the Learn More about Getting Started link, which takes you to the First Steps with WordPress article in the WordPress Codex, where you can read more about how to start using your new WordPress site.

# Site Health

The Site Health module of the Dashboard gives you details about the status of your WordPress and server configurations and settings. If you have a brand-new installation, you won't see anything listed in this module at all (refer to Figure 4-1). As time goes on, however, you may see notices about settings or configurations that require your attention, such as

>> Inactive or outdated plugins that should be removed or updated

>> Inactive or outdated themes that should be removed or updated

>> Notification that your site can't be indexed by search engines

Clicking the Visit the Site Health Screen link takes you to the Site Health screen of your Dashboard, where you see the information you need to improve the health of your site (see Figure 4-3).

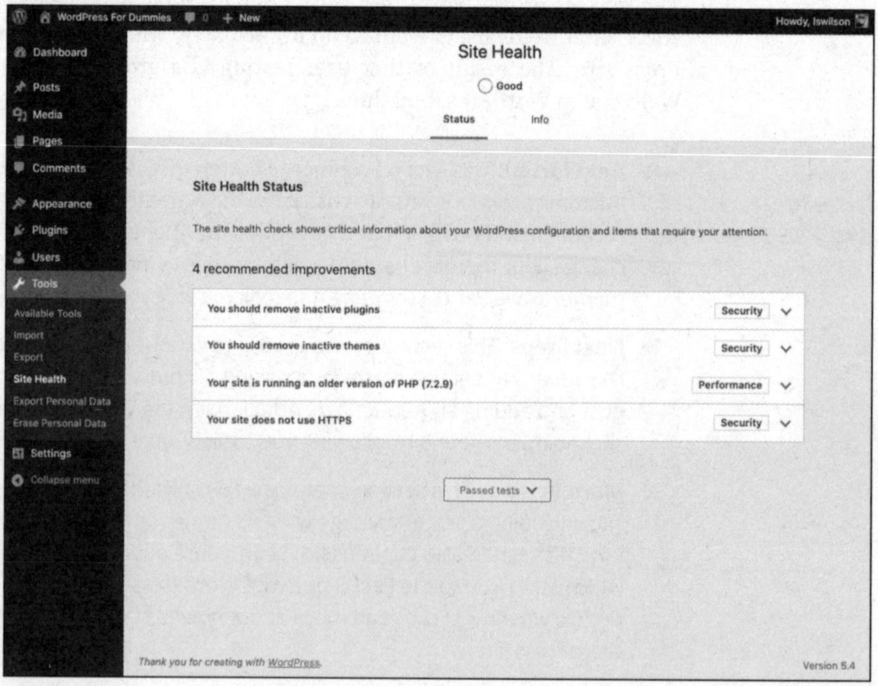

**FIGURE 4-3:**
The Site Health
screen.

# At a Glance

The At a Glance module of the Dashboard shows some details about what's going on in your website right now, this second! Figure 4-4 shows the expanded At a Glance module of my brand-spanking-new WordPress site.

The At a Glance module shows the following by default:

>> **The number of posts:** The total posts you currently have on your WordPress website. The website in Figure 4-4, for example, has one post. The link is blue, which means that it's clickable. When you click the link, you go to the Posts screen, where you can view and manage the posts on your website. (Chapter 5 covers managing posts.)

>> **The number of pages:** The number of pages on your website, which changes when you add or delete pages. (*Pages,* in this context, refers to the static pages you create in your website.) Figure 4-4 shows that the website has one page.

Clicking this link takes you to the Pages screen, where you can view, edit, and delete your current pages. (Find the difference between WordPress posts and pages in Chapter 5.)

>> **The number of comments:** The number of comments on your blog. Figure 4-4 shows that this website has one comment.

Clicking the Comments link takes you to the Comments screen, where you can manage the comments on your blog. Chapter 5 covers comments.

The last section of the Dashboard's At a Glance module shows the following information:

>> **Which WordPress theme you're using:** Figure 4-4 shows that the site is using the theme called Twenty Twenty. (Chapter 8 gives you information about WordPress themes.) The theme name is a link that, when clicked, takes you to the Manage Themes page, where you can view and activate themes on your blog.

>> **The version of WordPress you're using:** Figure 4-4 shows that this blog is using WordPress version 5.4. This version announcement changes if you're using an earlier version of WordPress. When WordPress software is upgraded, this statement tells you that you're using an outdated version of WordPress and encourages you to upgrade to the latest version.

## Activity

The module below the At a Glance module is Activity, shown in Figure 4-5.

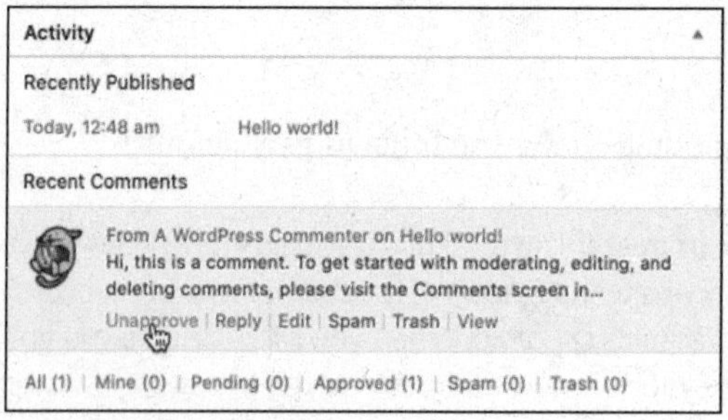

**FIGURE 4-5:**
The Activity module of the Dashboard.

Within this module, you find

>> **Posts most recently published:** WordPress displays a maximum five posts in this area. Each post's link is clickable and takes you to the Edit Post screen, where you can view and edit the post.

>> **Most recent comments published to your site:** WordPress displays a maximum of five comments in this area.

>> **The author of each comment:** The name of the person who left the comment appears above it. This section also displays the author's picture (or avatar), if she has one, or the default avatar if she doesn't.

>> **A link to the post the comment was left on:** The post title appears to the right of the commenter's name. Click the link to go to that post in the Dashboard.

>> **An excerpt of the comment:** This link is a snippet of the comment this person left on your site.

>> **Comment management links:** When you hover your mouse pointer over the comment, six links appear below it. These links give you the opportunity to manage those comments right from your Dashboard: The first link is Unapprove, which appears only if you have comment moderation turned on.

(Find out more about moderating comments in the "Comments" section later in this chapter.) The other five links are Reply, Edit, Spam, Trash, and View.

>> **View links:** These links — All, Mine, Pending, Approved, Spam, and Trash — appear at the bottom of the Recent Comments module.

You can find even more information on managing your comments in the "Comments" section later in this chapter.

## Quick Draft

The Quick Draft module, shown in Figure 4-6, is a handy form that allows you to write, save, and publish a blog post right from your WordPress Dashboard. The options are similar to the ones I cover in Chapter 5.

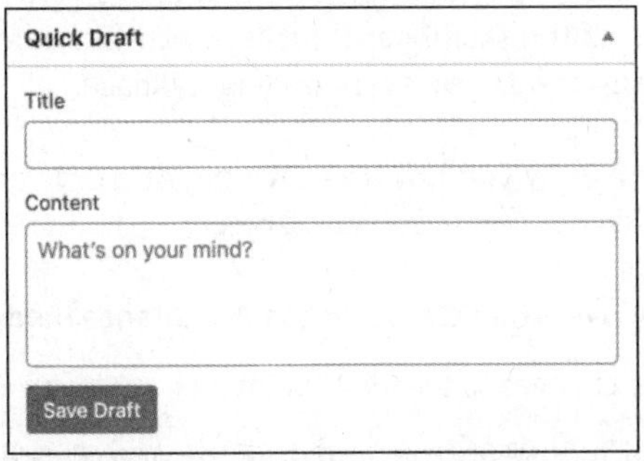

**FIGURE 4-6:** The Quick Draft module of the Dashboard.

If you're using a new WordPress blog and a new installation of WordPress, the Drafts list doesn't appear in the Quick Draft module, because you haven't written any posts that are set to Draft status. As time goes on, however, and you've written a few posts, you may save some of those posts as Drafts, to be edited and published at a later date. Those drafts show up in the Drafts sections of the Quick Draft module.

WordPress displays up to five drafts and displays the title of the post and the date it was last saved. Click the post title to go to the Edit Post page, where you can view, edit, and manage the draft post. Check out Chapter 5 for more information.

## WordPress Events and News

The WordPress Events and News module of the Dashboard pulls in posts from a site called WordPress Planet (https://planet.wordpress.org). In this module, you can stay in touch with several posts made by folks who are involved in WordPress development, design, and troubleshooting. You can find lots of interesting and useful tidbits if you keep this area intact. Quite often, I find great information about new plugins or themes, problem areas and support, troubleshooting, and new ideas, so I tend to stick with the default setting. You also see links to events like WordPress Meetups and WordCamps. If you click the pencil icon near the top of this module where it says Attend an Upcoming Event Near," a text box appears in which you can type the name of the city you live in, and this module will update with a listing of events near you.

# Arranging the Dashboard to Your Tastes

One feature of WordPress that I'm really quite fond of allows me to create my own workspace within the Dashboard. In the following sections, you find out how to customize your WordPress Dashboard to fit your needs, including modifying the layout, changing links and RSS feed information, and even rearranging the modules on different pages of the Dashboard. Armed with this information, you can open your Dashboard and create your very own workspace.

In the following steps, I show you how to move the At a Glance module so that it displays on the right side of your Dashboard screen:

**1.** **Hover your mouse over the title bar of the At a Glance module.**

Your mouse pointer changes to the Move pointer (a cross with arrows).

**2.** **Click and hold your mouse button, and drag the At a Glance module to the right side of the screen.**

As you drag the box, a light gray line with a dotted border appears on the right side of your screen. That gray line is a guide that shows you where you should drop the module (see Figure 4-7).

**3.** **Release the mouse button when you have the At a Glance module in place.**

The At a Glance module is now positioned on the right side of your Dashboard screen, at the top. The other modules on the right side of the Dashboard have shifted down, and the Activity module is the module in the top-left corner of the Dashboard screen.

**4.** **(Optional) Click the gray arrow to the right of the At a Glance title.**

The module collapses. Click the arrow again, and the module expands. You can keep that module opened or closed, based on your preference.

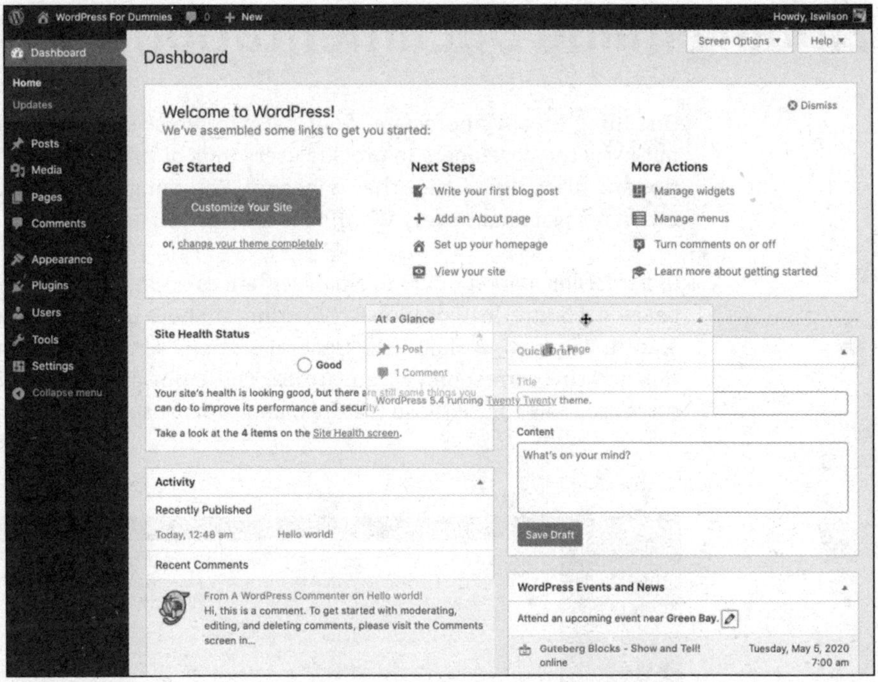

**FIGURE 4-7:**
A light gray line appears as a guide when you drag and drop modules in the WordPress Dashboard.

Repeat these steps with each module you see in the Dashboard so that the modules appear in the order you prefer.

**REMEMBER**

When you navigate away from the Dashboard, WordPress remembers the changes you've made. When you return, you still see your customized Dashboard and don't need to redo your changes.

If you find that your Dashboard contains a few modules that you never use, you can get rid of them by following these steps:

**1. Click the Screen Options button at the top of the Dashboard.**

The Screen Options drop-down menu opens, displaying the titles of the modules with a check box to the left of each title.

**2. Deselect the check box for the module you want to hide.**

The check mark disappears from the check box, and the module disappears from your Dashboard.

**TIP**

If you want a module that you hid to reappear, select that module's check box on the Screen Options drop-down menu.

# Finding Inline Documentation and Help

One thing I really appreciate about the WordPress software is the time and effort put in by the developers to provide users tons of inline documentation, as well as tips and hints right inside the Dashboard. You can generally find inline documentation for just about every WordPress feature you use.

*Inline documentation* refers to small sentences or phrases that you see alongside or below a feature in WordPress, providing a short but helpful explanation of the feature. Figure 4-8 shows the General Settings screen, where inline documentation and tips correspond with features. These tips can clue you into what the features are, how to use those features, and provide some recommended settings.

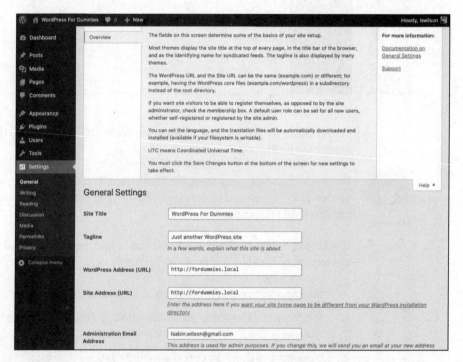

**FIGURE 4-8:** Inline documentation on the General Settings screen of the WordPress Dashboard.

In addition to the inline documentation that you find scattered throughout the Dashboard, a useful Help tab is located in the top-right corner of your Dashboard. Click this tab to open a panel containing help text that's relevant to the screen you're currently viewing in your Dashboard. If you're viewing the General Settings screen, for example, the Help tab displays documentation relevant to the General Settings screen. Likewise, if you're viewing the Add New Post screen, the

Help tab displays documentation with topics relevant to the settings and features you find on the Add New Post page of your Dashboard.

The inline documentation, topics, and text you find on the Help tab exist to assist and support you as you experience the WordPress platform, making it as easy to understand WordPress as possible. Another place you can find help and support is the WordPress Forums page at `https://wordpress.org/support`.

Throughout the pages of your WordPress Dashboard, you can apply the customization features that I cover for the main Dashboard page earlier in this chapter. Every section of the WordPress Dashboard is customizable with drag-and-drop modules, screen options, and inline help and documentation.

Have a look at Figure 4-9, which displays the Posts page of the WordPress Dashboard. (The Posts page is covered in greater detail in Chapter 5.) In the figure, the Screen Options menu shows your options for customization, including the following:

» Check boxes that you can select to display the Author, Categories, Tags, Comments, and Date of the posts listed on the Posts page

» A text field where you can input the number of posts you want to display on the Posts page

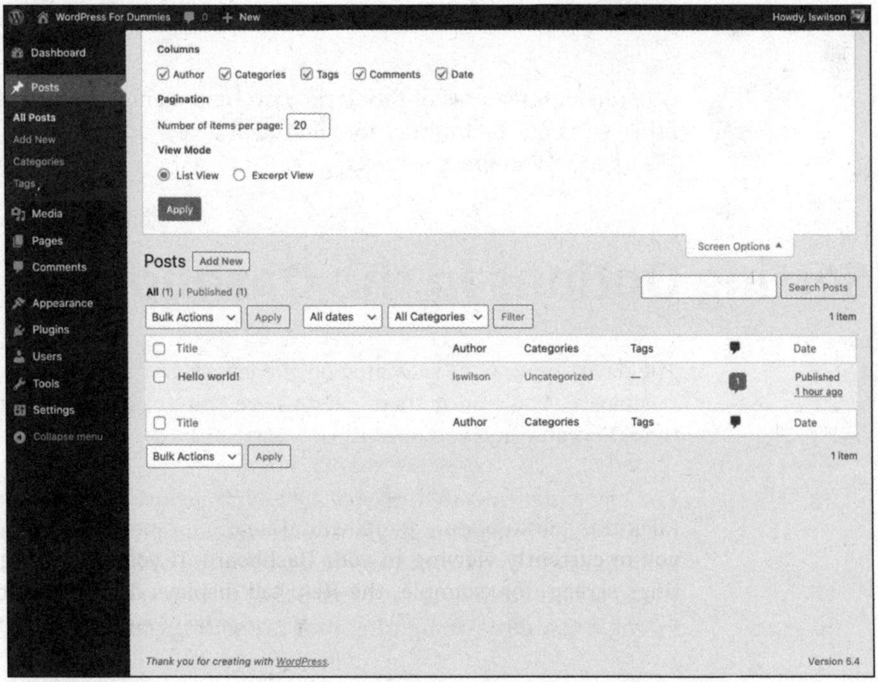

FIGURE 4-9: Screen Options on the Posts screen.

Figure 4-10 displays the Help topics on the Posts screen. Once you've clicked the Help tab at the top of the screen, the inline documentation for the page displays.

Other helpful features of the Dashboard Help menu are the links that lead you to other areas on the Internet for additional help, support topics, and resources on the various WordPress features.

# Setting Options in the Dashboard

The navigation menu is located on the left side of every screen of the WordPress Dashboard. You find it there everywhere you go; like a loyal friend, it's always there for you when you need it!

The navigation menu is divided into nine menus (not counting the Dashboard menu, which I mention previously). Hover your mouse over a menu, and another menu flies out to the right to reveal a submenu of items. The submenu items take you to areas within your Dashboard that allow you to perform tasks such as publishing a new post, configuring your site settings, and managing your comments.

The settings that allow you to personalize your site are the first ones that I cover in the next part of this chapter. Some of the menu items, such as creating and publishing new posts, are covered in detail in other chapters, but they're well worth a mention here as well so that you know what you're looking at. (Sections with additional information contain a cross-reference telling you where you can find more in-depth information on that topic in this book.)

# Configuring the Settings

At the bottom of the navigation menu, you find the Settings option. Hover over the Settings link. A menu appears, containing the following links, which I discuss in the sections that follow:

>> General

>> Writing

>> Reading

>> Discussion

>> Media

>> Permalinks

>> Privacy

## General

After you install the WordPress software and log in, you can put a personal stamp on your site by giving it a title and description, setting your contact email address, and identifying yourself as the author of the blog. You take care of these and other settings on the General Settings screen.

**REMEMBER**

Click the Save Changes button at the bottom of any page where you set new options. If you don't click Save Changes, your settings aren't saved, and Word-Press reverts to the preceding options. Each time you click the Save Changes button, WordPress reloads the current screen, displaying the new options that you just set.

To begin personalizing your site, follow these steps:

**1. Click the General link on the Settings menu.**

The General Settings screen appears (see Figure 4-11).

**2.** **Enter the name of your site in the Site Title text box.**

The title you enter here is the one that you've given your website to identify it as your own. I gave my new blog the title *WordPress For Dummies* (refer to Figure 4-11), which appears on my website as well as in the title bar of the viewer's web browser.

**TIP**

Give your website an interesting and identifiable name. You can use *Fried Green Tomatoes,* for example, if your website covers the book, or the movie, or even anything remotely related to the lovely Southern dish.

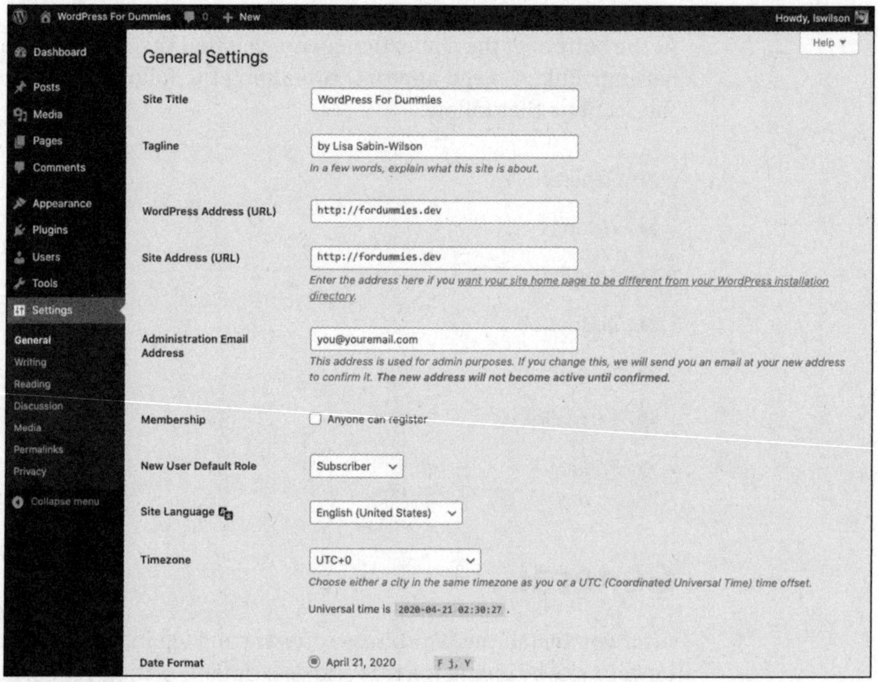

**FIGURE 4-11:**
Personalize the settings of your WordPress site on the General Settings screen.

**General Settings**

| | |
|---|---|
| Site Title | WordPress For Dummies |
| Tagline | by Lisa Sabin-Wilson |
| | *In a few words, explain what this site is about.* |
| WordPress Address (URL) | http://fordummies.dev |
| Site Address (URL) | http://fordummies.dev |
| | *Enter the address here if you want your site home page to be different from your WordPress installation directory.* |
| Administration Email Address | you@youremail.com |
| | *This address is used for admin purposes. If you change this, we will send you an email at your new address to confirm it. The new address will not become active until confirmed.* |
| Membership | ☐ Anyone can register |
| New User Default Role | Subscriber ▾ |
| Site Language | English (United States) ▾ |
| Timezone | UTC+0 ▾ |
| | *Choose either a city in the same timezone as you or a UTC (Coordinated Universal Time) time offset.* |
| | Universal time is 2020-04-21 02:30:27 . |
| Date Format | ⦿ April 21, 2020      F j, Y |

Dashboard • Posts • Media • Pages • Comments • Appearance • Plugins • Users • Tools • Settings • General • Writing • Reading • Discussion • Media • Permalinks • Privacy • Collapse menu

**3.** **In the Tagline text box, enter a five- to ten-word phrase that describes your blog.**

Figure 4-11 shows that my tagline is *by Lisa Sabin-Wilson*. So my website displays my site title followed by the tagline: *WordPress For Dummies by Lisa Sabin-Wilson.*

**REMEMBER**

The general Internet-surfing public can view your website title and tagline, which various search engines (such as Google, Yahoo!, and Bing) grab for indexing, so choose your words with this fact in mind.

4. **In the WordPress Address (URL) text box, enter the location where you installed the WordPress software.**

   Be sure to include the http:// portion of the URL and the entire path to your WordPress installation — for example, http://yourdomain.com. If you installed WordPress in a folder in your directory — inside a folder called wordpress, for example — you need to include it here. If I had installed WordPress in a folder called wordpress, the WordPress address would be http://yourdomain.com/wordpress.

5. **In the Site Address (URL) text box, enter the web address where people can find your blog by using their web browsers.**

   Typically, what you enter here is the same as your domain name (http://yourdomain.com). If you install WordPress in a subdirectory of your site, the WordPress installation URL is different from the blog URL. If you install WordPress at http://yourdomain.com/wordpress/ (WordPress URL), you need to tell WordPress that you want the blog to appear at http://yourdomain.com (the blog URL).

6. **Enter your email address in the Administration Email Address text box.**

   WordPress sends messages about the details of your website to this email address. When a new user registers for your site, for example, WordPress sends you an email alert.

7. **Select a Membership option.**

   Select the Anyone Can Register check box if you want to keep registration on your site open to anyone who wants to register. Leave the check box deselected if you'd rather not have open registration on your website.

8. **From the New User Default Role drop-down menu, choose the role that you want new users to have when they register for user accounts in your website.**

   You need to understand the differences among the user roles because each user role is assigned a different level of access to your website, as follows:

   - *Subscriber:* Subscriber is the default role. Assigning this role to new users is a good idea, particularly if you don't know who's registering. Subscribers are given access to the Dashboard screen, and they can view and change the options in their profiles on the Your Profile and Personal Options screen. (They don't have access to your account settings, however — only to their own.) Each user can change his username, email address, password, bio, and other descriptors in his user profile. Subscribers' profile information is stored in the WordPress database, and your site remembers them each time they visit so that they don't have to complete the profile information each time they leave comments on your website.

- *Contributor:* In addition to the access Subscribers have, Contributors can upload files and write, edit, and manage their own posts. Contributors can write posts, but they can't publish the posts; the administrator reviews all Contributor posts and decides whether to publish them. This setting is a nice way to moderate content written by new authors.

- *Author:* In addition to the access Contributors have, Authors can publish and edit their own posts.

- *Editor:* In addition to the access Authors have, Editors can moderate comments, manage categories, manage links, edit pages, and edit other Authors' posts.

- *Administrator:* Administrators can edit all the options and settings in the WordPress website. Simply put, Administrators have access to everything, so use caution when handing out Administrator access to your website.

**9.** **From the Site Language drop-down menu, choose your preferred language.**

The Site Language drop-down menu gives you several language options. The default setting is English, but at the time of this writing, the WordPress software is available in 112 languages. If your website should be in Spanish, use the drop-down menu to change it from English to Spanish.

**10.** **In the Timezone section, choose your UTC time from the drop-down menu.**

This setting refers to the number of hours that your local time differs from Coordinated Universal Time (UTC). This setting ensures that all your posts and comments left on your blog are time-stamped with the correct time. If you're lucky enough, as I am, to live on the frozen tundra of Wisconsin, which is in the Central time zone (CST), you would choose –6 from the drop-down menu because that time zone is six hours off UTC. WordPress also gives you the names of some major cities across the world to make the choice easier. Just select the name of the major city you live closest to; chances are that you're in the same time zone as that city.

TIP

If you're unsure what your UTC time is, you can find it at the Greenwich Mean Time website (https://greenwichmeantime.com). GMT is essentially the same thing as UTC.

**11.** **In the Date Format text box, enter the format in which you want the date to be displayed in your website.**

This setting determines the style of the date display. The default format is already selected and displayed for you: F j, Y (F = the full month name; j = the two-digit day; Y = the four-digit year), which gives you the date output. This default date format displays as September 16, 2020.

Select a different format by clicking the circle to the left of the option. You can also customize the date display by selecting the Custom option and entering your preferred format in the text box.

**TECHNICAL STUFF**

If you're feeling adventurous, you can find out how to customize the date format by clicking the Documentation on Date and Time Formatting link between the date and time options, which takes you to this page of the WordPress Codex: `https://wordpress.org/support/article/formatting-date-and-time`.

**12.** **In the Time Format text box (not shown in Figure 4-11), enter the format in which you want the time to be displayed in your site.**

This setting is the style of the time display. The default format is already inserted for you: g:i a (g = the two-digit hour; i = the two-digit minute; a = lowercase as a.m. or p.m.), which gives you the output `12:00 a.m.`

Select a different format by clicking the circle to the left of the option. You can also customize the date display by selecting the Custom option and entering your preferred format in the text box provided; find out how at `https://wordpress.org/support/article/formatting-date-and-time`.

**TIP**

You can format the time and date in several ways. Go to `https://www.php.net/manual/en/function.date.php` to find potential formats at the PHP website.

**13.** **From the Week Starts On drop-down menu (not shown in Figure 4-11), choose the day the week starts in your calendar.**

Displaying a calendar in the sidebar of your website is optional. If you choose to display a calendar, you can select the day of the week you want your calendar to start with.

# Writing

Choose Settings ⇨ Writing, and the Writing Settings screen opens (see Figure 4-12).

This screen of the Dashboard lets you set some basic options for writing your content. Table 4-1 gives you some information on choosing how your content looks and how WordPress handles some specific conditions.

After you set your options, be sure to click the Save Changes button; otherwise, the changes won't take effect.

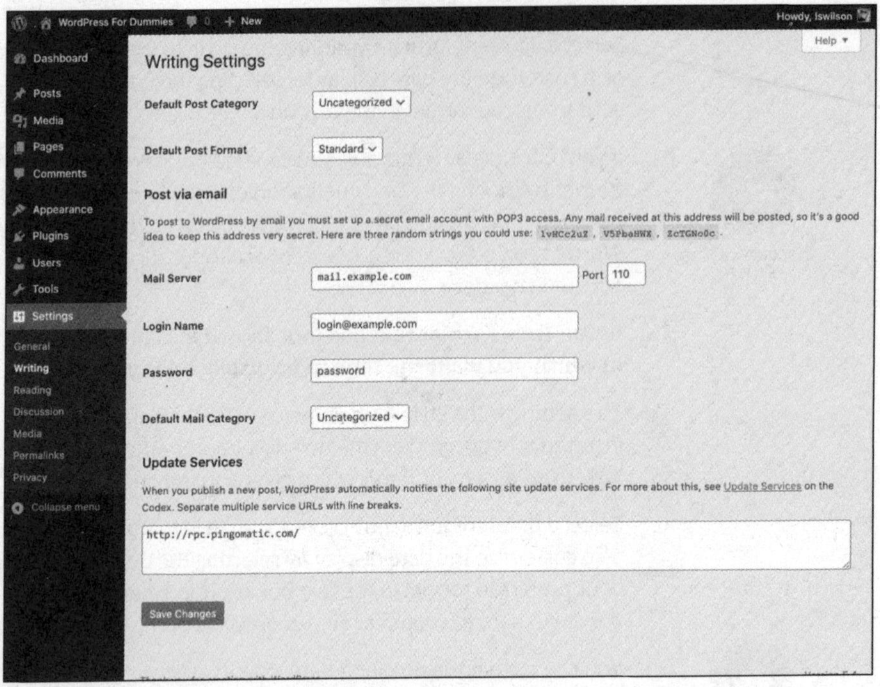

**FIGURE 4-12:**
The Writing Settings screen.

## TABLE 4-1     Writing Settings Options

| Option | Function | Default |
|---|---|---|
| Default Post Category | Lets you select the category that WordPress defaults to any time you forget to choose a category when you publish a post. | Uncategorized |
| Default Post Format | Select the format that WordPress defaults to any time you create a post and don't assign a post format. (This option is theme-specific; not all themes support post formats.) | Standard |
| Post via Email | Publish content from your email account by entering the mail server, port, login name, and password for the account you'll be using to send posts to your WordPress site. | N/A<br><br>You set the Mail Server, Login Name, Password according to your email settings, and then you select a default Mail Category. |
| Update Services<br><br>**Note:** This option is available only if you allow your site to be indexed by search engines (covered in the Reading Settings section). | Lets you indicate which ping service you want to use to notify the world that you've made updates or published new posts. The default, XML-RPC (`http://rpc.pingomatic.com`) updates all the popular services simultaneously. | `http://rpc.pingomatic.com` |

TIP

Go to `https://codex.wordpress.org/Update_Services` for comprehensive information on update services.

## Reading

The third item in the Settings drop-down menu is Reading. Choose Settings ⇨ Reading to open the Reading Settings screen (see Figure 4-13).

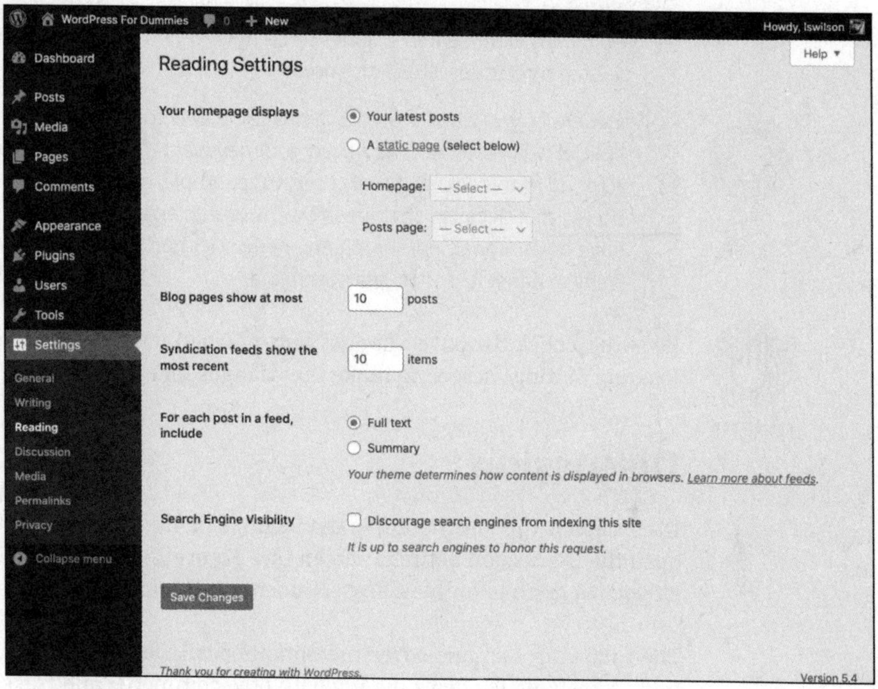

FIGURE 4-13:
The Reading
Settings screen.

You can set the following options in the Reading Settings screen:

>> **Your Homepage Displays:** Select the radio button to show a page instead of your latest posts on the front page of your site. You can find detailed information about using a static page for your front page in Chapter 12, including information on how to set it up by using the fields that appear after you select the radio button.

>> **Blog Pages Show at Most:** Type the maximum number of posts you want to display on each content page. The default is 10.

>> **Syndication Feeds Show the Most Recent:** In the Posts box, type the maximum number of posts you want to appear in your RSS feed at any time. The default is 10.

>> **For Each Post in a Feed, Include:** Select Full Text or Summary. Full Text publishes the entire post to your RSS feed, whereas Summary publishes an excerpt. By default, Full Text is selected. (Check out Chapter 5 for more information on WordPress RSS feeds.)

>> **Search Engine Visibility:** By default, the Discourage Search Engines from Indexing This Site option is deselected. If you're one of those rare bloggers who *don't* want search engines to be able to visit and index their websites in search directories, check the box.

TIP

Generally, you want search engines to be able to find your site. If you have special circumstances, however, you may want to enforce privacy on your site. A friend of mine has a family blog, for example, and she blocks search engine access to it because she doesn't want search engines to find it. When you have privacy enabled, search engines and other content bots can't find your website or list it in their search engines.

REMEMBER

Be sure to click the Save Changes button when you've set all your options in the Reading Settings screen to make the changes take effect.

# Discussion

Discussion is the fourth item in the Settings menu; choose Settings ⇨ Discussion to open the Discussion Settings screen (see Figure 4-14). The sections of this screen let you set options for handling comments and publishing posts to your site.

The following sections cover the options available to you in the Discussion Settings screen, which deals mainly with how comments and trackbacks are handled on your site.

## Default Post Settings

In the Default Post Settings section, you can tell WordPress how to handle post notifications. Here are your options:

>> **Attempt to Notify Any Blogs Linked To from the Post:** Select this check box, and your blog sends a notification (or *ping*) to any site you've linked to in your blog posts. This notification is commonly referred to as a *trackback*. (I discuss trackbacks in Chapter 2.) Deselect this check box if you don't want these notifications to be sent.

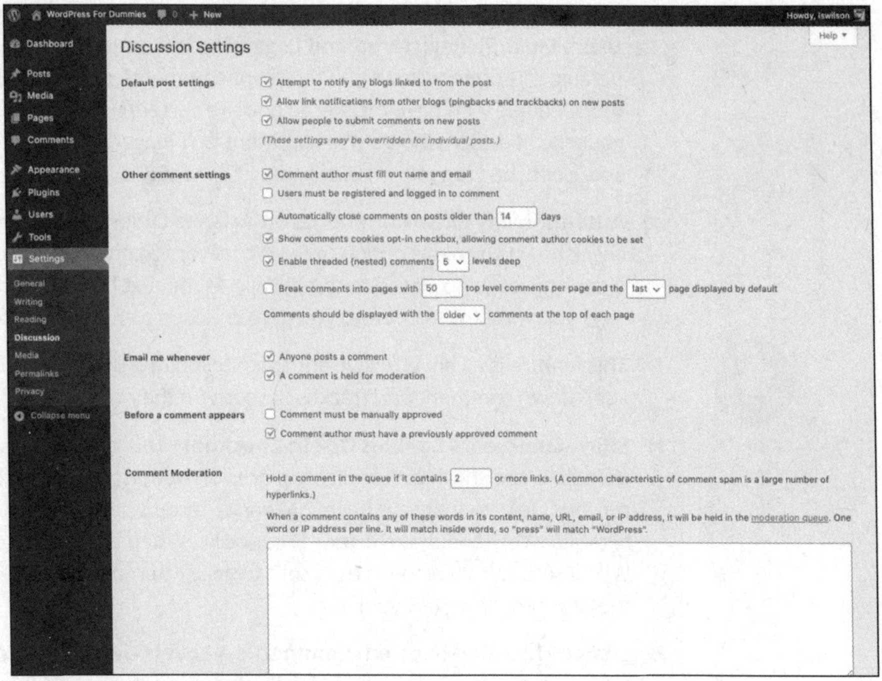

**FIGURE 4-14:**
The Discussion
Settings screen.

>> **Allow Link Notifications from Other Blogs (Pingbacks and Trackbacks) on New Articles:** By default, this check box is selected, and your site accepts notifications via pings or trackbacks from other sites that have linked to yours. Any trackbacks or pings sent to your site appear on your site in the Comments section of the post. If you deselect this check box, your site doesn't accept pingbacks or trackbacks from other sites.

>> **Allow People to Post Comments on New Posts:** By default, this check box is selected, and people can leave comments on your posts. If you deselect this check box, no one can leave comments on your content. (You can override these settings for individual articles; find more information in Chapter 5.)

## Other Comment Settings

The Other Comment Settings tell WordPress how to handle comments:

>> **Comment Author Must Fill Out Name and Email:** Enabled by default, this option requires all commenters on your blog to fill in the Name and Email fields when leaving a comment. This option is very helpful in combating comment spam. (See Chapters 2 and 7 for information on comment spam.) Deselect this check box to disable this option.

>> **Users Must Be Registered and Logged In to Comment:** Not enabled by default, this option allows you to accept comments on your site from only those people who have registered and are currently logged in as users on your site. If a user who tries to comment isn't logged in, she sees the message You must be logged in in order to leave a comment.

>> **Automatically Close Comments on Articles Older Than *X* Days:** Select the check box next to this option to tell WordPress that you want comments on older articles to be closed automatically. In the text box, enter the number of days you want to wait before WordPress closes comments on older articles.

This feature is a very effective antispam technique that many people use to keep down comment and trackback spam on their sites.

>> **Show Comments Cookies Opt-In Checkbox:** This check box is unselected by default. If you check it, WordPress places an option below the comments form on your website that says Save my name, email, and website in this browser for the next time I comment. When the user selects this box, WordPress sets a cookie in the user's browser to remember their information the next time they visit your site.

>> **Enable Threaded (Nested) Comments *X* Levels Deep:** The drop-down menu allows you to choose the level of threaded comments you'd like to have on your blog. The default is 5; you can choose up to 10 levels. Instead of displaying all comments on your site in chronological order (as they are by default), nesting them allows you and your readers to reply to a comment within the comment itself.

>> **Break Comments into Pages with *X* Top Level Comments Per Page and the Last/First Page Displayed by Default:** Fill in the text box with the number of comments you want to appear on one page (default: 50). This setting can really help sites that receive a large number of comments. It allows you to break a long string of comments into several pages, which makes the comments easier to read and helps speed the load time of your site because the page isn't loading such a large number of comments at the same time. If you want the last (most recent) or first page of comments to display, choose Last or First from the drop-down menu.

>> **Comments Should Be Displayed with the Older/Newer Comments at the Top of Each Page:** From the drop-down menu, choose Older or Newer. Older displays the comments on your site from oldest to newest. Newer does the opposite: It displays the comments on your site from newest to oldest.

## Email Me Whenever

The two options in the Email Me Whenever section are enabled by default:

» **Anyone Posts a Comment:** This option lets you receive an email notification whenever anyone leaves a comment on your site. Deselect the check box if you don't want to be notified by email about every new comment.

» **A Comment Is Held for Moderation:** This option lets you receive an email notification whenever a comment is awaiting your approval in the comment moderation queue. (See Chapter 5 for more information about the comment moderation queue.) Deselect this check box if you don't want this notification.

## Before a Comment Appears

The two options in the Before a Comment Appears section tell WordPress how you want WordPress to handle comments before they appear in your site:

» **Comment Must Be Manually Approved:** Disabled by default, this option keeps every single comment left on your blog in the moderation queue until you, the administrator, log in and approve it. Select this check box to enable this option.

» **Comment Author Must Have a Previously Approved Comment:** Enabled by default, this option requires comments posted by all first-time commenters to be sent to the comment moderation queue for approval by the administrator of the site. After comment authors have been approved for the first time, they remain approved for every comment thereafter, and this setting can't be changed. WordPress stores each comment author's email address in the database, and any future comments that match any stored emails are approved automatically. This feature is another measure built into WordPress to combat comment spam.

## Comment Moderation

In the Comment Moderation section (not shown in Figure 4-14), you can set options to specify what types of comments are held in the moderation queue to await your approval.

To prevent spammers from spamming your blog with a ton of links, select the check box titled Hold a Comment in the Queue If It Contains X or More Links. The default number of links allowed is 2. Give that setting a try, and if you find that you're getting lots of spam comments with multiple links, you may want to revisit this page and increase that number. Any comment with a higher number of links goes to the comment moderation area for approval.

The large text box in the Comment Moderation section lets you type keywords, URLs, email addresses, and IP addresses in comments that you want to be held in the moderation queue for your approval.

## Comment Blocklist

In this section (not shown in Figure 4-14), type a list of words, URLs, email addresses, and/or IP addresses that you want to flat-out ban from your site. Items placed here don't even make it into your comment moderation queue; the Word-Press system filters them as spam. Let me just say that the words I've placed in my blocklist aren't family-friendly and have no place in a nice book like this.

## Avatars

The final section of the Discussion Settings screen is Avatars (see Figure 4-15).

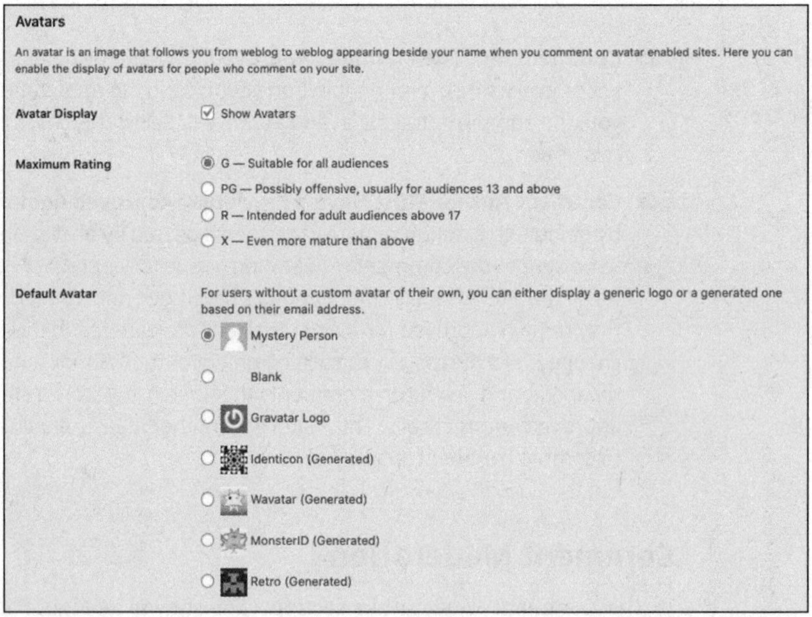

**Avatars**

An avatar is an image that follows you from weblog to weblog appearing beside your name when you comment on avatar enabled sites. Here you can enable the display of avatars for people who comment on your site.

Avatar Display ☑ Show Avatars

Maximum Rating
- ◉ G — Suitable for all audiences
- ○ PG — Possibly offensive, usually for audiences 13 and above
- ○ R — Intended for adult audiences above 17
- ○ X — Even more mature than above

Default Avatar
For users without a custom avatar of their own, you can either display a generic logo or a generated one based on their email address.
- ◉ Mystery Person
- ○ Blank
- ○ Gravatar Logo
- ○ Identicon (Generated)
- ○ Wavatar (Generated)
- ○ MonsterID (Generated)
- ○ Retro (Generated)

**FIGURE 4-15:** Default avatars you can display in your blog.

In this section, you can select settings for the use and display of avatars on your site, as follows:

1. **In the Avatar Display section, select the Show Avatars option if you want your site to display avatars beside comment authors' names.**

2. **In the Maximum Rating section, set the rating for the avatars that do display on your site.**

   This feature works similarly to the movie rating system you're used to. You can select G, PG, R, and X ratings for the avatars that appear on your site. If your site is family-friendly, you probably don't want it to display R- or X-rated avatars.

3. **Choose a default avatar in the Default Avatar section (refer to Figure 4-15):**

   - Mystery Person

   - Blank

   - Gravatar Logo

   - Identicon (Generated)

   - Wavatar (Generated)

   - MonsterID (Generated)

   - Retro (Generated)

4. **Click the Save Changes button.**

Avatars appear in a couple of places:

>> **The Comments screen of the Dashboard:** In Figure 4-16, the comment displays the commenter's avatar next to it.

>> **The comments on individual posts to your site:** Figure 4-17 shows a comment on my personal blog.

To enable the display of avatars in comments on your site, the Comments Template (comments.php) in your active theme has to contain the code to display them. Hop on over to Chapter 9 to find out how to do that.

**REMEMBER**

Click the Save Changes button after you've set all your options on the Discussion Settings screen to put the changes into effect.

See the nearby sidebar "Avatars and gravatars: How do they relate to WordPress?" for more information about avatars.

**FIGURE 4-16:**
Authors' avatars appear in the Comments screen of the WordPress Dashboard.

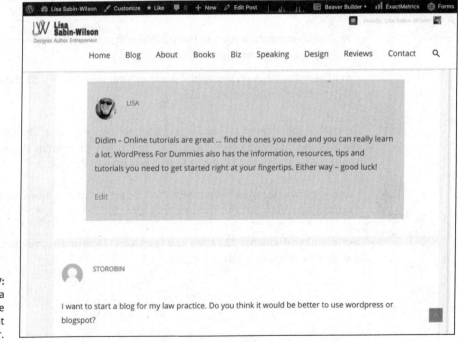

**FIGURE 4-17:**
Comments on a post, showing the comment author's avatar.

## AVATARS AND GRAVATARS: HOW DO THEY RELATE TO WordPress?

An *avatar* is an online graphical representation of a person. It's a small icon that people use to visually represent themselves on the web in areas where they participate in conversations, such as discussion forums and comments. *Gravatars* are globally recognized avatars; they're avatars that you can take with you wherever you go. A gravatar appears alongside comments, posts, and discussion forums as long as the site you are interacting with is gravatar-enabled.

In October 2007, Automattic, the core group behind the WordPress platform, purchased the Gravatar service and integrated it into WordPress so that everyone could enjoy and benefit from the service.

Gravatars aren't automatic; you need to sign up for an account with Gravatar before you can assign a gravatar to yourself, via your email address. You can find out more by visiting https://gravatar.com.

## Media

The next item on the Settings menu is Media. Choose Settings ⇨ Media to open the Media Settings screen (see Figure 4-18).

On the Media Settings screen, you can configure how your image files (graphics and photos) are resized for use on your site. The dimensions are referenced in pixels, first by width and then by height. (The setting 150 x 150, for example, means 150 pixels wide by 150 pixels high.)

The first set of options on the Media Settings page deals with images. WordPress automatically resizes your images for you in three sizes:

>> **Thumbnail Size:** The default is 150 x 150; enter the width and height of your choice. Select the Crop Thumbnail to Exact Dimensions check box to resize the thumbnail to the width and height you specified. Deselect this check box, and WordPress resizes the image proportionally.

>> **Medium Size:** The default is 300 x 300; enter the width and height of your choice.

>> **Large Size:** The default is 1024 x 1024; enter the width and height of your choice.

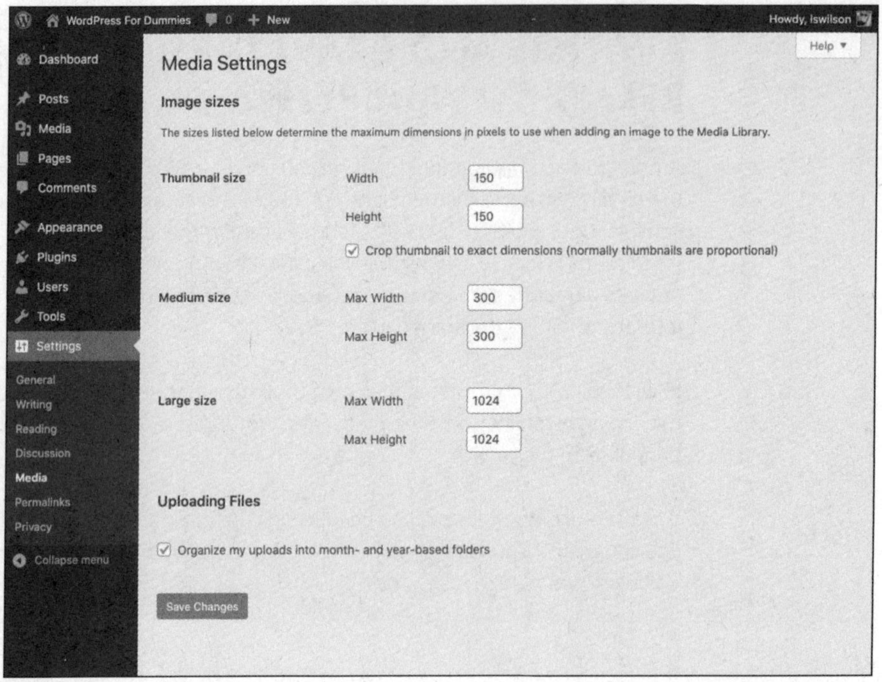

FIGURE 4-18:
The Media
Settings screen.

The last option on the Media Settings page is in the Uploading Files section. By default, the check box titled Organize My Uploads into Month- and Year-Based Folders is selected, directing WordPress to organize your uploaded files in folders by month and by year. Files that you upload in January 2021, for example, would be in the following folder: /wp-content/uploads/2021/01/. Deselect this check box if you don't want WordPress to organize your files by month and year.

Chapter 9 goes into detail about WordPress themes and templates, including how you can add image sizes other than these three. You can use these additional image sizes on your website, and you can also use a feature called Featured Image to create image thumbnails that get displayed in your posts, archive pages, and search result pages.

Be sure to click the Save Changes button to save your configurations!

## Permalinks

The next item on the Settings menu is Permalinks. Choose Settings ⇨ Permalinks to view the Permalink Settings screen (see Figure 4-19).

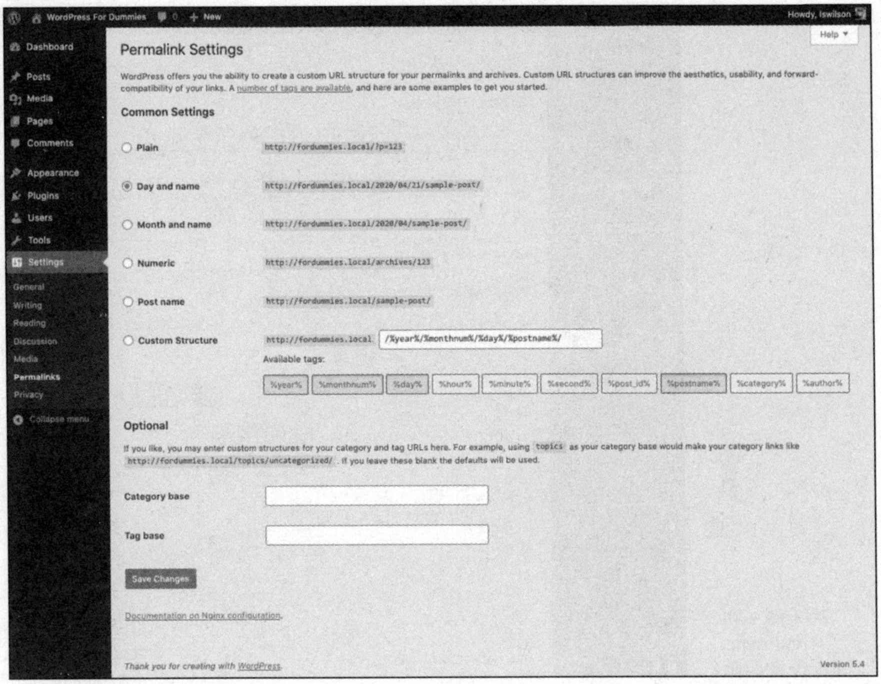

**FIGURE 4-19:**
The Permalink
Settings screen.

Each post you create on your site has a unique URL called a *permalink*, which is a permanent URL for all your website posts, pages, and archives. I cover permalinks extensively in Chapter 5 by explaining what they are, how you can use them, and how you set the options in this page.

# Privacy

The next link on the Settings menu is Privacy. Choose Settings ⟹ Privacy to view the Privacy Settings screen, shown in Figure 4-20.

On May 25, 2018, the European Union (EU) enacted a law called General Data Protection Regulation (GDPR for short). The GDPR is a set of rules designed to give European citizens more control of their personal data that's stored on the websites they browse on the Internet. As an owner of a website, you may be required to follow these laws to protect the privacy of your visitors' and users' data on your website. Part of this law requires you to publish a privacy page on your website that lays out your privacy policy, which helps your website remain in compliance with GDPR rules.

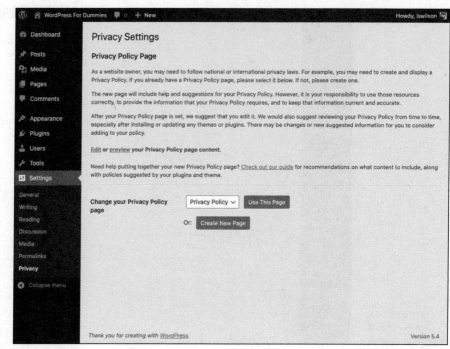

FIGURE 4-20:
The Privacy
Settings screen.

To help you, every new WordPress installation comes with a Privacy Policy page in draft form. The Privacy Settings screen displays this draft page in the Change Your Privacy Policy Page drop-down menu. Click the Use This Page button to use the default page that WordPress provided, or click the Create New Page button to create your own.

This is where I tell you that I'm not a lawyer and am not equipped to give you any legal advice on what your privacy policy page should contain to make your website compliant with the GDPR, but WordPress offers a handy guide. Click the Check Out Our Guide link on the Privacy Settings screen to view the Privacy Policy Guide screen in your Dashboard (see Figure 4-21). In addition to tips, the screen offers a template for a good privacy policy page.

TIP

Even though the GDPR was written to pertain to citizens of the EU, it has international implications, particularly if visitors to your website are located in the EU. Because you can't know whether an EU citizen is going to visit your site, it's a good rule of thumb for your website to have a basic privacy policy.

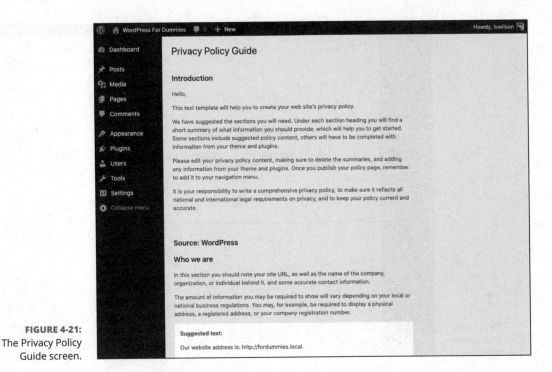

**FIGURE 4-21:**
The Privacy Policy
Guide screen.

# Creating Your Personal Profile

To personalize your blog, visit the Profile screen of your WordPress Dashboard.

To access the Profile screen, hover over the Users link on the Dashboard navigation menu, and click the Your Profile link. The Profile screen appears, as shown in Figure 4-22.

Here are the settings on this page:

>> **Personal Options:** The Personal Options section is where you can set four preferences for your site:

- *Visual Editor:* This selection enables you to use the Visual Editor when writing your posts. The Visual Editor gives you the formatting options you find in the Write Post screen (discussed in detail in Chapter 5). By default, the Visual Editor is on. To turn it off, select the Disable the Visual Editor When Writing check box.

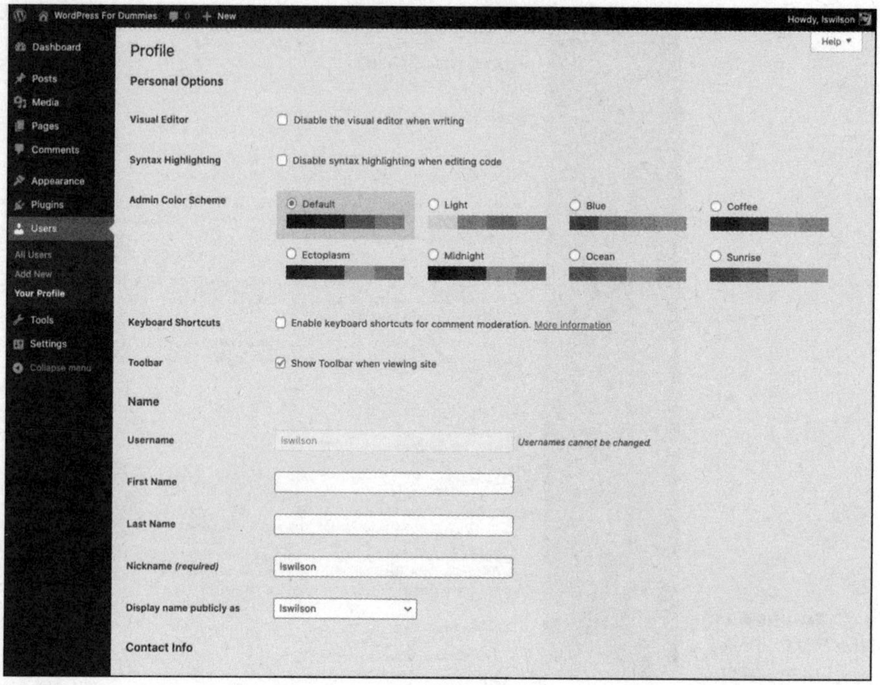

**FIGURE 4-22:**
Set your profile
details here.

- *Syntax Highlighting*: Select this check box to disable syntax highlighting. This option disables the feature in the Dashboard theme and plugin code editors that displays code in different colors and fonts, according to the type of code being used (HTML, PHP, CSS, and so on).

- *Admin Color Scheme:* These options set the colors for your Dashboard. The Default color scheme is automatically selected for you in a new installation, but you have other color options, including Light, Blue, Coffee, Ectoplasm, Midnight, Ocean, and Sunrise.

- *Keyboard Shortcuts:* This setting enables you to use keyboard shortcuts for comment moderation. To find out more about keyboard shortcuts, click the More Information link; you're taken to the Keyboard Shortcuts page (https://wordpress.org/support/article/keyboard-shortcuts) of the WordPress Codex.

- *Toolbar:* This setting allows you to control the location of the admin toolbar on your site. By default, the admin toolbar displays at the top of every page of your site when you're viewing it in your browser. It's important to understand that the admin toolbar appears only to users who are logged in. Regular visitors who aren't logged in to your site can't see the admin toolbar.

>> **Name:** Input personal information, such as your first name, last name, and nickname, and specify how you want your name to appear publicly. Fill in the text boxes with the requested information.

The rest of the options aren't shown in Figure 4-22; you have to scroll down to see them.

>> **Contact Info:** In this section, provide your email address and other contact information to tell your visitors who you are and where they can contact you. Your email address is the only required entry in this section. This address is the one WordPress uses to notify you when you have new comments or new user registrations on your blog. Make sure to use a real email address so that you get these notifications. You can also insert your website URL into the website text field.

>> **About Yourself:** This section is where you can provide a little bio and change the password for your website.

**WARNING**

When your profile is published to your website, not only can anyone view it, but it also gets picked up by search engines such as Bing and Google. Always be careful with the information in your profile. Think hard about the information you want to share with the rest of the world!

- *Biographical Info:* Type a short bio in the Biographical Info text box. This information can be shown publicly if you're using a theme that displays your bio, so be creative!

- *Profile Picture:* Display the current photo that you've set in your Gravatar account. You can set up a profile picture or change your existing one within your Gravatar account at https://gravatar.com.

- *Account Management:* Manage your password and user sessions.

- *New Password:* When you want to change the password for your website, type your new password in the first text box in the New Password section. To confirm your new password, type it again in the second text box.

Directly below the two text boxes is a little password helper where WordPress helps you create a secure password. WordPress alerts you if the password you've chosen is too short or not secure enough by calling it Very Weak, Weak, or Medium. When creating a new password, use a combination of letters, numbers, and symbols to make it hard for anyone to guess (such as *b@Fmn2quDtnSLQb1hm1%jexA*). When you create a password that WordPress thinks is a good one, it lets you know by calling it Strong.

**TIP**

**WARNING**

Change your password frequently. I can't recommend this practice strongly enough. Some people on the Internet make it their business to attempt to hijack blogs for their own malicious purposes. If you change your password monthly, you lower your risk by keeping hackers guessing.

- *Sessions:* If you're logged into your site on several devices, you can log yourself out of those locations by clicking the Log Out Everywhere Else button. This option keeps you logged in at your current location but logs you out of any other location where you may be logged in. If you're not logged in anywhere else, the button is inactive, and a message says `You are only logged in at this location`.

When you finish setting all the options on the Profile screen, click the Update Profile button to save your changes.

# Setting Your Site's Format

In addition to setting your personal settings in the Dashboard, you can manage the day-to-day maintenance of your site. This next section takes you through the links to these sections on the navigation menu, directly below the Dashboard link.

## Posts

Hover your mouse over the Posts menu, and a submenu appears with four links: All Posts, Add New, Categories, and Tags. Each link gives you the tools you need to publish content to your site:

>> **All Posts:** This link opens the Posts screen, where you see a list of all the saved posts you've written on your site. On this screen, you can search for posts by date, category, or keyword. You can view all posts, only posts that have been published, or just posts that have been saved but not yet published (drafts). You can also edit and delete posts from this page. Check out Chapter 5 for more information on editing posts on your site.

>> **Add New:** This link opens the Add New Post screen, where you compose your posts, set the options for posts (such as assigning a post to a category, or making it private or public), and publish the post to your site. You can find more information on posts, post options, and publishing in Chapter 5.

**TIP**

You can also get to the Add New Post screen by clicking the Add New button on the Posts screen or by clicking the +New link on the admin toolbar and selecting Post.

>> **Categories:** This link opens the Categories screen, where you can view, edit, add, and delete categories on your site. Find more information on categories in Chapter 5.

>> **Tags:** This link opens the Tags screen, where you can view, add, edit, and delete tags on your site. Chapter 5 provides more information about what tags are and why you might use them on your site.

## Media

Hover your mouse over the Media link on the navigation menu to reveal a submenu of two links:

>> **Library:** This link opens the Media Library screen. On this page, you view, search, and manage all the media files you've ever uploaded to your WordPress site.

>> **Add New:** This link opens the Upload New Media screen, where you can use the built-in uploader to transfer media files from your computer to the media directory in WordPress. Chapter 6 takes you through the details of uploading images, videos, audio, and other media files (such as Microsoft Word or PowerPoint documents) by using the WordPress upload feature.

**TIP**

You can also get to the Upload New Media screen by clicking the Add New button on the Media Library screen or by clicking the +New link on the admin toolbar and selecting Media.

## Pages

People use this feature to create pages on their sites such as About Me or Contact Me. Click the Pages menu to reveal these submenu links:

>> **All Pages:** This link opens the Pages screen, where you can search for, view, edit, and delete pages in your WordPress site.

>> **Add New:** This link opens the Add New Page screen, where you can compose, save, and publish a new page on your site. Table 4-2 describes the differences between a page and a post. The differences are subtle, but the two items are very different.

You can also get to the Add New Page screen by clicking the Add New button on the Pages screen or by clicking the +New link on the admin toolbar and selecting Page.

**TABLE 4-2**

## Differences Between Pages and Posts

| WordPress Options | Page | Post |
| --- | --- | --- |
| Appears in blog post listings | No | Yes |
| Appears as a static page | Yes | No |
| Appears in category archives | No | Yes |
| Appears in monthly archives | No | Yes |
| Appears in Recent Posts listings | No | Yes |
| Appears in site RSS feed | No | Yes |
| Appears in search results | Yes | Yes |

# Comments

Comments in the navigation menu don't have a submenu list of links. You simply click Comments to open the Comments screen, where WordPress gives you these options:

>> **All:** Shows all comments that currently exist on your site, including approved, pending, and spam comments

>> **Mine:** Shows all the comments you have created on your own website

>> **Pending:** Shows comments that you haven't yet approved and are pending in the moderation queue

>> **Approved:** Shows all comments that you've approved

>> **Spam:** Shows all the comments that are marked as spam

>> **Trash:** Shows comments that you've marked as Trash but not yet deleted from your blog

You can find information in Chapter 2 about the purpose of comments. In Chapter 5, I give you details on using the Comments section of your WordPress Dashboard.

# Appearance

When you hover your mouse over the Appearance link on the Dashboard navigation menu, a submenu appears, displaying the following links:

- » **Themes:** This link opens the Themes screen, where you can manage the themes available on your website. Check out Chapter 9 to find out how to use and manage themes on your WordPress blog.

- » **Customize:** Some themes have a Customize page, where you can configure different settings for the theme, such as the default Twenty Twenty theme. The Customize link appears on the Appearance menu only if the theme you're currently using has options available for configuration; if it doesn't, you won't see the Customize link here.

- » **Widgets:** The Widgets page allows you to add, delete, edit, and manage the widgets you use on your site.

- » **Menus:** This link opens the Menus screen, where you can build navigation menus to display on your site. Chapter 10 provides information on creating menus.

- » **Background:** This link opens the Background Image screen in the Customizer, where you can upload an image to use as the background of your WordPress website design. As with the Custom Header option, the Custom Background option exists in the Appearance menu only if you have the default Twenty Twenty theme (or any other theme that supports the Custom Background feature) activated. Not all WordPress themes use the Custom Background feature.

- » **Theme Editor:** This link opens the Edit Themes screen, where you can edit your theme templates. Chapters 9 through 11 have extensive information on themes and templates.

TIP

Customizing the background of your site helps you individualize the visual design of your website. You can find more information on tweaking and customizing your WordPress theme in Chapter 8. Chapter 9 gives you a great deal of information about using WordPress themes (including where to find, install, and activate them on your WordPress site), as well as detailed information on using WordPress widgets to display the content you want.

Part 4 provides information about WordPress themes and templates. You can dig deep into WordPress template tags and tweak an existing WordPress theme by using Cascading Style Sheets (CSS) to customize your theme a bit more to your liking.

# Plugins

The next item on the navigation menu is Plugins. Hover your mouse over the Plugins link to view the submenu links:

>> **Installed Plugins:** This link opens the Plugins screen, where you can view all the plugins installed on your website. On this page, you can activate, deactivate, and delete plugins.

>> **Add New:** This link opens the Add Plugins screen, where you can search for plugins in the official WordPress Plugins page (https://wordpress.org/plugins) by keyword, author, or tag. You can also install plugins directly to your site from the Plugins page.

>> **Editor:** Opens the Edit Plugins screen, where you can edit the plugin files in a text editor. Don't plan to edit plugin files unless you know what you're doing (meaning that you're familiar with PHP and WordPress functions).

**WARNING**

I strongly advise against editing plugin files unless you know exactly what you're doing — that is, you're familiar with PHP and WordPress functions.

See Chapter 7 for more on plugins.

# Users

The Users submenu has three links:

>> **All Users:** Click this link to go to the Users screen, where you can view, edit, and delete users on your WordPress site. Each user has a unique login name and password, as well as an email address assigned to her account. You can view and edit a user's information on the Users page.

>> **Add New:** This link opens the Add New User screen, where you can add new users to your WordPress site. Simply type the user's username, first name, last name, email (required), website, and a password in the fields provided and then click the Add User button. You can also specify whether you want WordPress to send login information to the new user by email. If you like, you can also assign a new role for the new user. Refer to the "General" section earlier in this chapter for more info about user roles.

>> **Your Profile:** Refer to the "Creating Your Personal Profile" section earlier in this chapter for more information about creating a profile.

# Tools

The last menu item on the navigation menu (and in this chapter!) is Tools. Hover your mouse over the Tools link to view the submenu list of links, which includes

>> **Available Tools:** This link opens the Tools screen of your Dashboard. WordPress comes packaged with two extra features that you can use on your site, if needed: Press This and Category/Tag Converter.

>> **Import:** Clicking this link opens the Import screen of your Dashboard. WordPress allows you to import content from a different publishing platform. I cover this feature in depth in Chapter 14.

>> **Export:** Clicking this link opens the Export screen of your Dashboard. WordPress allows you to export your content from WordPress so that you can import it to a different platform or another WordPress-powered site.

>> **Export Personal Data:** Opens the Export Personal Data screen. This screen gives you the opportunity to input the username or email address of a registered user on your site and obtain an authorization from the user to verify and approve the request to download personal data. When verification is complete, the user receives an email with a link to download their personal data in .zip format.

>> **Erase Personal Data:** Opens the Erase Personal Data screen in the Dashboard. This screen gives you the ability to input the user name or email address of a registered user to obtain verification from that user to erase his personal data from the site. After verification from the user has been completed, the user receives an email confirming that their personal data has been erased from the site.

**IN THIS CHAPTER**

» **Setting up categories**

» **Exploring permalinks**

» **Discovering RSS options**

» **Writing and editing posts and pages**

» **Using the block editor**

» **Managing and moderating comments**

Chapter **5**

# Establishing Your Publishing Routine

WordPress is a powerful publishing tool, especially when you use the full range of options available. With the basic settings configured (which I show you how to do in Chapter 4), now is the time to go forth and publish! You can skip to the "Writing Your First Entry" section in this chapter and jump right into creating new posts for your website. Or you can stay right here and discover some of the options you can set to make your website a bit more organized and logical from the get-go.

**TIP**

Content on your website can become unwieldy and disorganized, requiring you to revisit these next few features sometime in the near future so that you can get the beast under control. Why not do a little planning and get the work over with now? I promise that it won't take that long, and you'll thank me for it later.

## Staying on Topic with Categories

In WordPress, a *category* is what you determine to be the main topic of an individual piece of content on your site. Through the use of categories, you can file your posts into topics by subject. To improve your readers' experiences in navigating your site, WordPress organizes posts by the categories you assign to them. Visitors can click the categories they're interested in to see the posts you've written on those particular topics.

You should know ahead of time that the list of categories you set up is displayed on your site in a few places, including the following:

>> **Body of the post:** In most WordPress themes, you see the title followed by a statement such as `Filed In:` *Category 1, Category 2*. The reader can click the category name to go to a page that lists all the posts you've made in that particular category. You can assign a single post to more than one category.

>> **Navigation menu:** Almost all sites have a navigation menu that visitors can use to navigate your site. You can place links to categories on the navigation menu, particularly if you want to draw attention to particular categories.

>> **Sidebar of your theme:** You can place a full list of category titles in the sidebar. A reader can click any category and arrive at a page on your site that lists the posts you've made within that particular category.

*Subcategories* (also known as *category children*) can further refine the main category topic by listing specific topics related to the main (parent) category. In your WordPress Dashboard, on the Manage Categories page, subcategories are listed directly below the main category. Here's an example:

Books I Enjoy (main category)

Fiction (subcategory)

Nonfiction (subcategory)

Trashy Romance (subcategory)

Biographies (subcategory)

*For Dummies* (subcategory)

You can create as many levels of categories as you like. Biographies and *For Dummies* could be subcategories of Nonfiction, for example, which is a subcategory of the Books I Enjoy category. You aren't limited to the number of category levels you can create.

## Changing the name of a category

Upon installation, WordPress gives you one default category to get you started: Uncategorized. (See the Categories screen shown in Figure 5-1.) That category name is pretty generic, so you'll definitely want to change it to one that's more specific to you. (On my site, I changed it to Life in General. Although that name's still a bit on the generic side, it doesn't sound quite so . . . well, uncategorized.)

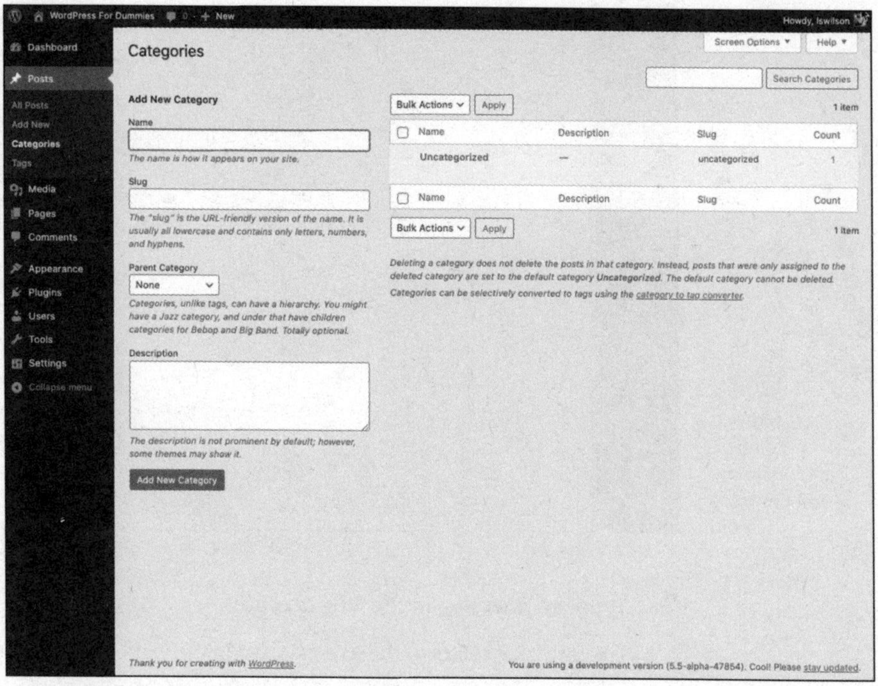

FIGURE 5-1:
The Categories
screen of a
brand-new site
shows the default
Uncategorized
category.

The default category also serves as a kind of fail-safe. If you publish a post to your site and don't assign that post to a category, the post is automatically assigned to the default category, no matter what you name the category.

**REMEMBER**

So how do you change the name of that default category? When you're logged in to your WordPress Dashboard, just follow these steps:

1. **Click the Categories link on the Posts submenu of the Dashboard navigation menu.**

   The Categories screen opens, containing all the tools you need to set up and edit category titles for your site.

2. **Click the title of the category you want to edit.**

   If you want to change the Uncategorized category, click the word *Uncategorized* to open the Edit Category screen (see Figure 5-2).

3. **Type the new name for the category in the Name text box.**

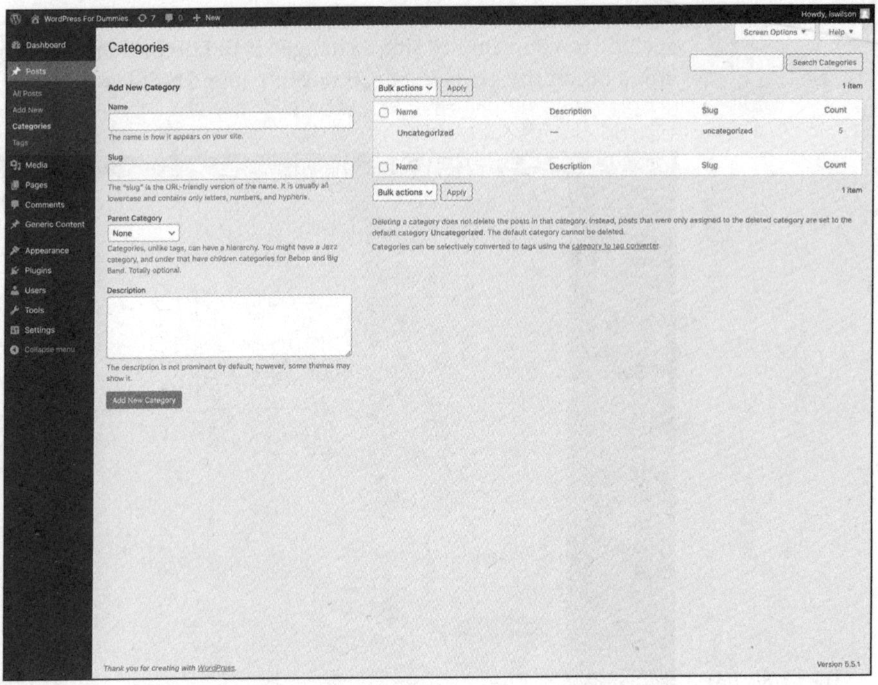

**FIGURE 5-2:**
Editing a category in the Edit Category screen.

4. **Type the new slug in the Slug text box.**

   The term *slug* refers to the word(s) used in the web address for the specific category. The category Books, for example, has a web address of http://yourdomain.com/category/books. If you change the slug to *Books I Like*, the web address is http://yourdomain.com/category/books-i-like. (WordPress automatically inserts hyphens between the slug words in the web address.)

5. **Choose a parent category from the Parent Category drop-down menu.**

   If you want this category to be a main category, not a subcategory, choose None.

6. **(Optional) Type a description of the category in the Description text box.**

   Use this description to remind yourself what your category is about. Some WordPress themes display the category description right on your site, too, which can be helpful for your visitors. (See Chapter 9 for more about themes.) You'll know that your theme is coded in this way if your site displays the category description on the category page(s).

7. **Click the Update button.**

   The information you just edited is saved, and the Categories screen reloads, showing your new category name.

**TIP**

If you want to edit a category's name only, you can click the Quick Edit link below the name on the Category screen, which you see when you hover your mouse over the Category name. Then you can do a quick name edit without having to load the Edit Category screen.

# Creating new categories and deleting others

Today, tomorrow, next month, next year — as your website grows in size and age, you'll continue adding new categories to further define and archive the history of your posts. You aren't limited in the number of categories and subcategories you can create.

Creating a new category is as easy as following these steps:

1. **Click the Categories link on the Posts submenu of the Dashboard navigation menu.**

   The Categories screen opens, displaying the Add New Category section on the left side (see Figure 5-3).

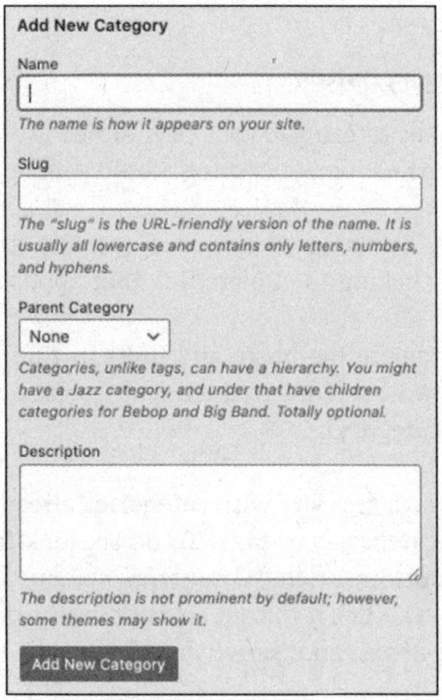

   **Add New Category**

   Name

   _The name is how it appears on your site._

   Slug

   _The "slug" is the URL-friendly version of the name. It is usually all lowercase and contains only letters, numbers, and hyphens._

   Parent Category

   None ⌄

   _Categories, unlike tags, can have a hierarchy. You might have a Jazz category, and under that have children categories for Bebop and Big Band. Totally optional._

   Description

   _The description is not prominent by default; however, some themes may show it._

   Add New Category

**FIGURE 5-3:**
Create a new category for your website.

2. **Type the name of your new category in the Name text box.**

   If you want to create a category in which you file all your posts about the books you read, for example, you might type **Books I Enjoy**.

3. **Type a name in the Slug text box.**

   The slug creates the link to the category page that lists all the posts you've made in this category. If you leave this field blank, WordPress automatically creates a slug based on the category name. If the category is Books I Enjoy, WordPress automatically creates a category slug like this: http://yourdomain.com/category/books-i-enjoy. If you want to shorten it, you can. Type **books** in the category Slug text box, and the link to the category becomes http://yourdomain.com/category/books.

4. **Choose the category's parent from the Parent Category drop-down menu.**

   Choose None if you want this new category to be a parent (or top-level) category. If you'd like this category to be a subcategory of another category, choose the category you want to be the parent of this one.

5. **(Optional) Type a description of the category in the Description text box.**

   Some WordPress templates are set up to display the category description directly below the category name (see Chapter 9). Providing a description helps you further define the category intent for your readers. The description can be as short or as long as you like.

6. **Click the Add New Category button.**

   That's it! You've added a new category to your blog. Armed with this information, you can add an unlimited number of categories to your new site.

You can delete a category by hovering your mouse over the title of the category you want to delete and then clicking the Delete link that appears below the title.

Deleting a category doesn't delete the posts and links in that category. Instead, posts in the deleted category are assigned to the Uncategorized category (or whatever you named the default category).

If you have an established WordPress site with categories already created, you can convert some or all of your categories to tags. To do so, look for the Category to Tag Converter link in the bottom of the Categories screen of your WordPress Dashboard (refer to Figure 5-1). Click it to convert your categories to tags. (See the nearby sidebar "What are tags, and how/why do I use them?" for more information.)

## WHAT ARE TAGS, AND HOW/WHY DO I USE THEM?

Tags are not to be confused with categories (as a lot of people do confuse them). *Tags* are clickable, comma-separated keywords that help you microcategorize a post by defining the topics in it. In contrast to WordPress categories, tags don't have a hierarchy; there are no parent tags and child tags. If you write a post about your dog, for example, you can put that post in the Pets category — but you can add some specific tags that let you get a whole lot more specific, such as `poodle` or `small dogs`. If someone clicks your `poodle` tag, he finds all the posts you've ever made that contain the `poodle` tag.

Another reason to use tags: Search-engine spiders harvest tags when they crawl your site, so tags help other people find your site when they search for specific words.

You can manage your tags in the WordPress Dashboard by clicking the Tags link on the Posts menu. The Tags screen opens, allowing you to view, edit, delete, and add new tags.

# Examining a Post's Address: Permalinks

Each WordPress post and page is assigned its own web page, and the address (or URL) of that page is called a *permalink*. Posts that you see in WordPress sites usually put their permalinks in any of four areas:

» The title of the blog post

» The Comments link below the post

» A separate permalink that appears (in most themes) below the post

» The titles of posts appearing in a Recent Posts sidebar

Permalinks are meant to be permanent links to your blog posts (which is where the *perma* part of that word comes from, in case you're wondering). Ideally, the permalink of a post never changes. WordPress creates the permalink automatically when you publish a new post.

By default, a permalink in WordPress looks like this:

```
http://yourdomain.com/?p=100/
```

The p stands for *post*, and 100 is the ID assigned to the individual post. You can leave the permalinks in this format if you don't mind letting WordPress associate each post with an ID number.

WordPress, however, lets you take your permalinks to the beauty salon for a bit of a makeover. I'll bet you didn't know that permalinks could be pretty, did you? They certainly can. Allow me to explain.

## Making your post links pretty

*Pretty permalinks* are links that are more pleasing to the eye than standard links and, ultimately, more pleasing to search-engine spiders. (See Chapter 12 for an explanation of why search engines like pretty permalinks.) Pretty permalinks look something like this:

```
http://yourdomain.com/2020/07/01/pretty-permalinks/
```

Break down that URL, and you see the date when the post was made, in year/month/day format. You also see the topic of the post.

To choose how your permalinks look, click Permalinks on the Settings menu. The Permalink Settings screen opens, as shown in Figure 5-4.

On this screen, you find several options for creating permalinks:

>> **Plain** (ugly permalinks): WordPress assigns an ID number to each post and creates the URL in this format: `http://yourdomain.com/?p=123`.

>> **Day and Name** (pretty permalinks): For each post, WordPress generates a permalink URL that includes the year, month, day, and post slug/title: `http://yourdomain.com/2020/07/01/sample-post/`.

>> **Month and Name** (also pretty permalinks): For each post, WordPress generates a permalink URL that includes the year, month, and post slug/title: `http://yourdomain.com/2020/07/sample-post/`.

>> **Numeric** (not so pretty): WordPress assigns a numerical value to the permalink. The URL is created in this format: `http://yourdomain.com/archives/123`.

>> **Post Name** (my preferred option): WordPress takes the title of your post or page and generates the permalink URL from those words. The name of the page that contains my bibliography, for example, is called simply Books; therefore, for this permalink structure, WordPress creates the permalink URL `http://lisasabin-wilson.com/books`. Likewise, a post titled WordPress Is Awesome gets a permalink URL like this: `http://lisasabin-wilson.com/wordpress-is-awesome`.

>> **Custom Structure:** WordPress creates permalinks in the format you choose. You can create a custom permalink structure by using tags or variables, as I discuss in the next section.

To create a pretty-permalink structure, select the Post Name radio button; then click the Save Changes button at the bottom of the page.

## Customizing your permalinks

A *custom permalink structure* is one that lets you define which variables you want to see in your permalinks by using the tags listed in Table 5-1.

**TIP**

One nifty feature of WordPress is that it remembers when you change your permalink structure and automatically writes an internal redirect from the old permalink structure to the new one.

If you want your permalink to show the year, month, day, category, and post name, select the Custom Structure radio button on the Permalink Settings page, and type the following tags in the Custom Structure text box:

```
/%year%/%monthnum%/%day%/%category%/%postname%/
```

**TABLE 5-1** **Custom Permalinks**

| Permalink Tag | Results |
|---|---|
| `%year%` | 4-digit year (such as 2020) |
| `%monthnum%` | 2-digit month (such as 07 for July) |
| `%day%` | 2-digit day (such as 20) |
| `%hour%` | 2-digit hour of the day (such as 15 for 3 p.m.) |
| `%minute%` | 2-digit minute (such as 45) |
| `%second%` | 2-digit second (such as 10) |
| `%postname%` | Text — usually, the post name — separated by hyphens (such as `making-pretty-permalinks`) |
| `%post_id%` | The unique numerical ID of the post (such as 123) |
| `%category%` | The text of the category name that you filed the post in (such as `books-i-read`) |
| `%author%` | The text of the post author's name (such as `lisa-sabin-wilson`) |

If you use this permalink format, a link for a post made on July 1, 2020, called WordPress For Dummies and filed in the Books I Read category, would look like this:

```
http://yourdomain.com/2020/07/01/books-i-read/wordpress-for-dummies/
```

**REMEMBER**

Be sure to include slashes (/) before tags, between tags, and at the very end of the string of tags. This format ensures that WordPress creates correct, working permalinks by using the correct `rewrite` rules located in the `.htaccess` file for your site. (See "Making sure that your permalinks work with your server" later in this chapter for information on `rewrite` rules and `.htaccess` files.)

**WARNING**

Changing the structure of your permalinks in the future affects the permalinks for all the posts on your blog — new and old. Keep this fact in mind if you decide to change the permalink structure. An especially important reason: Search engines (such as Google and Bing) index the posts on your site by their permalinks, so changing the permalink structure makes all those indexed links obsolete.

Don't forget to click the Save Changes button at the bottom of the Permalink Settings page; otherwise, your permalink changes won't be saved!

# Making sure that your permalinks work with your server

After you set the format for the permalinks for your site by using any options other than the default, WordPress writes specific rules, or directives, to the .htaccess file on your web server. The .htaccess file in turn communicates to your web server how it should serve up the permalinks, according to the permalink structure you've chosen to use. To use an .htaccess file, you need to know the answers to two questions:

» Does your web server configuration use and give you access to the .htaccess file?

» Does your web server run Apache with the mod_rewrite module?

If you don't know the answers, contact your hosting provider to find out.

If the answer to both questions is yes, continue with the following steps. If the answer is no to both questions, skip to the sidebar "Working with servers that don't use Apache mod_rewrite" later in this chapter.

You and WordPress work together in glorious harmony to create the .htaccess file that lets you use a pretty-permalink structure on your website. The file works like this:

1. **Locate the .htaccess file on your web server, or create one, and put it there.**

TIP

   If .htaccess already exists, you can find it in the root of your directory on your web server — that is, the same directory where you find your wp-config.php file. If you don't see the file in the root directory, try changing the options of your SFTP (Secure File Transfer Protocol) client to show hidden files. Because the .htaccess file starts with a period (.), it may not be visible until you configure your SFTP client to show hidden files.

   If you need to create the file and put it on your web server, follow these steps:

   (a) *Using a plain-text editor (such as Notepad for Windows or TextEdit for Mac), create a blank file, and name it htaccess.txt.*

   (b) *Upload htaccess.txt to your web server via SFTP. (See Chapter 3 for more information about SFTP.)*

   (c) *Rename the file .htaccess (notice the period at the beginning), and make sure that it's writable by the server by changing permissions to 755 or 777. (See Chapter 3 for information on changing permissions on server files.)*

2. **Create the permalink structure on the Permalink Settings screen of your WordPress Dashboard.**

3. **Click the Save Changes button at the bottom of the Permalink Settings screen.**

   WordPress inserts into the `.htaccess` file the specific rules necessary for making the permalink structure functional in your blog.

If you followed these steps correctly, you have an `.htaccess` file on your web server that has the correct permissions set so that WordPress can write the correct rules to it. Your pretty-permalink structure works flawlessly. Kudos!

If you open the `.htaccess` file and look at it now, you see that it's no longer blank. It should contain code called *rewrite rules*, which looks something like this:

```
# BEGIN WordPress
<IfModule mod_rewrite.c>
RewriteEngine On
RewriteBase /
RewriteCond %{REQUEST_FILENAME} !-f
RewriteCond %{REQUEST_FILENAME} !-d
RewriteRule . /index.php [L]
</IfModule>
# END WordPress
```

## WORKING WITH SERVERS THAT DON'T USE APACHE MOD_REWRITE

Using permalink structures requires that your web-hosting provider have a specific Apache module option called `mod_rewrite` activated on its servers. If your web-hosting provider doesn't have this item activated on its servers, or if you're hosting your site on a Windows server, the custom permalinks work only if you type **index.php** in front of any custom permalink tags.

Create the custom permalink tags like this:

```
/index.php/%year%/%monthnum%/%day%/%postname%/
```

This format creates a permalink like this:

```
http://yourdomain.com/index.php/2020/07/01/wordpress-for-dummies
```

You don't need an `.htaccess` file to use this permalink structure.

I could delve deeply into `.htaccess` and all the things you can do with this file, but I'm restricting this section to how it applies to WordPress permalink structures. If you'd like to unlock more mysteries of `.htaccess`, check out "Comprehensive Guide to .htaccess" at `www.javascriptkit.com/howto/htaccess.shtml`.

# Discovering the Many WordPress RSS Options

For your readers to stay updated with the latest and greatest content you post to your site, they need to subscribe to your RSS feed, which is like a web-based subscription that lets your readers know when you've posted new content.

RSS feeds come in different flavors, including RSS 0.92, RDF/RSS 1.0, RSS 2.0, and Atom. The differences lie within the base code that makes up the functionality of the syndication feed. What's important is that WordPress supports all versions of RSS — which means that anyone can subscribe to your RSS feed with any type of feed reader available.

I mention many times throughout this book that WordPress is very intuitive, and this section on RSS feeds is a shining example of a feature that WordPress automates. WordPress has a built-in feed generator that works behind the scenes to create feeds for you. This feed generator creates feeds from your posts, comments, and even categories.

The RSS feed for your posts is *autodiscoverable,* which means that almost all RSS feed readers and most browsers (Firefox, Chrome, Edge, and Safari, for example) automatically detect the RSS feed URL for a WordPress blog. Table 5-2 gives you some good guidelines on how to find the RSS feed URLs for the different sections of your site.

**TIP**

If you're using custom permalinks (see "Making your post links pretty" earlier in this chapter), you can simply add `/feed` to the end of any URL on your blog to find the RSS feed. Some of your links will look similar to these:

» `http://yourdomain.com/feed` — your main RSS feed

» `http://yourdomain.com/comments/feed` — your comments RSS feed

» `http://yourdomain.com/category/cat-name/feed` — RSS feed for a category

**TABLE 5-2** **URLs for Built-In WordPress Feeds**

| Feed Type | Example Feed URL |
|---|---|
| RSS 0.92 | `http://yourdomain.com/wp-rss.php` or `http://yourdomain.com/?feed=rss` |
| RDF/RSS 1.0 | `http://yourdomain.com/wp-rdf.php` or `http://yourdomain.com/?feed=rdf` |
| RSS 2.0 | `http://yourdomain.com/wp-rss2.php` or `http://yourdomain.com/?feed=rss2` |
| Atom | `http://yourdomain.com/wp-atom.php` or `http://yourdomain.com/?feed=atom` |
| Comments RSS | `http://yourdomain.com/?feed=rss&p=50` p stands for *post,* and 50 is the post ID. You can find the post ID in the Dashboard by clicking the Posts link. Locate a post, and hover the mouse over the title to find the ID in the URL that displays in your browser status bar. |
| Category RSS | `http://yourdomain.com/wp-rss2.php?cat=50` cat stands for *category,* and 50 is the category ID. You can find the category ID in the Dashboard by clicking the Categories link. Locate a category, and hover the mouse over the title to find the ID in the URL that displays in your browser's status bar. |

Try this technique with any URL on your site. Add `/feed` at the end, and you'll have the RSS feed for that page.

RSS feeds are important parts of delivering content from your blog to your readers. RSS feeds are expected these days, so the fact that WordPress takes care of everything for you — provides the feeds for you, complies with all RSS formats, and offers many internal feeds — gives the software a huge advantage over any other content management system.

# Writing Your First Entry

It's finally time to write your first post on your new WordPress site! The topic you choose to write about and the writing techniques you use to get your message across are all on you; I have my hands full writing this book! I *can* tell you, however, how to write the wonderful passages that can bring you blog fame. Ready?

Composing a post is a lot like typing an email: You give it a title, you write the message, and you click a button to send your words into the world. By using the different options that WordPress provides — content blocks, discussion options, categories, and tags, for example — you can configure each post the way you like. This section, however, covers the minimal steps you take to compose and publish a post on your site.

**TIP**

When you're writing (or editing) a blog post, the editor window adopts the visual look and feel of the WordPress theme you're using. In this chapter, I'm using the default Twenty Twenty theme; the post editor looks like the theme, so I know how it'll look on my live website when I publish the post.

Follow these steps to write a basic post:

**1. Click the Add New link on the Posts menu of the Dashboard.**

The Edit Post screen opens, as shown in Figure 5-5.

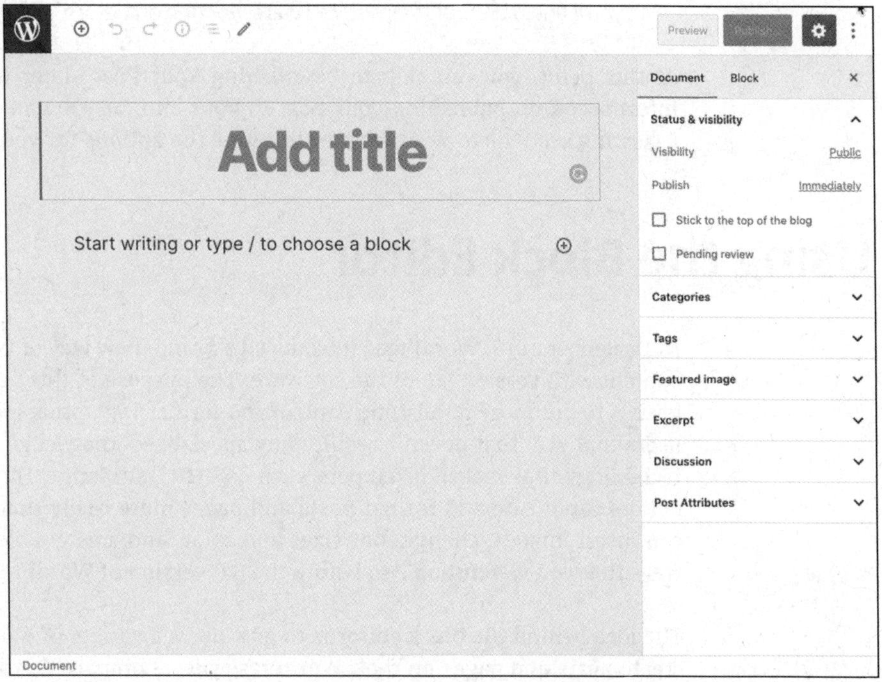

**FIGURE 5-5:**
Give your post a title, and write your post body.

**2. Type the title of your post in the Add Title text field at the top of the Edit Post screen.**

**3. Type the content of your post in the area below the Add Title field.**

The first time you visit the Edit Post screen, this area displays a message that says Start writing or type / to choose a new block. I cover blocks in "Using the Block Editor" later in this chapter. For the purposes of this section, you'll type the text of your new post in this area.

4. **Click the Save Draft link, located in the top-right corner of the Edit Post screen.**

   The Save Draft link changes to a message that says Saved (not visible in Figure 5-5).

   WordPress has a built-in autosave feature to make sure that your content is saved and protected from being lost. Imagine spending an hour writing a long post and then the power goes out due to a storm in your area! You don't need to worry about all your work being lost because WordPress thoughtfully saved it for you. The default interval is 10 seconds, so if you don't click the Save Draft link, WordPress saves your post for you automatically. When the storm is over, you'll find a draft of the post you were working on before lightning struck.

At this point, you can skip to "Publishing Your Post" later in this chapter for information on publishing your post to your site, or you can continue with the following sections to discover how to refine the options for your post.

# Using the Block Editor

In December 2018, WordPress introduced a brand-new way of writing and editing content with version 5.0 of the software. The purpose of this new editing experience is to put more publishing control and formatting options in the hands of the users in a way that doesn't require any specialized knowledge or training in the technology that makes it happen, such as PHP, JavaScript, HTML, or CSS. Now, editors can create and format posts and pages more easily than ever before. You can insert images, change font sizes and color, and create tables and columns in ways that you weren't able to before the 5.0 version of WordPress.

The idea behind the block editor is to give users a variety of blocks with which to create posts and pages on their WordPress sites. Compare WordPress blocks with the blocks you played with as a child; you were able to take one block and stack it on top of the next block and the next to build a tower of blocks to the moon. Each block within the WordPress editor gets filled with content (text, images, video, and so on) and is stacked atop another block with more content. Once you have a full page of content created with blocks, you can configure options to control formatting and display and move around on the page to create the experience you want for your readers.

This section of the chapter takes you through the new block editor in WordPress.

**TIP**

WordPress named the new block editor Gutenberg after Johannes Gutenberg, the inventor of the printing press. Generically, everyone refers to it as the block editor, but if you hear or see the term *Gutenberg* tossed around in the WordPress support forums or at a WordPress meetup or WordCamp, you'll know that people are talking about the editor in WordPress.

## Discovering available blocks

By default, when you first load the Edit Post screen in your Dashboard (refer to Figure 5-5), the area where you type the text of your post consists of standard paragraph blocks. Hover over this area with your mouse, and you see that the block is outlined, as shown in Figure 5-6.

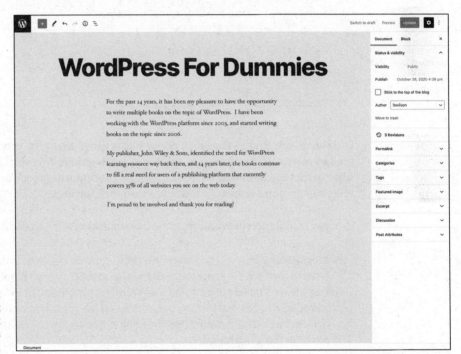

**FIGURE 5-6:**
Default paragraph block in the WordPress block editor.

You can use the standard paragraph blocks to write your posts, and leave it at that. But the block editor has many features that give you a variety of options for formatting your content by using a variety of blocks.

You can discover the different types of blocks available by clicking the small plus sign in the top-left corner of the Edit Post (or Edit Post) screen. Clicking that icon displays a drop-down menu of blocks, as shown in Figure 5-7.

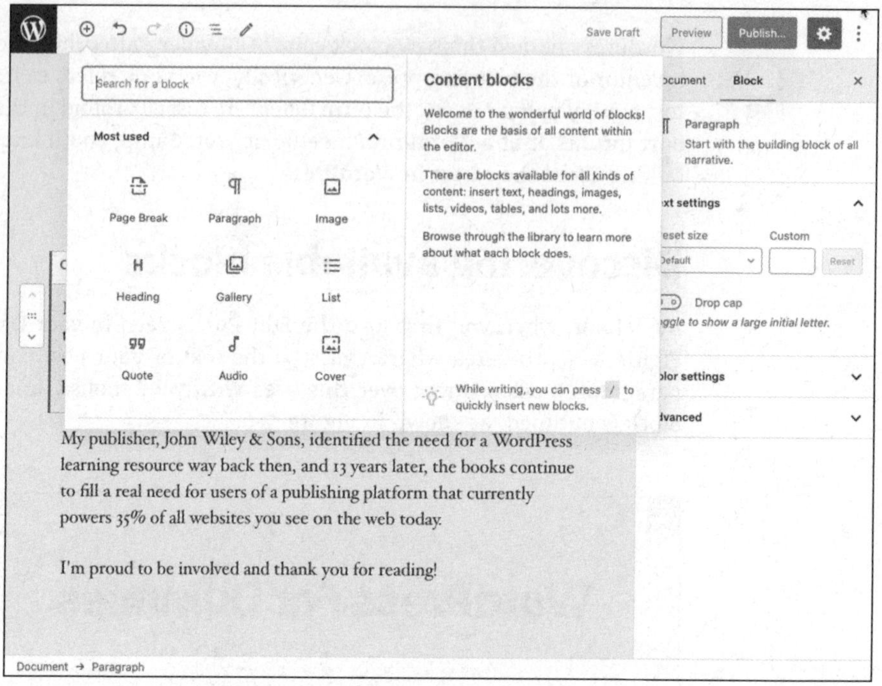

**FIGURE 5-7:**
Menu of available blocks in the WordPress block editor.

Each block you find in the block editor has settings that give you display options such as color settings, font size and color, and width. In this section, you discover what blocks are available for you to use. In "Configuring block settings" later in this chapter, you see how to configure the block settings.

In a new installation of WordPress, the following blocks are available for you to use:

>> **Most Used:** When you first install WordPress, the blocks listed in the Most Used section are the same as the ones listed in the Common Blocks section (see the next bullet point). As you use WordPress more and more, the software learns your habits and lists the blocks that you use most when writing and editing posts. Click the Most Used heading to display a drop-down list of available blocks in this section.

>> **Common Blocks:** The blocks included in this section are common, which means that most users use these blocks when writing and editing posts. Click the Common Blocks heading to display a drop-down list of available blocks in this section:

- *Paragraph:* Clicking this option inserts a block that allows you to create a standard paragraph of text.

- *Image:* Clicking this option inserts a block that allows you to insert an image into your post or page.

- *Heading:* Clicking this option inserts a block that allows you to insert header text with H1, H2, H3, H4, H5, and H6 tags to help your visitors (and search engines) understand the structure of your content.

- *Gallery:* Clicking this option inserts a block that allows you to upload and display multiple images in your post or page.

- *List:* Clicking this option inserts a block that allows you to create a bulleted or numbered list.

- *Quote:* Clicking this option inserts a block that allows you to enter a quote with a citation that's stylized for visual emphasis.

- *Audio:* Clicking this option inserts a block that allows you to upload an audio file and embed it in an audio player in your post or page.

- *Cover:* Clicking this option inserts a block that allows you to upload an image and add it to your post or page with text overlaid on it.

- *File:* Clicking this option inserts a block that allows you to upload a file (such as `.doc` or `.pdf`) and add it to your post or page for your visitors to download.

- *Video:* Clicking this block inserts a block that allows you to upload a video file and embed it in a video player in your post or page.

>> **Formatting:** The blocks in this section are for special formatting needs such as code, HTML, and quotes. Not everyone is going to use these formatting blocks, but for those who do (such as programmers), these blocks are extremely helpful for formatting specialized text. Click the Formatting heading to display a drop-down list of available blocks in this section:

- *Code:* Clicking this option inserts a block that allows you to insert and display code snippets that respect standard code-formatting rules and prevents the application from executing the code you've written.

- *Classic:* Clicking this option inserts a block that users of earlier WordPress versions are used to and may be more comfortable with.

- *Custom HTML:* Clicking this option inserts a block with an HTML editor that allows you to write HTML code and preview it as you type and edit.

- *Preformatted:* Clicking this option inserts a block that allows you to add text that gets displayed exactly as it is intended in code or HTML format. The text is typically displayed in a monospace font (such as Courier), and the Preformatted editor respects your spacing and tabs, keeping them in place. Preformatted text is helpful for people who include code samples within their posts or pages. Also, this option prevents the application from executing the code.

- *Pullquote:* Clicking this option inserts a block that allows you to enter a quote, with citation, that gives special visual emphasis to the text. The `Code is Poetry. – WordPress` example in Figure 5-8 is a Pullquote.

- *Table:* Clicking this option inserts a table editor that allows you to include a table of rows and columns, much as you'd do in a standard word processing program such as Microsoft Word.

- *Verse:* Clicking this option inserts a verse editor that allows you to enter a verse of poetry or song lyrics or to quote a small number of lyrics or lines of poetry. The text gets special formatting and spacing to give it visual emphasis. Figure 5-9 shows an example of a verse from a popular sports song.

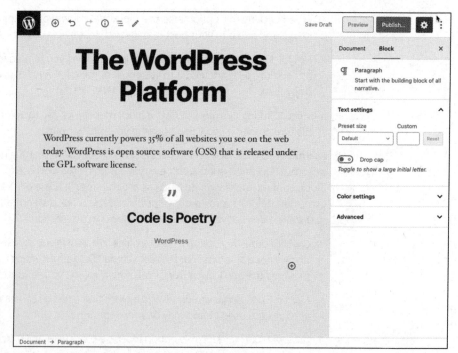

**FIGURE 5-8:**
A Pullquote block in the WordPress block editor.

>> **Layout Elements:** The blocks in this section allow you to create different layouts for your content on every post or page. The layouts include tables, columns, and buttons. These elements enable you to create pages and posts that look different or the same, so you can be as creative as you want to be. Click the Layout Elements option to display a drop-down menu of available blocks in this section:

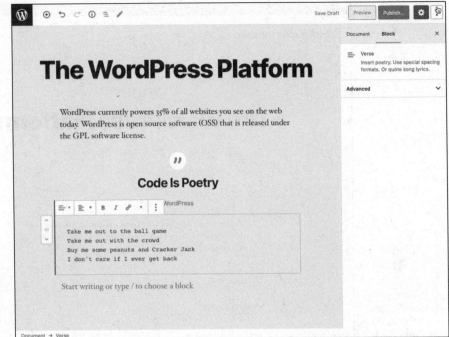

FIGURE 5-9:
An example of
the Verse block in
the WordPress
block editor.

- *Page Break:* Clicking this option inserts a block that serves as a marker point for a page break, allowing you to create a post or a page that has a multi-page experience. The content that appears above this block is displayed on the page with navigation links, prompting the reader to navigate to page 2 of your post or page to read the rest. Figure 5-10 shows a post that uses the Page Break block to create a post that spans two pages.

- *Button:* Clicking this option inserts a clickable button within your page or post. People usually use a button to link to a section of their site that they feel is important, such as a contact form or an online shop.

- *Columns:* Clicking this option inserts a column editor that allows you to create a section of text in side-by-side columns. Figure 5-11 shows the use of the Columns block to create three columns of text with headers in a post. The options for this block allow you to define the number of columns as well as the colors of the columns and text.

- *Group*: A group block can contain multiple blocks within one container. Another way to think of it is a parent block that has child blocks inside. The options for this block allow you to set color options for the text and background color.

**FIGURE 5-10:**
Example of a post that uses the Page Break block.

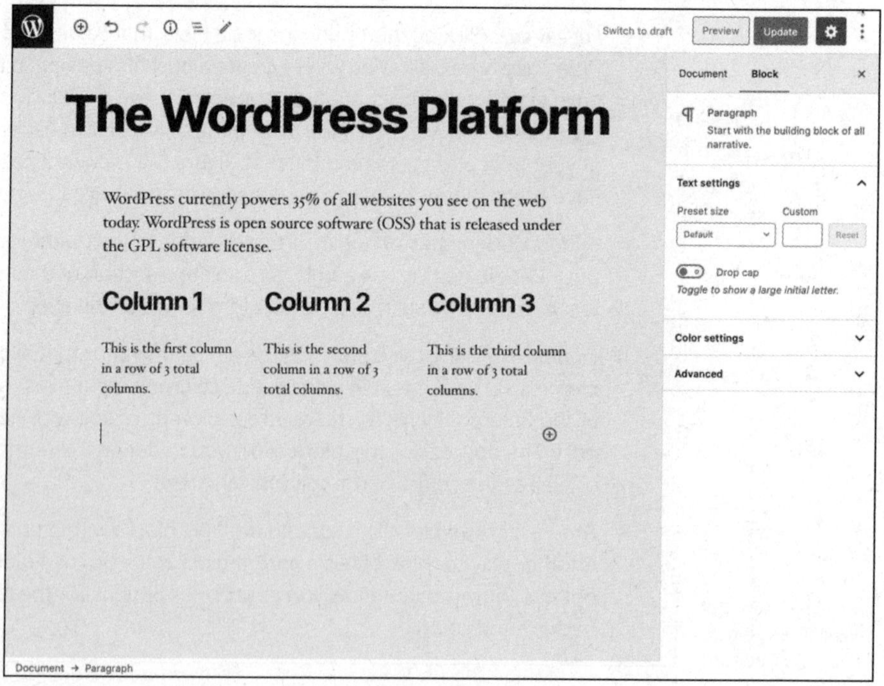

**FIGURE 5-11:**
An example of the Columns block in the WordPress block editor.

- *Media & Text:* Clicking this option inserts a two-column block that allows you to display media (image or video) and text side by side. The options for this block allow you to align the media to the right or the left of the text block. You also can define the width of the block, making it the width of the reader's computer screen or the width of the rest of the content on your page. Figure 5-12 displays the Media & Text block in use on a web page on which the media is displayed to the left of the text.

- *More:* Clicking this option inserts a block that serves as a marker point for your post or page excerpt. The content that appears above this block is shown as the excerpt on pages such as archive and search results pages.

- *Separator:* Clicking this option inserts a block that creates a break between sections of content by using a horizontal separator or line. Options for this block allow you to determine the style of the separator line: Short Line, Wide Line, or Dots.

- *Spacer:* Clicking this option inserts a block that creates white space within your post or page. This block doesn't get filled with any kind of content; rather, it allows you to create a space between content blocks at a height that you can define in the block options.

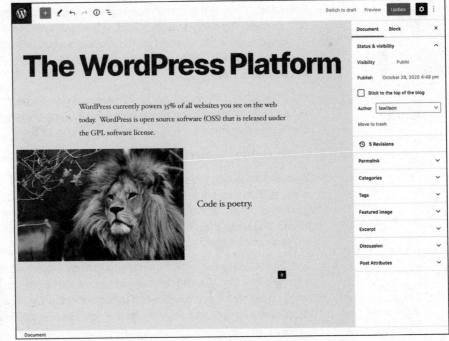

**FIGURE 5-12:** An example of the Media & Text block in the WordPress block editor.

>> **Widgets:** I cover the use of WordPress widgets on your website in Chapter 12, so see that chapter to find out what widgets are and what they do. Some widgets in WordPress are available in the block editor to allow you to insert blocks with predefined content. Click the Widgets option to display a drop-down list of available blocks in this section:

- *Shortcode:* Clicking this option inserts a block with a text field that allows you to include a shortcode for adding custom elements to your WordPress page or post.

- *Archives:* Clicking this option inserts the Archives widget, which displays a list of monthly archives of your posts.

- *Calendar:* Clicking this option inserts a block with a calendar of posts on your site in a month-by-month calendar format.

- *Categories:* Clicking this option inserts the Category widget, which displays a list of categories on your site.

- *Latest Comments:* Clicking this option inserts the Latest Comments widget into your post or page. This widget displays a list of the most recent comments left on your site. Options for this block allow you to define the number of comments you want to display and to toggle the display of the comment avatar, date, and excerpt.

- *Latest Posts:* Clicking this option inserts the Recent Posts widget into your post or page. This widget displays a list of the most recent posts you've published to your site. Options for this block allow you to sort the post list by newest to oldest, oldest to newest, and ascending or descending in alphabetical order.

- *RSS*: Clicking this option inserts the RSS widget into your post or page. This widget gives you the option to insert an RSS URL into a text field that will display the content of the RSS feed inside your post or page.

- *Search*: Clicking this option inserts a search form in your post or page.

- *Social Icons*: Clicking this option inserts a block with various social media icons, such as Facebook, Twitter, and Instagram. After you add this block, you can click each icon and type the URL to the corresponding social media profile. The social icons are displayed on the front end of your website, and readers can click them to visit your social media profiles.

- *Tag Cloud*: Clicking this option inserts a block with a tag cloud. A *tag cloud* is a listing of the tags you use on your website, with the font size of each tag being based on the prominence of a tag on your site. Tags that you use often have a larger font size, and tags that you use less often have a smaller font size.

» **Embeds:** This block allows you to embed content from various services on the web, such as a video from YouTube, a specific tweet from Twitter, or a photo from Instagram. Currently, the WordPress block editor allows you to embed content from 34 services. The services in this embed list are likely to change over time, with new ones being added and old ones removed as services on the web continue to evolve. Click the Embeds option to display a drop-down menu of available blocks in this section.

## Inserting new blocks

WordPress gives you a variety of ways to add a new block to your post or page. As you work with the block editor more and more, you'll develop a favorite method of adding new blocks based on your preferences and writing style.

In a brand-new post, the Edit Post screen gives you the title field to type your page title in and a Paragraph block to write the content of your page. From there, you can add new blocks to add different types of content to your page. You can insert a new block into your page by using any of the following methods:

» **Use the Top Block Inserter.** This method is covered in "Discovering available blocks" earlier in this chapter and illustrated in Figure 5-6.

» **Use the Editor Block Inserter.** When you've added a block, put content in it, and are ready to add a new block, hover your mouse over the existing block. You see an icon that looks like a plus sign. Click that icon to display a list of available blocks that you can add to your page (see Figure 5-13). Choosing a block from that list inserts the block directly below the existing one.

» **Use More Options.** Every block has a small toolbar of styling options for that block (discussed in "Configuring block settings" later in this chapter). The item on the right side of the toolbar menu is labeled More Options; click that item to display a drop-down list of additional settings (see Figure 5-14). Two of those options are labeled Insert Before and Insert After. Click either option to insert a standard Paragraph block above or below the block you're currently working in. Then you can configure the block type by using the Editor Block Inserter (refer to Figure 5-13).

» **Use shortcut icons.** On the Edit Post screen, three icons allow you to add three commonly used blocks: Image, Code, and Quote (see Figure 5-8 for the Quote block). Click one of those icons to add the corresponding block to your page.

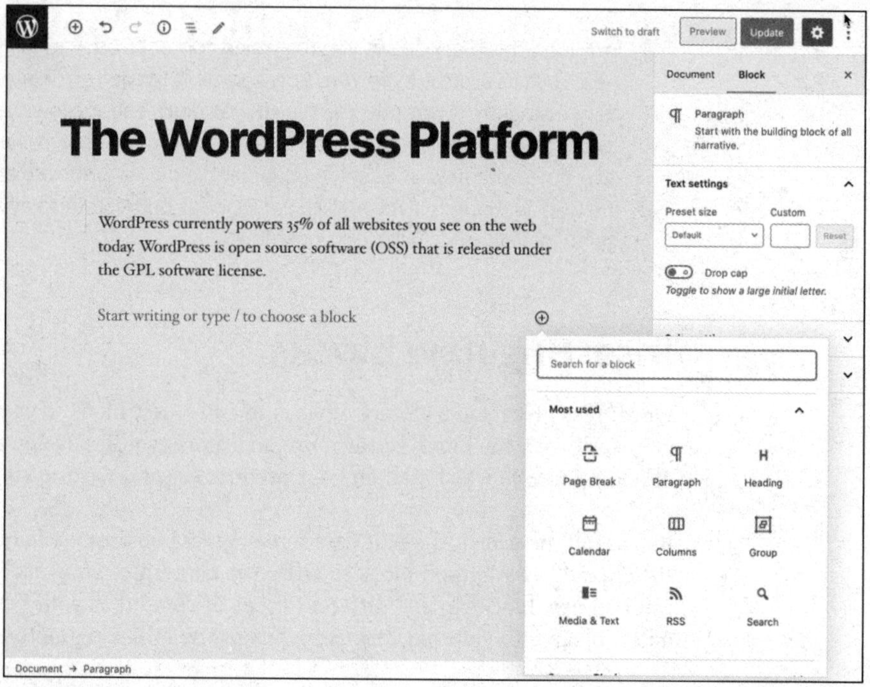

**FIGURE 5-13:**
The Editor Block
Inserter.

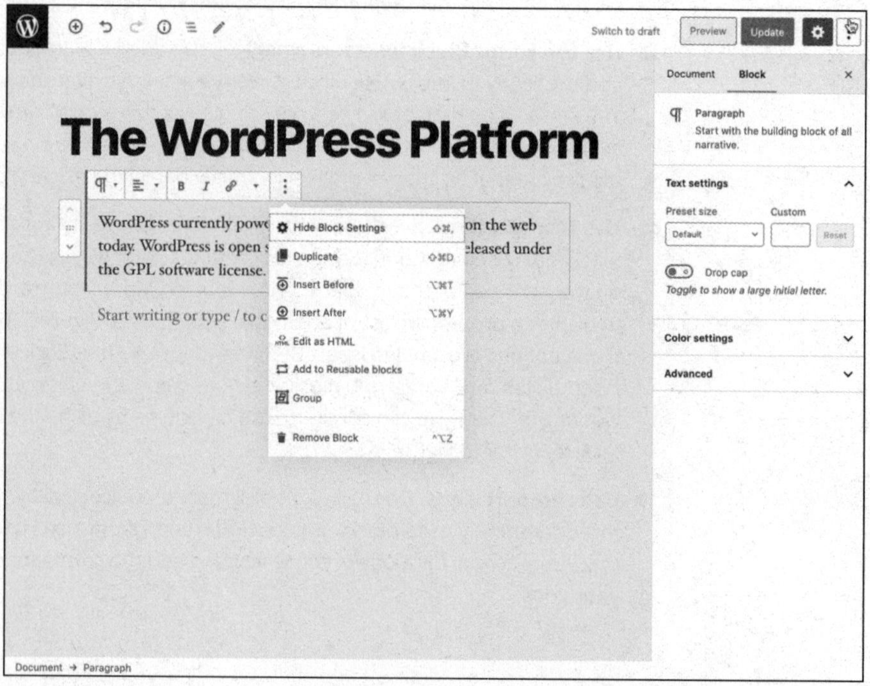

**FIGURE 5-14:**
More Options
menu in the block
editor.

>> **Press Enter (or Return, on a Mac).** Press the Enter key on your keyboard when you're in a standard Paragraph block to insert a new Paragraph block into your page. Then you can continue using the Paragraph block or use the Editor Block Inserter (refer to Figure 5-13) to change the block type.

>> **Use slash commands.** When you click inside a standard Paragraph block and press the slash (/) key on your keyboard, a list of blocks appears. This list enables you to add a block without moving your hands away from the keyboard.

Figure 5-15 shows the blocks that are available when you press the slash key. You can navigate to the block you need by pressing the down-arrow key and then pressing Enter to select it, or you can finish typing the name of the block and then press the Enter key to insert it into your page. This editing experience is intended to be mouseless; you can keep typing away and adding blocks from your keyboard.

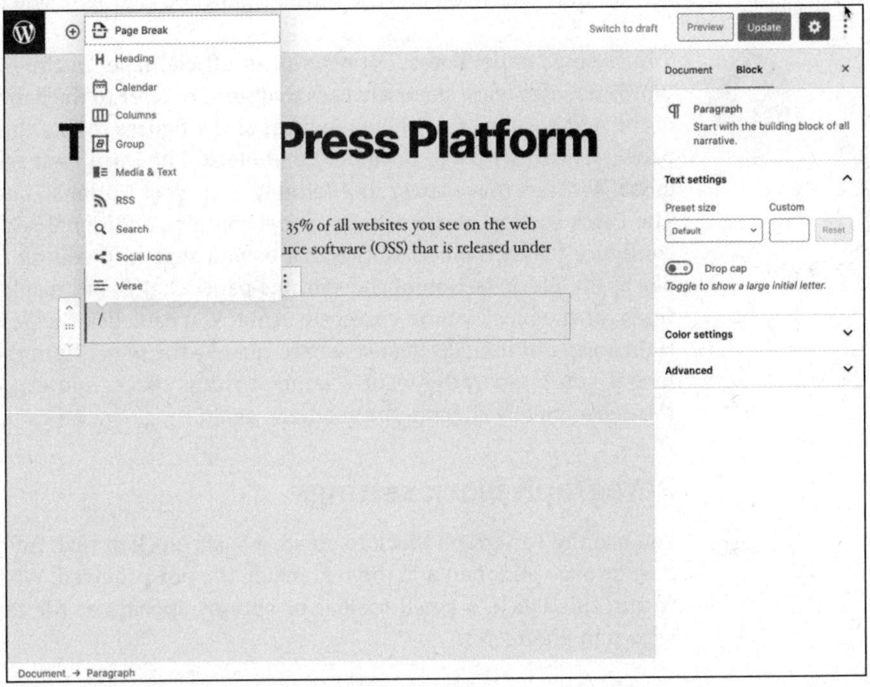

**FIGURE 5-15:**
Slash commands give you a mouseless experience when adding blocks.

# Configuring block settings

Each block on the block editor menus has options that let you configure display settings for your content, such as font size, font color, background color, and block width and/or height. Each block has its own set of options.

In this section, you discover how to configure the settings and options for four of the most commonly used blocks:

» Paragraph

» Image

» Media & Text

» Pullquote

You configure options for the block you're using in two areas of the Edit Post screen: the block itself and the settings panel on the right side of the screen.

TIP

The settings panel doesn't appear as an official label in the Dashboard, but the WordPress dev team currently uses that term to refer to the panel on the right side of the Add Post page, as shown in most of the figures in this chapter. The settings panel has two sections: Document and Block. The Document section contains the global settings that I cover in "Refining Your Post Options" later in this chapter. The Block section of the settings panel contains additional settings that you can configure for each block. Notice that as you switch to editing a different kind of block, the Block section of the settings panel changes to display the unique settings for the block you're currently using. You can click the gear icon in the top-right corner of the Edit Post screen to remove the panel from the screen, which is nice if you temporarily want a larger writing space. You can easily restore the settings panel by clicking the gear icon again.

## Paragraph block settings

You use the Paragraph Block to create a basic block of text. Add the block to your post or page and then add the text inside the box provided. When you're working within this block, a small toolbar of options appears at the top of the block, as shown in Figure 5-16.

Paragraph block toolbar

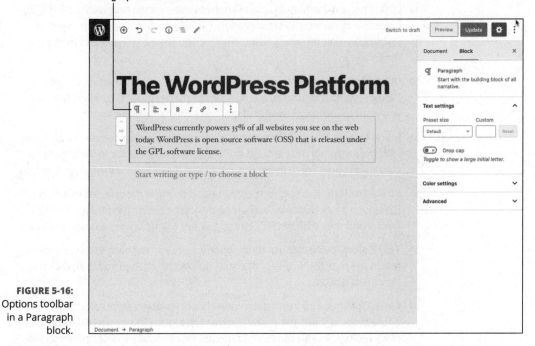

**FIGURE 5-16:**
Options toolbar
in a Paragraph
block.

This toolbar provides a variety of options, including the following (from left to right):

>> **Change Block Type:** Clicking this option allows you to change the type of block you're using. If you want to change from a Paragraph to a Quote block, for example, click the Change Block Type icon and then choose the Quote block to swap it. The block types you can change to from a Paragraph block include Quote, Verse, Heading, List, and Preformatted.

>> **Align Text Left:** This block-formatting option positions the text within a Paragraph block to the left side of the page.

>> **Align Text Center:** This block-formatting option positions the text within a Paragraph block to the center of the page.

>> **Align Text Right:** This block-formatting option positions the text within a Paragraph block to the right side of the page.

>> **Bold:** This text-formatting option changes the selected text within a Paragraph block to a bold (darker) font. Select the text you want to format and then click the Bold icon. Example: **bold text.**

>> **Italic:** This text-formatting option changes the selected text within a Paragraph block to an italic (slanted) font. Select the text you want to format and then click the Italic icon. Example: *italic text.*

» **Link:** This text-formatting option changes the selected text within a Paragraph block to a hyperlink (text that readers can click to visit a new web pages or website, in their browser). Select the text you want to format and then click the Link icon.

» **Inline Code:** This text-formatting option changes the formatting within the Paragraph block to display as code, rather than regular text. Select the text you want to format as code and then click the Inline Code option.

» **Inline Image:** This text-formatting option allows you to insert an image inline with the paragraph text around it. Place your mouse cursor in the area in the paragraph you'd like to insert the image and then click the Inline Image option.

» **Strikethrough:** This text-formatting option changes the selected text within a Paragraph block to display with a line through it. Select the text you want to format and then click the Strikethrough icon. Example: `strikethrough text`.

» **Text Color:** This text-formatting option allows you to change the color of the text in the paragraph. Select the text you want to change and then click the Text Color option.

» **More Options:** Clicking this icon reveals a drop-down menu of options for the entire block, not just the content within the block. (This option exists on every block toolbar, so you can refer to this list for other blocks later in this section.) Click the More Options icon on the block toolbar to reveal a drop-down list of options:

   • *Hide Block Settings:* Clicking this option removes the settings panel on the right side of the Edit Post screen, giving you a larger space in which to create your content.

   • *Duplicate:* Clicking this option duplicates the block you're currently using and inserts it below the current block.

   • *Insert Before:* Clicking this option inserts a blank default block directly above the block you're currently using.

   • *Insert After:* Clicking this option inserts a blank default block directly below the block you're currently using.

   • *Edit As HTML:* Clicking this option changes the block editor to an HTML editor so you can view and create content in HTML code.

   • *Add to Reusable Blocks:* Click this option to save the block you're currently using to a library of blocks that you can reuse in other places on your site. This feature is helpful when you create blocks on one page and want to use the same blocks on other pages. Saving a block as a Reusable block makes it available for use on other pages, exactly as it appeared when you saved it. Editing the block applies those changes everywhere the block is used on your site.

# USING KEYBOARD SHORTCUTS

Some people appreciate the ability to use keyboard shortcuts to accomplish things that normally take a mouse click to do. Keyboard shortcuts can make quick work of common editing actions by reducing dependence on the computer mouse, which can also benefit those who have carpal tunnel syndrome or other ailments associated with joint or muscle pain and strain. WordPress provides a variety of keyboard shortcuts to accomplish things like making text bold or italic, saving a post, and undoing changes. In the top-right corner of the Edit Post screen is an icon that looks like three dots stacked on top of one another. Hover your mouse over this icon, and you see the label Show More Tools & Options. Click this icon to open a menu of options. Choose Keyboard Shortcuts to open a small window that shows you the shortcuts you can use when editing and creating posts and pages in WordPress.

**TIP**

When you create a Reusable block, it's stored on the Blocks menu (refer to Figure 5-7) in a new section labeled Reusable.

- *Group:* Click this option to create a group block.

- *Remove Block:* Clicking this option removes the block from the Edit Post screen. Use this feature carefully, because when you remove a block, it's gone (unless you've saved it as a Reusable block).

Additional options for the Paragraph block are available in the settings panel on the right side of the Edit Post screen, as shown in Figure 5-17.

Those options include

>> **Text Settings:** This section allows you to set font size and drop cap options in the Paragraph block.

- *Font Size:* You have two ways to define the size of the font you're using in the block that you're editing. In the Text Settings section of the settings panel, you can choose Small, Normal, Large, or Huge from the Font Size drop-down menu to adjust the text sizes that are predefined by WordPress. Alternatively, you can enter a specific number in the text box to the right of the Font Size menu if you want to use a specific-size font. When you adjust the Text Settings in the settings panel, the changes immediately occur in the block you're editing.

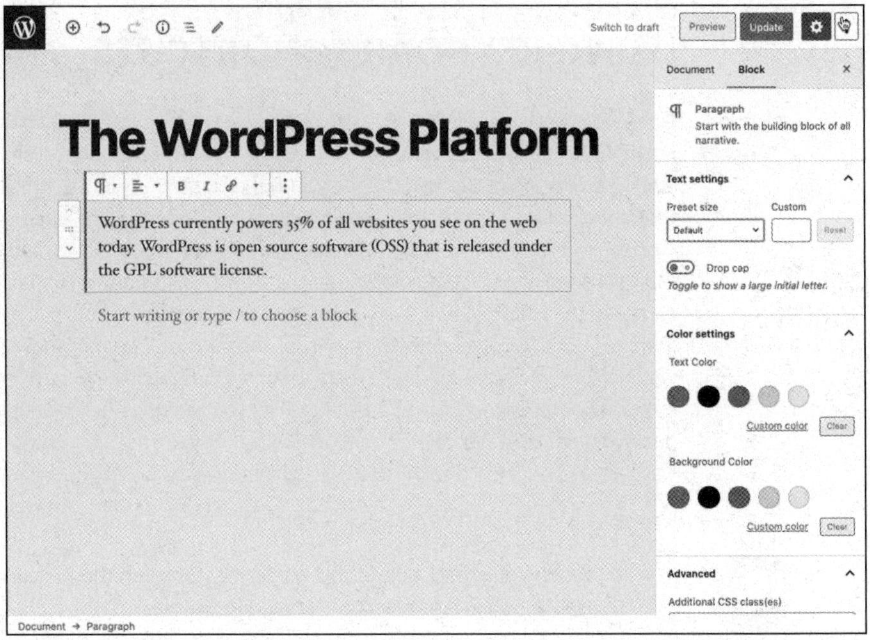

**FIGURE 5-17:**
Options in the settings panel for the Paragraph block.

- *Drop Cap:* A drop cap can be applied to a paragraph of text to make the first letter of the paragraph much larger and bolder than the rest of the text within the block. You often see this method used in magazine and newspaper articles; it can have a dramatic effect on articles that appear on the web. Figure 5-18 shows an example of two paragraphs that use the Drop Cap method. To enable it, click the Drop Cap toggle button.

» **Color Settings:** This section allows you to set the text color and background color of the Paragraph block.

- *Text Color:* In the Color Settings section of the block settings panel, you can change the color of the text used in the Paragraph block you're currently using. Choose one of five preselected colors or click the custom color picker icon to select another color. Figure 5-19 shows a Paragraph block with a black background and white text.

- *Background Color:* Within the Color Settings section of the settings panel, you can select a background color for the Paragraph block you're currently using. Choose one of five preselected colors or click the custom color picker icon to select another color (see Figure 5-20).

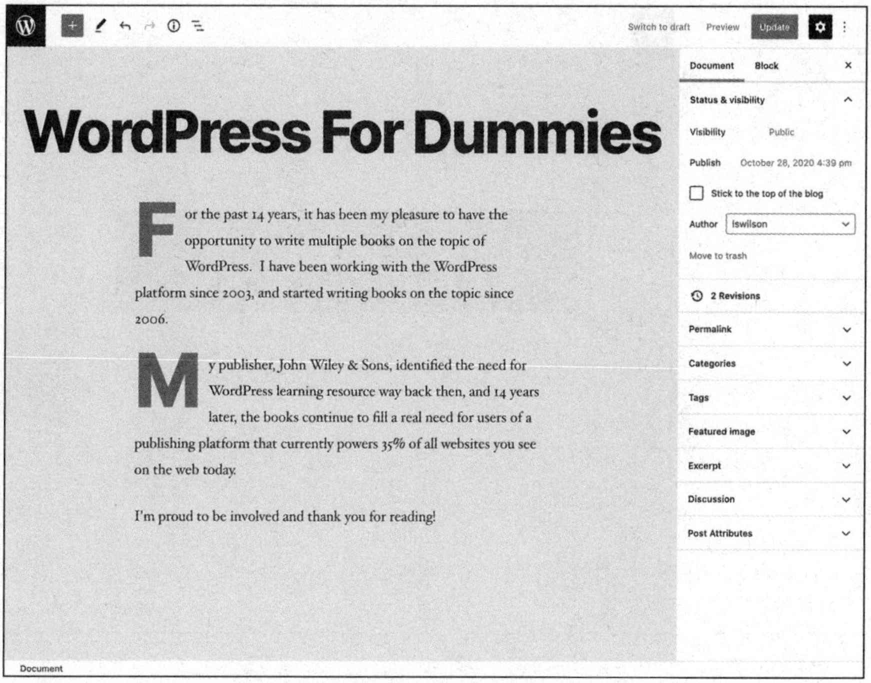

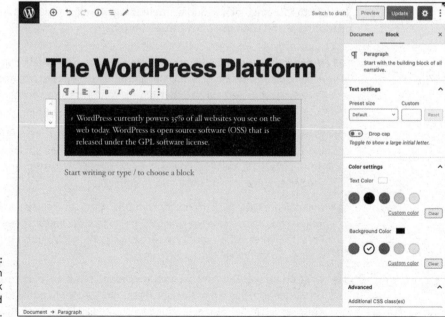

**FIGURE 5-18:**
Use of the Drop
Cap option.

**FIGURE 5-19:**
A Paragraph
block with a black
background and
white text.

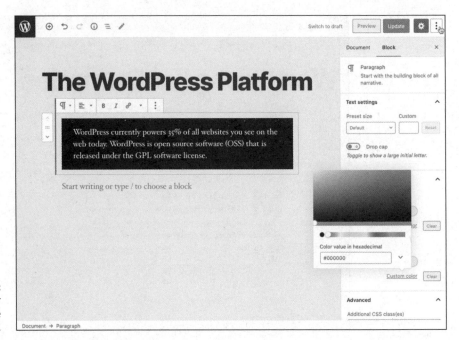

**FIGURE 5-20:**
The custom color
picker tool in the
settings panel.

## Image block settings

You use the Image block to add a single image to your post. Add the block to your post or page and then use one of these options:

>> **Upload:** Click the Upload button to select an image from your computer. This action uploads the image from your computer to your website and inserts the image into your post via an Image block. WordPress also adds this image to your Media Library so you can reuse this image in the future.

>> **Media Library:** Click the Media Library button to choose an image from the WordPress Media Library. When the Select or Upload Media screen opens (see Figure 5-21), select an image in the Media Library section and then click the Select button to add the image to your post.

>> **Insert from URL:** Click the Insert from URL button to display a small text box where you can paste or type the URL (or link) for the image you want to use. Press the Enter key on your keyboard or click the Apply button to insert the image into the Image block you're using.

>> **Drag an Image:** This cool option allows you to select an image from your computer and drag it into the WordPress block editor to add it to a block. The dragged image also gets added to the Media Library for future use.

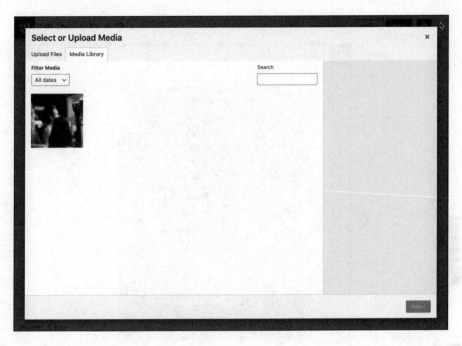

**FIGURE 5-21:**
Clicking the
Media Library
button in an
Image block
opens the Select
or Upload Media
screen and
displays images
in the WordPress
Media Library.

>> **Write Caption:** When you've added an image to an Image block, you see an optional field directly below it labeled Write Caption. This field is optional; if you do nothing with it, nothing displays on your site. If you type words in this field, however, those words appear below the image as a caption.

When you're working within this block, notice the small toolbar of options that appears at the top of the block, as shown in Figure 5-22. This toolbar provides a variety of options for the Image block, including the following (from left to right):

>> **Change Block Type:** Click this option to change the type of block you're using. If you want to change from an Image block to a Cover block, for example, click the Change Block Type icon and then select the Cover block to swap it. The block types you can change from an Image block include Media & Text, Gallery, Cover, and File.

>> **Align Left:** This block-formatting option positions the image within the Image block against the left margin of the page.

>> **Align Center:** This block-formatting option positions the image within the Image block in the center of the page.

>> **Align Right:** This block-formatting option positions the image within the Image block against the right margin of the page.

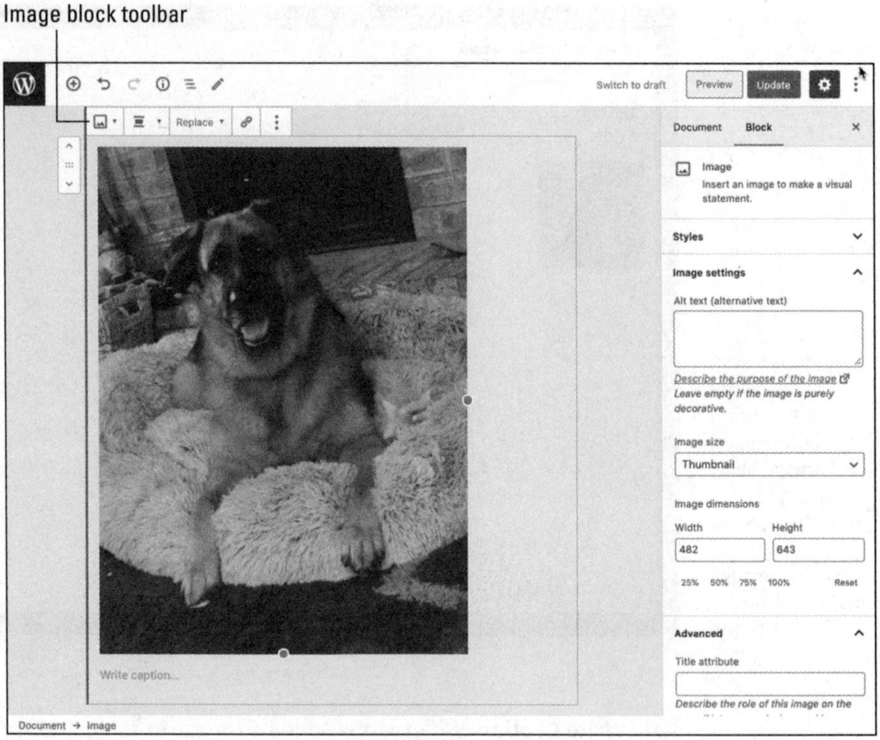


**FIGURE 5-22:** Options toolbar in the Image block.

- **Wide Width:** Click this option to set the width of the image to the width of the content on the page.
- **Full Width:** Click this option to increase the width of the image to the width of the screen you're viewing the content on. In Figure 5-23, you see a post that I created with a full-width image. Notice that the left and right edges of the image extend all the way to the left and right sides of the viewing screen.

  TIP

  If you don't select Wide Width or Full Width on the Image block toolbar, you can set the desired width of the image in the Image block settings panel, as covered in the next section.

- **Replace:** This option allows you to replace the image and opens the Media Library when you click it.
- **Link:** This option allows you to set a URL for the image, which makes the image clickable. When a visitor to your site clicks the image, they're taken to the URL you specified.
- **More Options:** The settings here are the same as the ones discussed in "Paragraph block settings" earlier in this chapter.

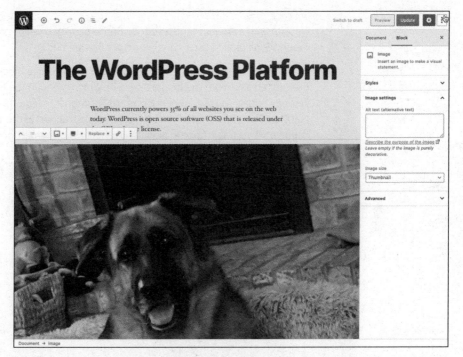

**FIGURE 5-23:**
Example of an
Image block with
the Full Width
setting.

Additional options are available for the Image block in the settings panel on the
right side of the Edit Post screen, as shown in Figure 5-24.

The options include

» **Styles**: The options in this section allow you to set the style of the image using
the Image block.

- *Default*: The Default style is set automatically and displays the image in its
original state.

- *Rounded*: The Rounded style can be set to display the image with rounded
corners.

» **Image Settings:** The options in this section allow you to set the alternative
text, size, and dimensions of the image in the Image block.

- *Alt Text:* Enter descriptive text in the Alt Text field to describe the image
you're using in the Image block. Also referred to as *alternative text*, this
description helps people who can't see the image on your site; when the
image doesn't load on your site, the alt text is displayed, providing context
for the missing image. This option is also an accessibility feature that helps
those who use screen readers to browse the web. (Screen readers allow
visually impaired users to understand the text that's displayed on the

website with a speech synthesizer or Braille display.) Additionally, alt text descriptions assist in search engine optimization (SEO).

- *Image Size:* To set the size of the image used in the Image block, choose a predefined size from the Image Size drop-down menu in the Image block settings panel. The available options are Thumbnail, Medium, Large, and Full Size. The dimensions for these options are defined in the WordPress settings on your Dashboard (see Chapter 3) to see them, choose Settings ➪ Media.

- *Image Dimensions:* To set a specific width and height for the image used in the Image block, enter numbers in the Width and Height fields in the Image Dimensions section of the Image block settings panel.

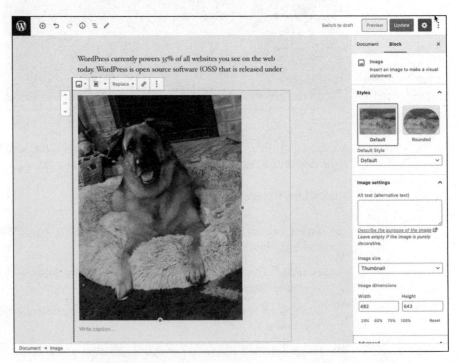

**FIGURE 5-24:**
Options in the settings panel for the Image block.

**TIP**

When you're creating content inside of any block on a page, and you find that you've made a mistake, the Undo button at the top of the Edit Post screen undoes the last action you took on the screen. Alternatively, a Redo button allows you to redo the last undone action.

## Media & Text block settings

You use the Media & Text block to insert a two-column block that displays media (image or video) and text side by side. When you add the block to your post, it adds a block with the image settings on the left and text settings on the right, as shown in Figure 5-25.

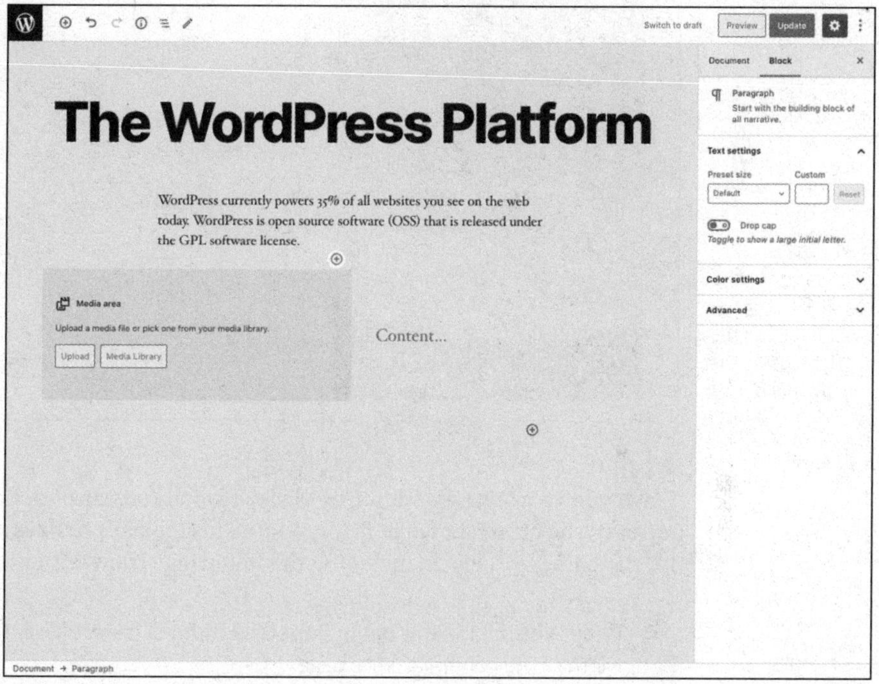

**FIGURE 5-25:**
The Media & Text block.

In this block, you add the media in the Media Area section on the left, using the method covered in "Image block settings" earlier in this chapter and adding your text in the Content section on the right, as shown in Figure 5-26.

**TIP**

When you enter text in the Content section on the right side, a toolbar of text-formatting options appears. This toolbar has the same options that I cover in "Paragraph block settings" earlier in this chapter.

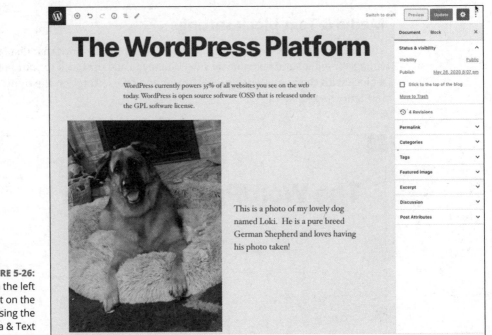

When you're working within this block, a small toolbar of options appears at the top of the block, as shown in Figure 5-27. This toolbar provides a variety of options for the paragraph block, including the following (from left to right):

>> **Wide Width:** Click this option to set the width of the Media & Text block to the width of the content on the page.

>> **Full Width:** Click this option to increase the width of the image to the width of the screen you're viewing the content on. The left and right edges of the block extend all the way to the left and right sides of the viewing screen.

>> **Show Media on Left:** Click this option to set the media on the left and the text on the right. This option is the default setting.

>> **Show Media on Right:** Click this option to set the text on the left and the media on the right.

>> **Vertical Align Top**: Click this option to align the text with the top of the image.

>> **Vertical Align Middle**: Click this option to align the text with the middle of the image.

>> **Vertical Align Bottom**: Click this option to align the text with the bottom of the image.

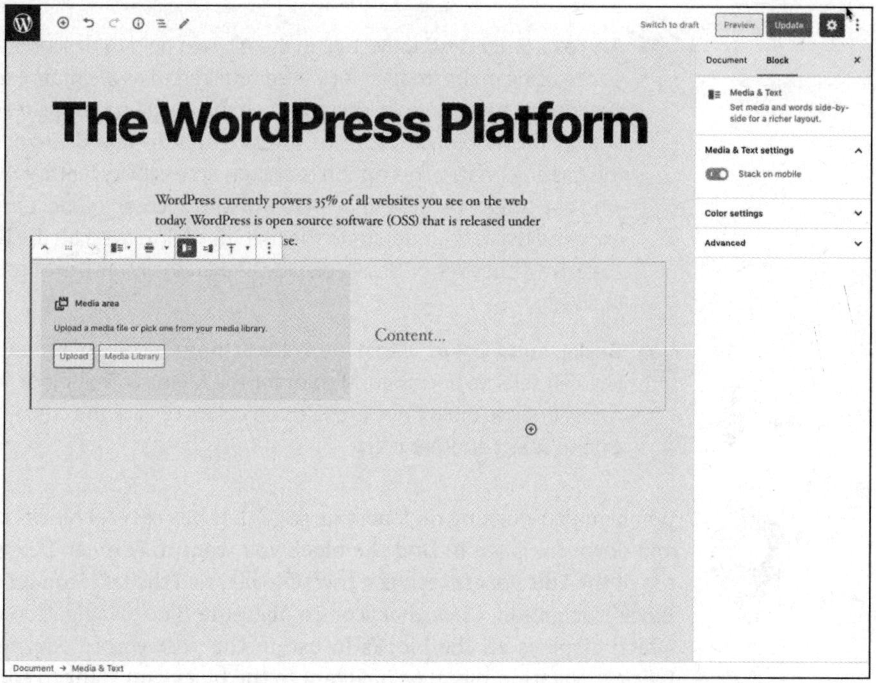

FIGURE 5-27: Options in the settings panel for the Media & Text block.

>> **Link:** This option allows you to set a URL for the image, which makes the image clickable. When a visitor to your site clicks the image, they're taken to the URL you specified.

>> **(Replace) Open Media Library**: Click. this option to replace the image with one in your Media Library.

>> **(Replace) Upload**: Click this option to replace the image with one you'd like to upload from your computer.

>> **More Options:** The settings here are the same as the ones I discuss in "Paragraph block settings" earlier in this chapter.

Additional options for the Media & Text block available in the settings panel on the right side of the Edit Post screen include

>> **Stack on Mobile:** Click this option to define the display of the Media & Text block on mobile devices. If you leave this option deselected, the block spans the width of the mobile screen. Depending on the image and text you've included in the block, the block could be difficult to read. Select the Stack on Mobile option to display the content within Media & Text blocks stacked on top of one another. If you have the media set to display on the left with the text on the right, for example, on mobile devices, the media displays above the text.

>> **Alt Text:** Enter descriptive text in the Alt Text field to describe the image you're using in the Image block. Also referred to as *alternative text*, this description helps people who can't see the image on your site; when the image doesn't load on your site, the alt text is displayed, providing context for the missing image. This option is also an accessibility feature that helps those who use screen readers to browse the web. (Screen readers allow visually impaired users to understand the text that's displayed on the website with a speech synthesizer or Braille display.) Additionally, Alt Text descriptions assist in SEO.

>> **Background Color:** Within the Color Settings section of the settings panel, you can select a background color for the Media & Text block you're currently using. Choose one of five preselected colors or click the custom color picker icon to select another color.

TIP

When you're working on a post or page that has several blocks in use, scrolling up and down the page to find the block you want to edit can be cumbersome. At the top of the Edit Post screen is a row of icons, and the last icon on the right is labeled Block Navigation. Click that icon to open the Block Navigation drop-down menu, which displays all the blocks in use in the post you're editing. You can use the Block Navigation menu to navigate to the block you want to edit.

TIP

If you're ever curious about the structure of the content on a page you're creating, see the small icon at the top of the Edit Post screen that looks like a lowercase *i* with a circle around it. (It's the fourth icon from the left.) When you hover over this icon, you see the label Content Structure. Click the Content Structure icon to open a small window with some details about the post or page you're editing.

If you're a total geek like me and like to work with code, or if you're curious to see the underlying HTML code for the blocks you're creating, WordPress provides a way to do that. In the top-right corner of the Edit Post screen is an icon that looks like three dots stacked on top of one another. Hover your mouse over this icon, and you see the label Show More Tools & Options. Click the icon to open a menu of options. To view the code versions of the blocks you've created on your site, choose the Code Editor option on the Editor section of the menu. This action changes the display of your post content to code rather than the default visual editor. Figure 5-28 shows a post in code.

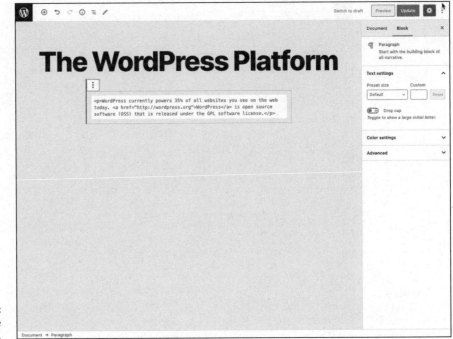

The WordPress Platform

<p>WordPress currently powers 35% of all websites you see on the web today. <a href="http://wordpress.org">WordPress</a> is open source software (OSS) that is released under the GPL software license.</p>

Document → Paragraph

**FIGURE 5-28:**
Using the Code
Editor on a post.

# Refining Your Post Options

After you write the post, you can choose a few extra options before you publish it for the entire world to see. On the right side of the Edit Post screen is the settings panel, which you should be familiar with from earlier sections of this chapter. Click the Document link at the top to view the options you can set for the post globally, shown in Figure 5-29. Unlike settings for individual blocks, the Document settings pertain to the entire post.

**TIP**

You'll see a lot of options and settings on the screens where you add new and edit existing posts and pages. Should you find all those options, links, and menus distracting, WordPress allows you to write in distraction-free mode. In the top-right corner of the Edit Post screen is an icon that looks like three dots stacked on top of one another. Hover your mouse over this icon, and you see the label Show More Tools & Options. Click this icon to open a menu of options. Choose the Full Screen Mode option in the View section of the settings panel to change the screen view to full screen, which removes all the distractions of those pesky links, menus, and settings. You can restore the screen to normal by performing the same action and deselecting Full Screen Mode.

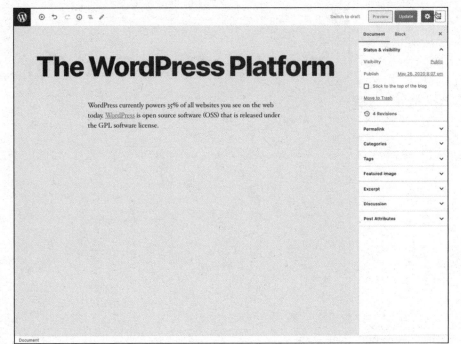

**FIGURE 5-29:**
The Document
section of the
settings panel.

The Document settings include the following:

>> **Status and Visibility:** By default, the visibility status of your post is set to
Public. You can select the status option by clicking the Public link in the Status
and Visibility panel:

- *Public:* Select this option to make the post viewable by everyone who visits
your site.

- *Private:* Select this option to make the post viewable only by site administrators and editors. Saving a post with this status prevents anyone else
from viewing the post on your site.

- *Password Protected:* Select this option to create a password for your post.
By assigning a password to a post, you can publish a post that only you can
see. You also can share the post password with a friend who can see the
content of the post after she enters the password. But why would anyone
want to password-protect a post? Suppose that you just ate dinner at your
mother-in-law's house, and she made the *worst* pot roast you've ever
eaten. You can write all about it! Protect it with a password and give the
password to your trusted friends so that they can read about it without
offending your mother-in-law.

- *Publish:* By default, WordPress assigns the publish date and time as the exact date and time when you originally published the post to your site. If you want to future-publish this post, you can set the time and date for any time in the future. If you have a vacation planned and don't want your site to go without updates while you're gone, for example, you can write a few posts, and set the date for a time in the future. Those posts are published to your site while you're somewhere tropical, diving with the fish. Click the date and time displayed, and a date and time picker appears. Use this picker to set the date and time when you'd like this post to publish to your site.

- *Stick to the Front Page:* Select this check box to have WordPress publish the post to your site and keep it at the top of all posts until you change this setting; this type of post is known as a *sticky post.* Typically, posts are displayed in chronological order on your site, with the most recent post at the top. If you make a post sticky, it remains at the top no matter how many other posts you make after it. When you want to unstick the post, deselect the Front Page check box.

- *Pending Review:* Select this check box to save the post as a draft with the status of Pending Review. This option alerts the administrator of the site that a contributor created a post that's waiting for administrator review and approval. (This feature is helpful for sites that have multiple authors.) Generally, only contributors use the Pending Review option. Note that this option is available only for new posts. You won't see it in the settings panel of posts that have already been published.

- *Author:* If you're running a multiauthor blog, you can choose the name of the author you want to assign to the post you're editing. By default, your name is selected as the author name in the Author drop-down list. (You don't see this option in Figure 5-29 because my site has only one author.)

- *Move to Trash:* Click this button to delete the post you've been working on. This action doesn't permanently delete the post, however. You can find and restore that post by visiting the Posts screen in your Dashboard (choose Posts ➪ All Posts) and clicking the Trash link.

» **Revisions:** In "Writing Your First Entry" earlier in this chapter, I talk about the autosave feature. Its function is to automatically save the work you've done on posts you're creating so that you don't lose any of it. Each time you edit a post, WordPress automatically saves the old version of your post and stores it as a revision, making it available for you to access later. This section gives you an indication of how many revisions a post has. When you click the Revisions link, you see the Compare Revisions screen, where you can review and restore revisions of your post.

>> **Permalink:** A *permalink* is the direct link, or URL, to the post you're about to publish (see the earlier section "Making your post links pretty"). Although you can't change the domain portion of this URL (`https://domain.com`), you can adjust the part of the URL that appears in the link after the final slash at the end of your domain. For a post titled WordPress Tips, WordPress automatically creates a URL from that title like `http://domain.com/wordpress-tips`. Use the URL field in the Permalink section of the settings panel to set different words for your post (or page) from the ones that WordPress automatically creates for you. You could shorten the slug for the post title WordPress Tips to wordpress so that the URL for the post is `http://domain.com/wordpress`.

>> **Categories:** You can file your posts in different categories to organize them by subject. (See more about organizing your posts by category in "Staying on Topic with Categories" earlier in this chapter.) Select the check box to the left of the category you want to use.

Don't see the category you need in the Category section? Click the Add New Category link, and add a new category right there on the page you're using to create or edit your post.

>> **Tags:** Type your desired tags in the Add New Tag text box. Be sure to separate tags with commas so that WordPress knows where each tag begins and ends. Cats, Kittens, Feline represents three different tags, but if you enter the tags without the commas, WordPress would consider those three words to be one tag.

>> **Featured Image:** Some WordPress themes are configured to use an image (photo) to represent each post on your site. The image can appear on the home/ front page, blog page, archives page, search results page, or anywhere within the content displayed on your website. If you're using a theme that has this option, you can easily define it by clicking Set Featured Image in the Featured Image section of the settings panel. This action opens a window that allows you to upload a new image or select an existing image from the Media Library.

>> **Excerpt:** *Excerpts* are short summaries of your posts. Many authors use snippets to show teasers of their posts on their website, thereby encouraging readers to click the Read More links to read the posts in their entirety. By default, WordPress automatically creates an excerpt based on the text contained in the first paragraph of your post. But if you want to control what text is displayed for the excerpt of your post, type your desired text in the Write an Excerpt box, which is displayed when you click the Excerpt section in the setting panel. Excerpts can be any length in terms of words, but the point is to keep them short and sweet to tease your readers into clicking the Read More link.

>> **Discussion:** Decide whether to let readers submit comments through the comment system by selecting Allow Comments in the Discussion section of the settings panel. Additionally, you can enable pingbacks and trackbacks by

clicking the check box in the Discussion section labeled Allow Pingbacks & Trackbacks. By default, both options are selected for posts you create on your site. For more on trackbacks, see Chapter 2.

# Publishing Your Post

You've given your new post a title and written the content of the post by assembling all the content blocks you need to create the post you desire. Maybe you've even added an image or other type of media file to the post (see Chapter 6), and you've configured the tags, categories, and other options in the settings panel. Now the question is this: To publish or not to publish (yet)?

WordPress gives you three options for saving or publishing your post when you're done writing it. These options are located in the top-right corner of the Add New (or Edit) Post screen. Figure 5-30 shows the available options: Save Draft, Preview, and Publish.

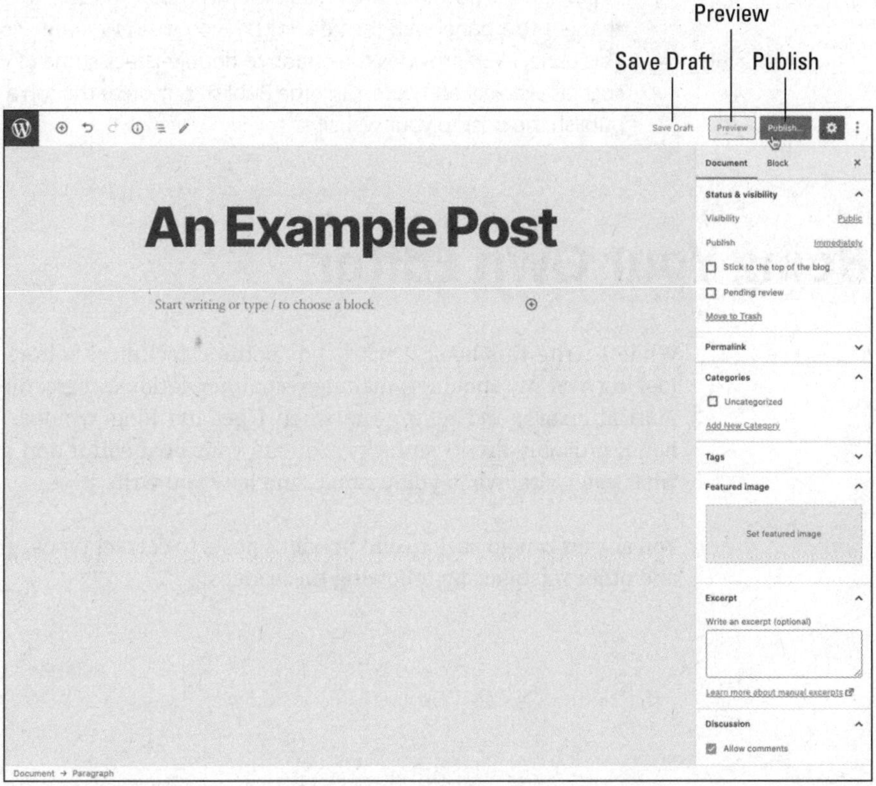

**FIGURE 5-30:** The publish options for your posts and pages.

The options for saving or publishing your post include

>> **Save Draft:** Click this link to save your post as a draft. The Save Draft link refreshes with a message that says Saved, indicating that your post has been successfully saved as a draft. The action of saving as a draft also saves all the post options you've set for the post, including blocks, categories, tags, and featured images. You can continue editing now, tomorrow, the next day, or next year; the post is saved as a draft until you decide to publish it or delete it. Posts saved as drafts can't be seen by visitors to your site. To access your draft posts in your Dashboard, visit the Posts screen (choose Posts ⇨ All Posts), and click the Drafts link on the top menu.

>> **Preview:** Click the Preview button to view your post in a new window, as it would appear on your live site if you'd published it. Previewing the post doesn't publish it to your site yet. Previewing simply gives you the opportunity to view the post on your site and check it for any formatting or content changes you want to make.

>> **Publish:** Click the Publish button when you're ready to publish your post or page to your website and allow your visitors to view it when they visit. WordPress puts a small fail-safe feature in place to make sure that you want to publish the post live; when you click the Publish button, the settings panel changes to a panel with the heading Do you really want to publish? This panel even provides the option to double-check some of your settings, such as visibility and date. Click the Publish button at the top a second time to publish the post to your website.

# Being Your Own Editor

While I write this book, I have copy editors, technical editors, and proofreaders looking over my shoulder, making recommendations, correcting typos and grammatical errors, and telling me when I get too long-winded. You, on the other hand, probably aren't so lucky! You are your own editor and have full control of what you write, when you write it, and how you write it.

You always can go back to edit previous posts to correct typos, grammatical errors, and other mistakes by following these steps:

1. **Find the post that you want to edit by clicking the All Posts link on the Posts menu of the Dashboard.**

   The Posts screen opens, listing the 20 most recent posts you've created.

**TIP**

   To filter that listing of posts by date, choose a date from the All Dates drop-down menu at the top of the Posts screen (choose Dashboard ⇨ Posts). If you choose January 2021, the Posts page reloads, displaying only those posts that were published in January 2021.

   You also can filter the post listing by category. Choose your desired category from the All Categories drop-down list.

2. **When you find the post you need, click its title.**

   Alternatively, you can click the Edit link that appears below the post title when you hover your mouse over it.

   The Edit Post screen opens. In this screen, you can edit the post and/or any of its options.

**TIP**

   If you need to edit only the post options, click the Quick Edit link that appears below the post title when you hover your mouse over it. A Quick Edit menu appears, displaying the post options that you can configure, such as title, status, password, categories, tags, comments, and time stamp. Click the Update button to save your changes.

3. **Edit your post; then click the Update Post button.**

   The Edit Post screen displays a message that the post has been updated.

# Look Who's Talking on Your Site

The feature that really catapulted blogging into the limelight is the comments feature, which lets visitors interact with the authors of sites. Comments and trackbacks are great ways for readers to interact with site owners, and vice versa. I cover both comments and trackbacks in Chapter 2.

## Managing comments and trackbacks

To find your comments, click the Comments link on the Dashboard navigation menu. The Comments page opens (see Figure 5-31).

FIGURE 5-31:
Clicking the
Comments menu
shows you the
Comments page,
with all the
comments and
trackbacks on
your site.

When you hover your mouse pointer over your comments, several links appear that give you the opportunity to manage those comments:

>> **Unapprove:** This link appears only if you have comment moderation turned on, and only for approved comments. The comment is placed in the moderation queue, which you get to by clicking the Awaiting Moderation link that appears below the Manage Comments header. The moderation queue is kind of a holding area for comments that haven't yet been published to your blog. (See "Moderating comments and trackbacks" later in this chapter for more on the moderation queue.)

>> **Reply:** Click this link, and a text box drops down. In this text box, you can type and submit your reply to the person who commented. This feature eliminates the need to load your live site to reply to a comment.

>> **Quick Edit:** Click this link to open the comment options without leaving the Comments page. You can configure post options such as name, email, URL, and comment content. Click the Save button to save your changes.

>> **Edit:** Click this link to open the Edit Comment page, where you can edit fields such as name, email, URL, and comment content (see Figure 5-32).

>> **Spam:** Click this link to mark the comment as spam and toss it into the spam bin, where it will never be heard from again!

>> **Trash:** This link does exactly what it says: sends the comment to the Trash and deletes it from your blog.

TIP

If you have a lot of comments listed in the Comments page and want to bulk-manage them, select the boxes to the left of all the comments you want to manage. Then choose one of the following from the Bulk Actions drop-down menu in the top-left corner: Approve, Mark As Spam, Unapprove, or Delete.

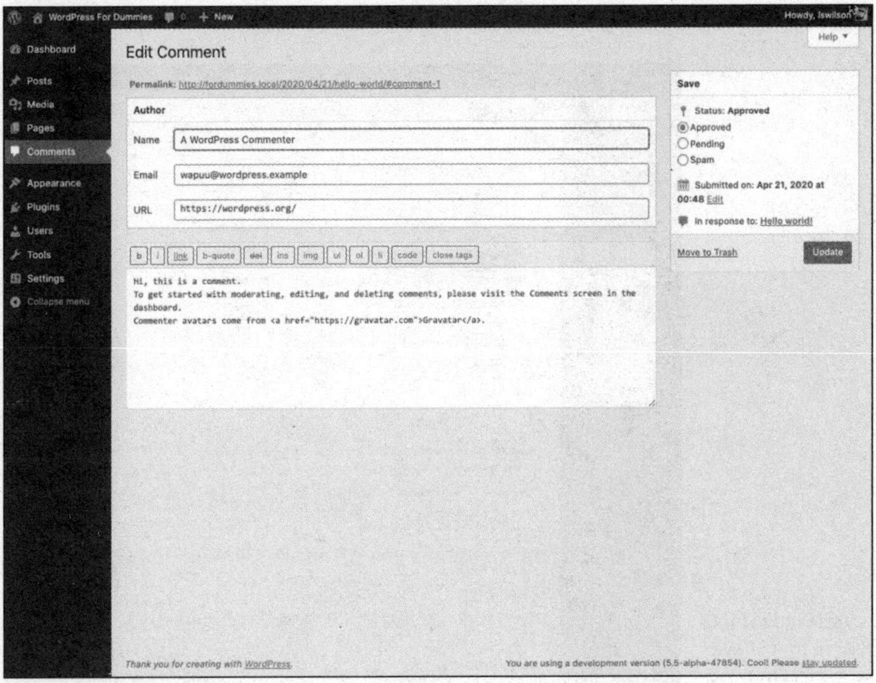

FIGURE 5-32:
Edit a user's
comment in the
Edit Comment
page.

## Moderating comments and trackbacks

If you have your options set so that comments aren't published to your site until you approve them, you can approve comments from the Comments screen as well. Just click the Pending link on the Comments screen, and you go to the Edit Comments page. If you have comments and/or trackbacks awaiting moderation, you see them on this page, where you can approve them, mark them as spam, or delete them.

A nice feature of WordPress is that it immediately notifies you of any comments sitting in the moderation queue, awaiting your action. This notification appears as a small circle to the right of the Comments menu on the left navigation menu of every single page. Figure 5-33 shows my Dashboard page with an indicator in the Comments menu showing a comment awaiting moderation. (An indicator also appears on the top toolbar.)

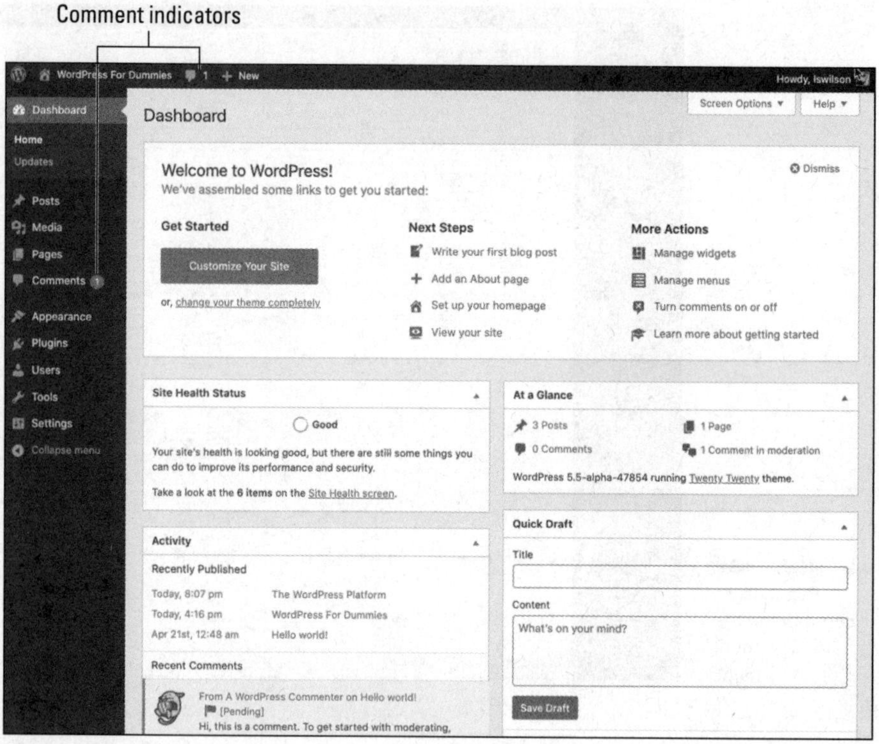

Comment indicators

FIGURE 5-33:
These indicators
tell me that I have
a comment
awaiting
moderation.

# Tackling spam with Akismet

I touch on Akismet a few times throughout this book because it's my humble opinion that Akismet is the mother of all plugins and that no WordPress site is complete without a fully activated version of Akismet running on it.

Apparently, WordPress agrees, because the plugin has been packaged in every WordPress software release beginning with version 2.0. Akismet was created by the folks at Automattic.

I've been blogging since 2002 when I started blogging with the Movable Type blogging platform. I moved to WordPress in 2003. As blogging became more and more popular, comment and trackback spam became more and more of a nuisance. One morning in 2004, I found that 2,300 pieces of disgusting comment spam had been published to my blog. Something had to be done! The folks at Automattic did a fine thing with Akismet. Since the emergence of Akismet, I've barely had to think about comment or trackback spam except for the few times a month I check my Akismet spam queue.

I talk in greater detail about plugin use in WordPress in Chapter 7, where you find out how to activate Akismet and make sure that it's protecting your blog from trackback and comment spam.

# 3

# Flexing and Extending WordPress

**IN THIS PART . . .**

Learn how to use the WordPress Media Library to manage images and media.

Embed videos in your content to provide media for your readers to interact with.

Discover, download, and install plugins for WordPress to extend the features available on your website.

Find and install WordPress themes that give your website a great visual look.

# Chapter **6**

# Media Management: Images, Audio, and Video

A dding images and photos to your posts and pages can really dress up the content. By using images and photos, you give your content a dimension that you can't express in plain text. Through visual imagery, you can call attention to your content and improve the delivery of the message by adding depth to it.

The same goes for adding video and audio files to your posts and pages. Video lets you provide entertainment through moving, talking (or singing!), and streaming video. Audio files let you talk to your visitors and add a personal touch. Many website owners use video and audio to report news and to broadcast Internet radio and television shows. The possibilities are endless!

In this chapter, you discover how to enhance your website by adding images, video, and audio to your content. You even find out how to run a full-fledged photo gallery on your site, all through the WordPress.org software and its integrated Media Library.

REMEMBER

You add these extras to your site in the Upload/Insert area of the Add New Post page. You can add them as you're writing your post or come back and add them later. The choice is yours!

# Inserting Images into Your Content

Adding an image to a post is easy with the WordPress image uploader. Jump right in and give it a go. From the Dashboard, click the Add New link on the Posts menu, and the Edit Post screen loads in your browser. On the Edit Post screen, click the Add Block icon to open the block selector, and click the Image icon to add an Image block to your post, as shown in Figure 6-1.

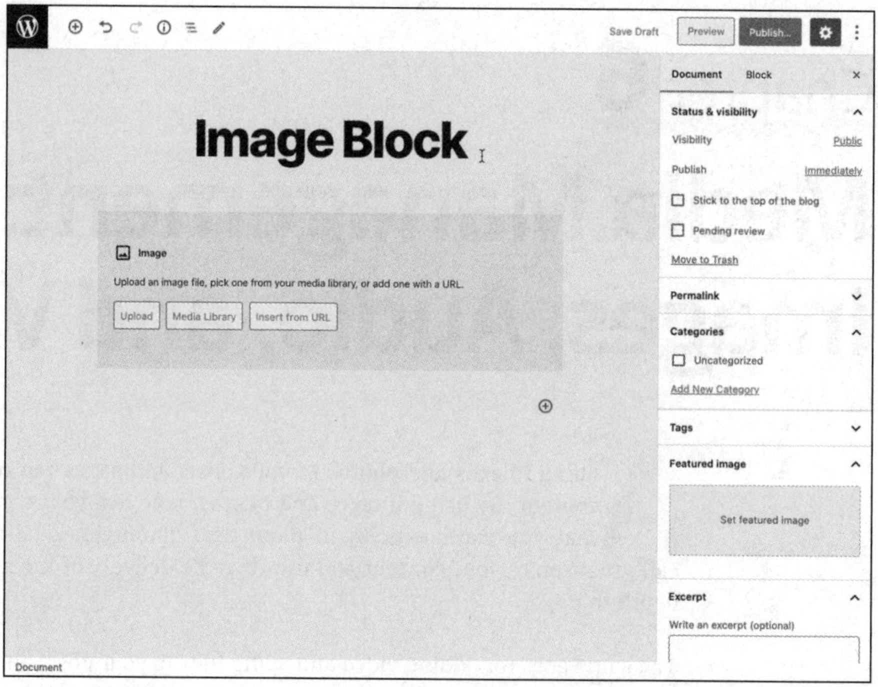

**FIGURE 6-1:**
An Image block in
the WordPress
block editor.

The Image block gives you four ways to add an image to your post, which are described in detail in the following sections of this chapter:

>> **Upload:** Select an image from your computer and upload it to your site.

>> **Media Library:** Select an image from the WordPress Media Library.

>> **Insert from URL:** Use an image from a different source by adding an image URL.

>> **Drag an image:** Drag an image from your computer to your WordPress site.

# Uploading an image from your computer

After you've added the Image block to your post, you can add an image from your computer's hard drive by following these steps:

1. **Click the Upload button in the Image block.**

   A dialog box opens, allowing you to select an image (or multiple images) from your computer's hard drive. (See Figure 6-2.)

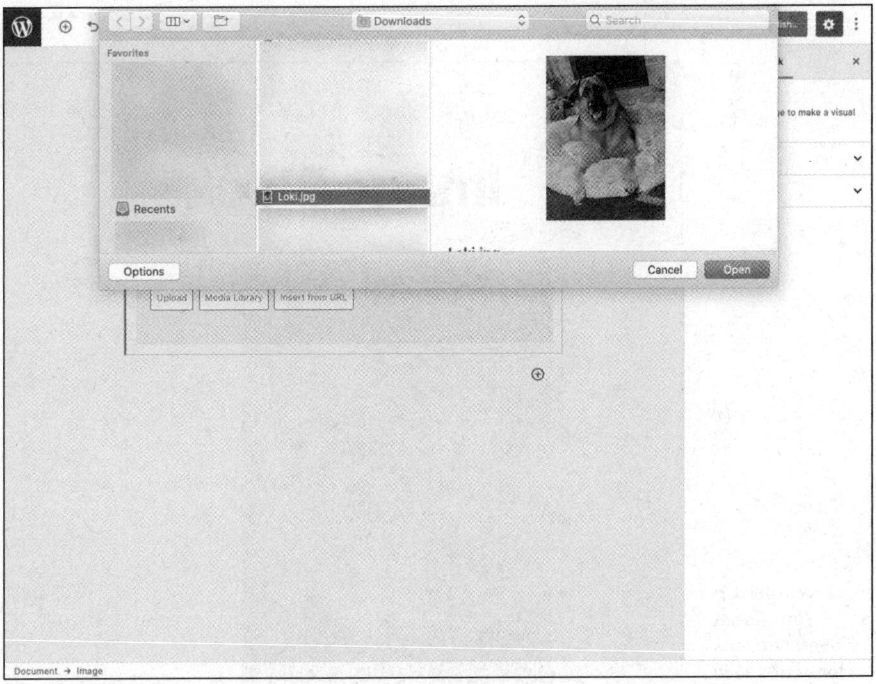

**FIGURE 6-2:**
Uploading an image from your computer.

2. **Select your image(s) from your hard drive and then click the Open button.**

   The image is uploaded from your computer to your website, and the Edit Post screen displays your uploaded image ready for editing, if needed.

3. **Edit the details for the image in the Image Settings section of the settings panel on the right side of the Edit Post screen (see Figure 6-3).**

   The Image Settings section provides several image options, which are covered in Chapter 5:

   - Alt Text
   - Image Size

- Image Dimensions

- Link To

**TIP**

WordPress automatically creates small and medium-size versions of the original images you upload through the built-in image uploader. A thumbnail is a smaller version of the original file. You can edit the size of the thumbnail by clicking the Settings link and then clicking the Media menu link. In the Image Sizes section of the Media Settings page, designate the desired height and width of the small and medium thumbnail images generated by WordPress.

**FIGURE 6-3:**
You can set several options for your image after you add it to your post.

4. **Use the Image block toolbar to set the display options for the image.**

   Figure 6-4 shows the toolbar for the Image block. In Chapter 5, I cover how to work with the toolbar options for the Image block at length.

5. **Continue adding content to your post, or publish it.**

Image block toolbar

**Image Block**

**FIGURE 6-4:**
When you click
the Image block,
the toolbar
appears at the
top of the Edit
Post screen.

## Inserting an image from the Media Library

The WordPress Media Library (see Chapter 5) contains all the images you've ever uploaded to your website, making those images available for use in any post or page you create on your site. After you've added an Image block to your post, you can add an image from the Media Library by following these steps:

1.  **Click the Media Library button in the Image block (refer to Figure 6-1).**

    The Select or Upload Media screen opens, with the Media Library section displayed (see Figure 6-5).

2.  **Select the image you want to use by clicking it.**

3.  **Click the Select button.**

    The Select or Upload Media screen closes, and the Edit Post screen reappears. WordPress inserts the image you've chosen into the post you're creating.

4.  **Set the options for the image.**

    Complete steps 3 and 4 of "Uploading an image from your computer" earlier in this chapter.

5.  **Continue adding content to your post, or publish it.**

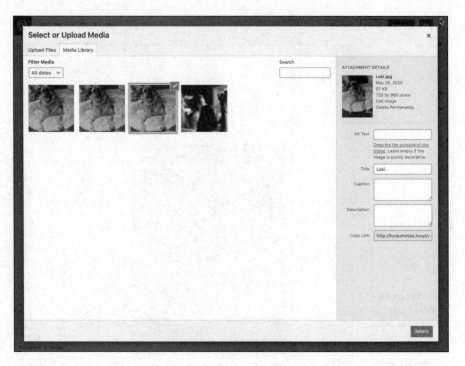

**FIGURE 6-5:**
The Select or
Upload Media
screen.

You can also insert an image into your post by using a URL or by dragging an image from your computer to your WordPress website. These techniques are covered in Chapter 5.

**TIP**

# Using the Columns Block to Insert Multiple Images in a Row

The Layout Elements section of the WordPress block editor has a block called Columns. You can use this block to add a row of columns to your post and then insert images into those columns to create a row of multiple images for display in your post. Figure 6-6 displays a post with a grid of images created by using a variety of columns and images. I used the standard Image block for the first image you see; then I used a Columns block to create the two side-by-side images on the page.

To add and configure a Columns block with images in your post, follow these steps:

FIGURE 6-6:
A grid of images
in a blog post
using the
Columns and
Image blocks.

1. **Add the Columns block with 2 Columns in your post on the Edit Post screen.**

   You can find the methods for adding blocks to your post in Chapter 5. Figure 6-7 shows the Columns block inserted into a post with different variations to choose among. Select the two-column variation.

2. **Click the Add Block icon to add an Image block in the left column.**

   When you hover your mouse in the first (left) column, a small plus icon appears. When you click that icon, the WordPress block-selector window appears, and you can select the Image block and insert it into the first column, as shown in Figure 6-8.

3. **Add your image to the left column, using the Image block options.**

   Follow the steps in "Inserting an Image into Your Content" earlier in this chapter.

4. **Repeat steps 2 and 3 for the column on the right side of the Columns block to insert an image into the right column.**

5. **Continue adding content to your post, or publish it.**

   When you're done, your Edit Post screen looks like Figure 6-9, with your two selected images displayed next to each other. (If you want to create a grid like the one in Figure 6-6, add an Image block above the Columns block in the Edit Post screen to add a single image, for a grid of three images.)

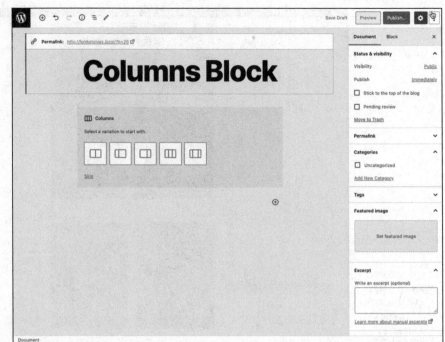

**FIGURE 6-7:**
The Columns
block in the
WordPress
Block Editor.

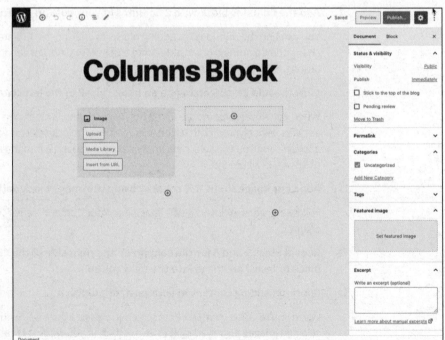

**FIGURE 6-8:**
Adding the
Image block to
one of the
columns in the
Columns block.

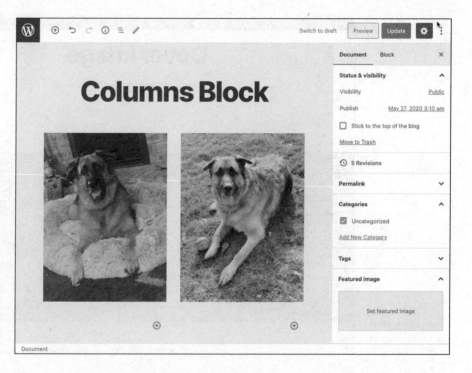

**FIGURE 6-9:**
The Columns
block on the Edit
Post screen.

**TIP**

You can add more than two columns by using the Columns block. In the settings panel for the Columns block (on the right side of the Edit Post screen shown in Figure 5-7), you can define how many columns you want the Columns block to have. You can have three or four images displayed in a row, for example. You could get creative, creating grids of images by stacking multiple Columns blocks on top of one another, each with differing numbers of columns. You might have four rows of images in which the first row has two columns, the second row has four columns, the third row has two columns, and the fourth row has one column. That arrangement would create an interesting image grid.

# Inserting a Cover Image into a Post

The WordPress Block Editor has a Cover block, which you can use in place of a heading to add additional emphasis to a section of your site. The Cover block allows you to display a short line of text on top of an image of your choice. Figure 6-10 displays a post I created about my dogs.

Using the Cover block is a nice way to separate different sections of your post content. You can use it in place of a regular heading to add more visual emphasis and appeal to your content, for example.

**FIGURE 6-10:**
An example of
the Cover block in
use on my site.

To add and configure a Cover block in your post, follow these steps:

1. **Add the Cover block in your post on the Edit Post screen.**

   You can find the methods for adding blocks to your post in Chapter 5. Figure 6-11 shows the Cover block inserted into a post.

2. **Click the Upload button to select an image from your computer or the Media Library button to select an image from the Media Library.**

   The steps for these two methods are listed in "Uploading an image from your computer" and "Inserting an image from the Media Library," earlier in this chapter. When you're done, the image appears in your content on the Edit Post screen.

3. **Add your desired text in the Cover block.**

   Click the text `Write title` in the Cover block, and type your own text over it. In Figure 6-12, I added text that displays on top of the image I added in the Cover block.

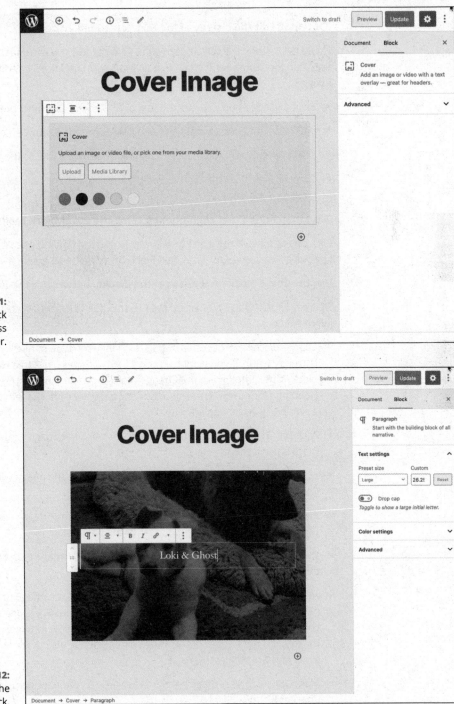

**FIGURE 6-11:**
The Cover block in the WordPress Block Editor.

**FIGURE 6-12:**
Adding text in the Cover block.

4. **(Optional) Edit the display of the text you added in step 3.**

Figure 6-12 shows the text added to the Cover block. You can use the small four-icon toolbar above the text to adjust the format of the text added to the image:

- *Bold:* Click this icon to make the text bold (darker). Example: **bold text.**

- *Italic:* Click this icon to make the text italic (slanted). Example: *italic text.*

- *Link:* Click this icon to create a hyperlink by adding a URL to the text.

- *Strikethrough:* Click this icon to apply a line through the text. Example: ~~strikethrough text.~~

5. **(Optional) Adjust text alignment by using the Cover block toolbar.**

Click one of these three icons to align the text in the Cover block: Align Text Left, Align Text Center, or Align Text Right. By default, text is center-aligned.

6. **Use the Cover block toolbar to set the display options for the image.**

Figure 6-13 shows the toolbar for the Cover block. The options include (from left to right)

- *Change Block Type:* Click this option to change the type of block you're currently using. If you want to change from a Cover block to a Heading block, for example, click the Change Block Type icon and then select the Heading block to swap it. The only block types you can change to from a Cover block are Image and Heading.

- *Align Left:* This block-formatting option positions the image within the Cover block against the left margin of the page.

- *Align Center:* This block-formatting option positions the image within the Cover block in the center of the page.

- *Align Right:* This block-formatting option positions the image within the Cover block against the right margin of the page.

- *Wide Width:* Click this option to set the width of the image to the width of the content on the page.

- *Full Width:* Click this option to increase the width of the image to the width of the screen you're viewing the content on. In Figure 6-10 earlier in this chapter, you see a post I created with a Full Width cover image. Notice that the left and right edges of the image extend all the way to the left and right sides of the viewing screen.

- *More Options:* The settings here are the same as the ones discussed in Chapter 5.

Cover block toolbar

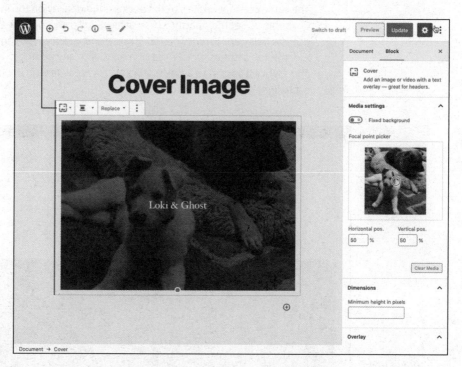

**FIGURE 6-13:**
The Cover block
toolbar.

**7.** **(Optional) Set the Cover block settings to a fixed background.**

In the settings panel on the right of the Edit Post screen, you see settings for the Cover block. In the Cover Settings section is a toggle setting called Fixed Background. By default, this setting is set to off. Click the toggle button to set the background image to fixed. A fixed background means that the image is locked in place and doesn't move as your visitors scroll down the page of your website. It's a neat effect; give it a try! If you don't like the effect, set the Fixed Background option to off.

**8.** **(Optional) Set a color overlay for the image in the Cover block.**

In the settings panel, you see a section called Overlay, with these options:

● *Overlay Color:* You can set the color for the background of the image as one of the five predefined colors in the settings panel, or click the custom color picker icon to set a specific image. The color you choose is overlaid on the image.

● *Background Opacity:* When you've set the overlay color, you can set the opacity of that color as well. *Opacity* refers to the transparency of the color. Suppose that the image you're using is hanging on a wall in your house, and you want to hang a curtain in front of it. If that curtain is solid black, you could say that the curtain has 100 percent opacity (or zero transparency) because you can't see the image on the wall through the curtain.

The same concept applies to the background opacity on the image in the Cover block. Set the opacity to 0 to achieve full transparency; set it to 50 to achieve half transparency or 100 for no transparency. You can use the slider in the settings panel to set the opacity to any point between 0 and 100. Figure 6-14 shows a Cover block on my website with the overlay color set to light gray and 80 percent opacity.

**FIGURE 6-14:** The Cover block used on a site with an image that has a light gray overlay color set to 80 percent opacity.

# Inserting a Photo Gallery

You can also use the WordPress Block Editor to insert a full photo gallery into your posts. Upload all your images; then, instead of adding an Image block, use the Gallery block.

Follow these steps to insert a photo gallery into a blog post:

**1. Add the Gallery block to your post on the Edit Post screen.**

You can find the methods for adding blocks to your post in Chapter 5. Figure 6-15 shows the Gallery block inserted into a post.

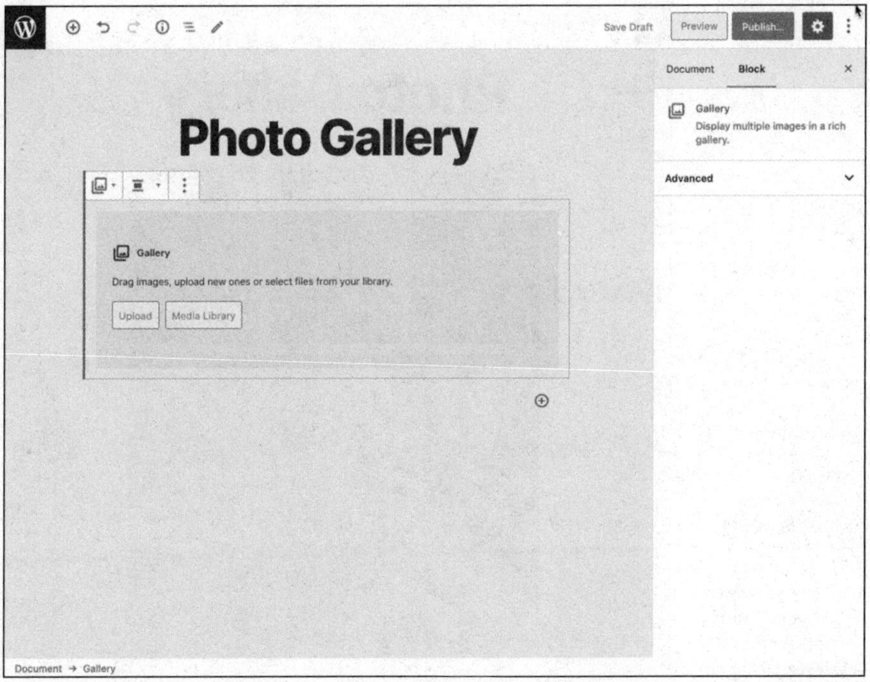

FIGURE 6-15:
The Gallery block
in the WordPress
Block Editor.

2. **Click the Upload button to select an image from your computer or the Media Library button to select an image from the Media Library.**

The Create Gallery screen opens. The steps for these two methods appear in "Uploading an image from your computer" and "Inserting an image from the Media Library" earlier in this chapter. The only difference is that you can select multiple images to include in the Gallery block; when selected, the images appear at the bottom of the Create Gallery screen.

3. **Click the Create a New Gallery button.**

The Edit Gallery screen opens, displaying all the images you selected in step 2.

4. **(Optional) Add a caption for each image by clicking the Caption This Image area and typing a caption or short description for the image.**

5. **(Optional) Set the order in which the images appear in the gallery by using the drag-and-drop option on the Edit Gallery page.**

Drag and drop images to change their order.

6. **Click the Insert Gallery button.**

WordPress inserts the selected images into your post in the Gallery block (see Figure 6-16).

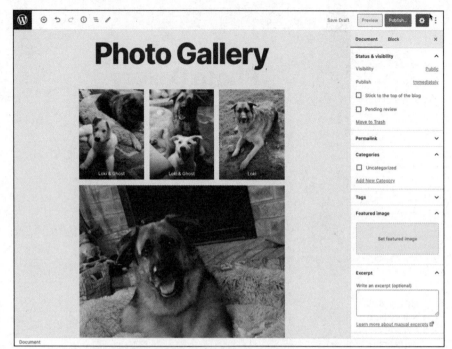

**FIGURE 6-16:**
The Gallery block
populated with
selected images.

7. **Use the Gallery block toolbar to set the display options for the gallery.**

   Figure 6-17 shows the toolbar for the Gallery block. The options include (from left to right)

   - *Change Block Type:* Click this option to change the type of block you're currently using. If you want to change from a Gallery block to an Image block, for example, click the Change Block Type icon and then select the Image block to swap it. The only block type you can change to from a Gallery block is an Image block.

   - *Align Left:* This block-formatting option positions the image within the Gallery block against the left margin of the page.

   - *Align Center:* This block-formatting option positions the image within the Gallery block in the center of the page.

   - *Align Right:* This block-formatting option positions the image within the Gallery block against the right margin of the page.

   - *Wide Width:* Click this option to set the width of the image to the width of the content on the page.

Gallery block toolbar

FIGURE 6-17:
The Gallery block
toolbar.

- *Full Width:* Click this option to increase the width of the Gallery to the full width of the screen your visitors are viewing the content on. In Figure 6-18, you see a post I created with a full-width gallery. Notice that the left and right edges of the image extend all the way to the left and right sides of the viewing screen.

- *More Options:* These settings are the same ones discussed in Chapter 5.

8. **Use the Gallery block options in the settings panel on the right side of the Edit Post page to configure options for your gallery:**

- *Columns:* Select how many columns of images you want to appear in your gallery.

- *Crop Images:* By default, this option creates image thumbnails that are cropped to the same size so that they align evenly. You can turn this setting off if you prefer different-size thumbnails in your gallery.

- *Link To:* Select Attachment Page, Media File, or None to tell WordPress what you'd like the images in the gallery to link to.

- *Image Sizes:* Select Thumbnail, Medium, or Full size to tell WordPress what size images you want to use in your gallery.

**FIGURE 6-18:**
A photo gallery on my website using the Gallery block.

Check out the "WordPress image and gallery plugins" sidebar for a few interesting and helpful plugins that can help you create beautiful image galleries and photo albums on your website.

TIP

Matt Mullenweg, co-founder of WordPress, created a blog that uses only photos and images in a grid format. Check out his photo blog at `https://matt.blog`.

# Inserting Video Files into Your Posts

Many website owners want to go beyond offering written content for the consumption of their visitors by offering different types of media, including audio and video files. WordPress makes it easy to include these different types of media files in your posts and pages by using the built-in file-upload feature.

The audio files you add to your site can include music or voice in formats such as `.mp3`, `.midi`, or `.wav` (to name just a few). Some website owners produce their own audio files in regular episodes, called *podcasts*, to create an Internet radio show. Often, you can find these audio files available for syndication on a variety of streaming services, such as iTunes and Spotify.

You can include videos in posts or pages by embedding code offered by popular third-party video providers such as YouTube (`https://www.youtube.com`) and Vimeo (`https://vimeo.com`). Website owners can also produce and upload their own video shows, an activity known as *vlogging* (video blogging).

TIP

Check out a good example of a video blog at `https://www.tmz.com/videos`. TMZ is a popular celebrity news website that produces and displays videos for the web and for mobile devices.

WARNING

When dealing with video and audio files on your site, remember to upload and use only media that you own or have permission to use. Copyright violation is a very serious offense, especially on the Internet, and using media that you don't have permission to use can have serious consequences; results can include having your website taken down, facing heavy fines, and even going to jail. I'd really hate to see that happen to you. So play it safe, and use only those media files that you have permission to use.

Whether you're producing your own videos for publication or embedding other people's videos, placing a video file in a post or page has never been easier with WordPress.

# Adding a link to a video from the web

Adding a video from the web, in these steps, adds a hyperlink to the video. Use these steps if all you want to do is provide a text link to a page that has the video on it, rather than embed the video in your post or page (covered in the "Adding video using the Embed Block" section later in this chapter.)

To add a link to a video from the web, follow these steps:

1. **Add a Paragraph block to your post, and type your content in it.**

2. **Select the text you'd like to link.**

   The Insert from URL page appears, as shown in Figure 6-19.

3. **Click the Link icon on the Paragraph Block toolbar.**

   A small text box opens.

4. **Type the URL (Internet address) of the video in the text box.**

   Type the full URL, including the http:// and www portions of the address, as shown in Figure 6-19. Video providers, such as YouTube, usually list the direct links for the video files on their sites; you can copy and paste one of those links into the text box.

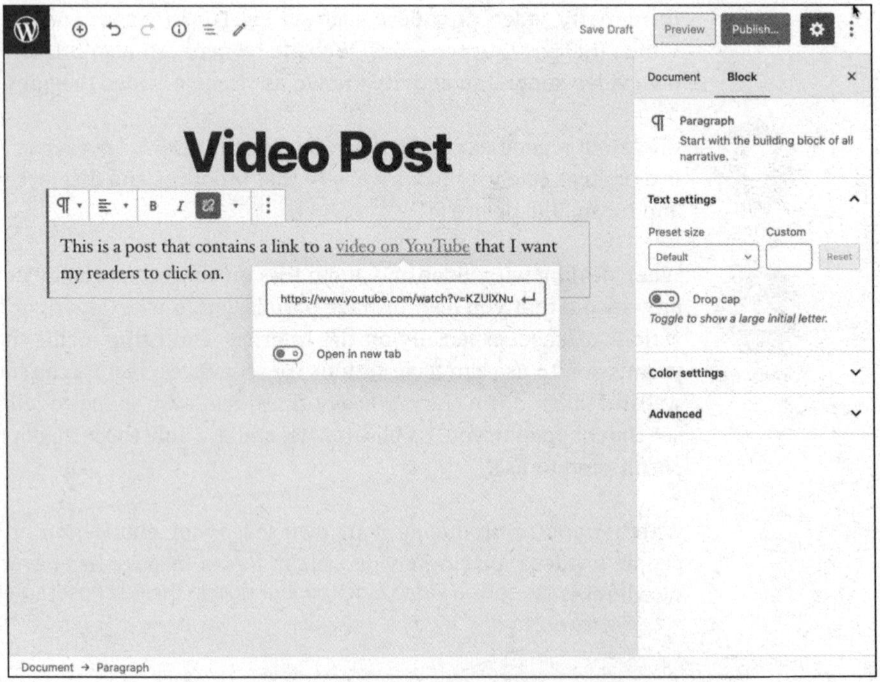

**FIGURE 6-19:**
Add a video by linking to a URL.

5. **Press Enter.**

   A link to the video is inserted into your post, as shown in Figure 6-20. WordPress doesn't embed the actual video in the post; it inserts only a link to the video. Your site visitors click the link to load another page on which the video plays.

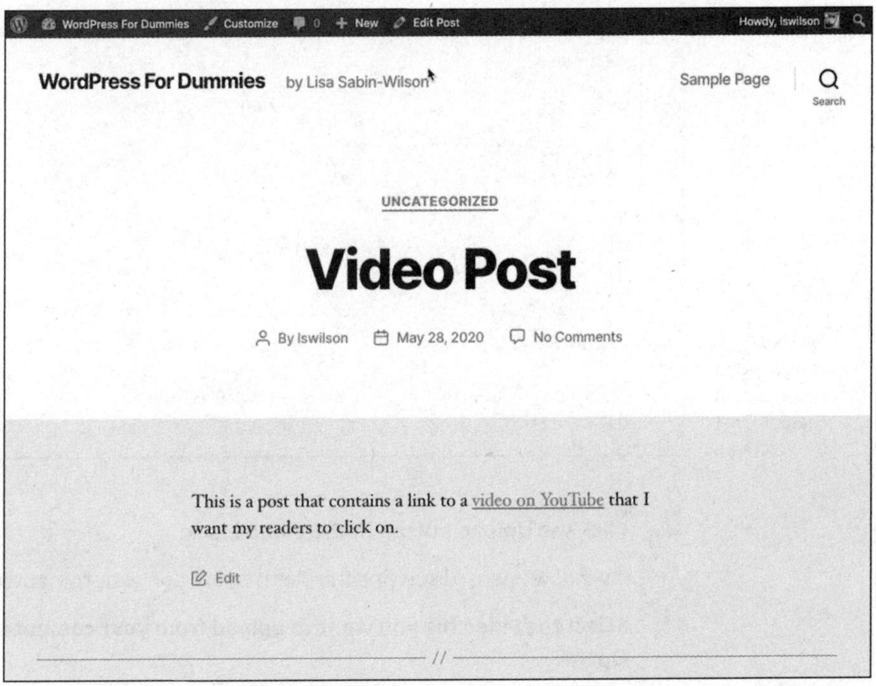

WordPress For Dummies
Customize
0
New
Edit Post
Howdy, lswilson
WordPress For Dummies
by Lisa Sabin-Wilson
Sample Page
Search
UNCATEGORIZED
Video Post
By lswilson
May 28, 2020
No Comments
This is a post that contains a link to a video on YouTube that I want my readers to click on.
Edit


**FIGURE 6-20:** A link to a video in my blog post.

**TIP**

I'm using the Link icon in the Paragraph block to illustrate how you can link to a web page or service that contains a video, but you can use the Link icon to link to any URL on the web, not just videos. Keep this method in mind when you want to insert a hyperlink into your post content.

## Adding video from your computer

To upload and post to your blog a video from your computer, follow these steps:

1. **Click the Add Block icon at the top left of the Edit Post screen, and select the Video block in the Common Blocks section.**

   WordPress adds a Video block to your post in the Edit Post screen, as shown in Figure 6-21.

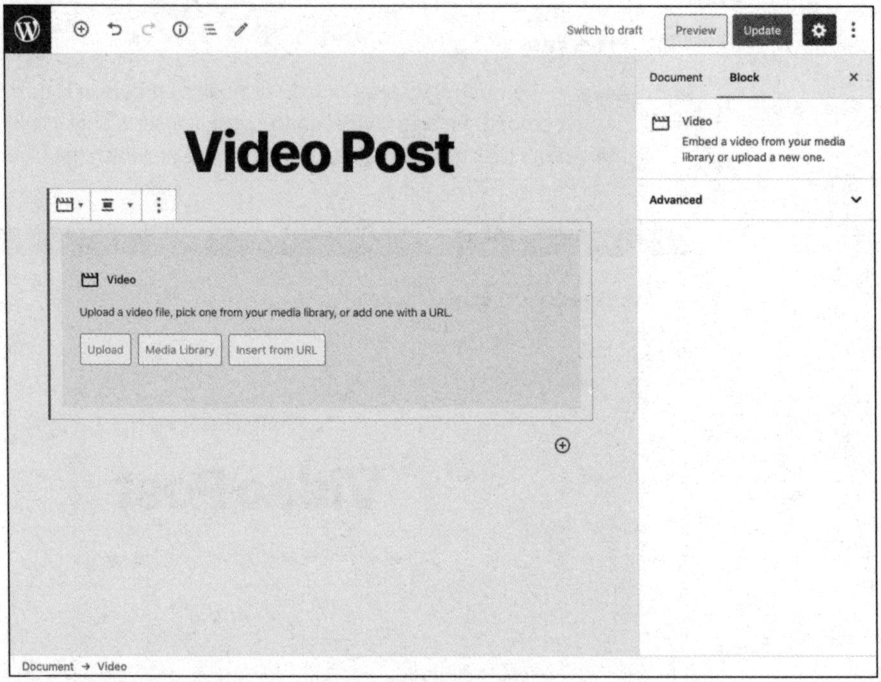

**FIGURE 6-21:**
Adding a Video
block to a post in
the Edit Post
screen.

2. **Click the Upload button in the Video block.**

   A window opens, displaying the files that exist on your computer.

3. **Select the video file you want to upload from your computer, and click Open.**

   The video is uploaded from your computer to your web server and gets inserted into the Video block on the Edit Post screen.

4. **(Optional) Add a caption in the text field below the video player, if desired.**

5. **Use the Video block's toolbar to set display options for the video.**

   The toolbar provides the following options:

   - *Change Block Type*: Clicking this option allows you to change the type of block you're using. If you want to change this Video block to a Cover block, for example, click the Change Block Type icon and then select the Cover block to swap it. For the Video block, the block types you can change to include Cover, Media & Text, and File.

   - *Align Left*: This block-formatting option positions the video within the Video block on the left side of the page.

- *Align Center*: This block-formatting option positions the video within the Video block in the center of the page.

- *Align Right*: This block-formatting option positions the video within the Video block on the right side of the page.

- *Wide Width:* Click this option to set the width of the video to span the width of the content on the page.

- *Full Width*: Click this option to increase the width of the video to span the full width of the screen on which you're viewing the content. The left and right edges of the video extend all the way to the left and right sides of the viewing screen.

- *Replace*: Select Open Media Library, Upload, or Insert from URL to replace the video with a different one.

- *More Options*: The settings here are the same as the ones discussed in Chapter 5.

6. **In the settings panel on the right side of the Edit Post screen, configure the options for the video you uploaded in step 3:**

   - *Autoplay:* By default, this option is set to off. Turn it on to set the video to start playing automatically when your visitors load this post in their browser.

   - *Loop:* By default, this option is set to off. Turn it on to set the video to play, on repeat, when your visitors load this post in their browser.

   - *Muted:* By Default, this option is set to off. Turn it on to mute the sound of the video automatically when it plays; your visitors will have to toggle the sound on when they play your video.

   - *Playback Controls:* This option is on by default; toggle it off to remove playback controls (such as the Play and Pause buttons) from your video.

     This option may seem like an odd one to disable. After all, why put a video on your page if no one can click the Play button to view it? Well, if you use this in conjunction with the Autoplay option, you can set a video to automatically play for your visitors without giving them a way to pause or mute the video. However, please don't do this — it is a horrible experience for visitors on your site because it autoplays the video and your visitors may not be expecting it. Always give your visitors the ability to pause, play, and mute videos. People of the internet will thank you.

   - *Play Inline:* By default, this option is off; turn it on to ensure that the video doesn't automatically enter full-screen mode when playing on smartphones.

   - *Preload:* Preload is an HTML5 attribute that tells the web browser how much of the video data it should fetch and cache (or store) when a web page with a video is visited. Preload can reduce the amount of lag time it could take to load a video, especially if the video file is very large. This attribute is used when a video is served from the same server that hosts the website — not to embed third-party videos. The options are Auto (fetches the entire video), Metadata (fetches only the metadata, such as video dimensions and length), and None (fetches none of the video data). The default attribute for the Video block is Metadata; to change it, make a different choice from the Preload drop-down menu.

- *Poster Image:* When you embed a video in a post, by default, the first thing your visitor sees is the first frame of that video. Click the Select Poster Image button to upload an image, or select an image from the Media Library, to use as the video poster image on your site.

  Figure 6-22 shows a video of a lion I uploaded to a post. You see that the video displays a black screen of the video before visitors click the Play button. Using the Poster Image, I uploaded an image of a lion that replaces the first frame and gives the video a nicer appearance with an image of people at a concert. Figure 6-23 shows the video on my site after I applied the Poster Image to the video.

**7.** **Save and publish your post, or add more content and publish the post later.**

**FIGURE 6-22:**
A video displayed without a poster image defined.

# Video Post

By lswilson · May 28, 2020 · No Comments

▶ 0:00 / 0.37

Edit

**FIGURE 6-23:**
A video displayed
with a poster
image defined.

**TIP**

I don't recommend uploading your own videos directly to your WordPress site, if you can help it. Many video service providers, such as YouTube and Vimeo, give you free storage for your videos. Embedding videos in a WordPress page or post from one of those services is easy, and by using those services, you're not using your own storage space or bandwidth limitations. Additionally, if many people visit your site and view your video at the same time, loading it from a third-party site such as YouTube will make visitors' experience much faster.

## Adding video using the Embed block

The preceding steps enable you to insert a video from your computer that is hosted on your own web hosting account. If you use the Embed block in the WordPress Block Editor, WordPress automatically embeds the video(s) in a video player within your posts and pages.

With this feature, WordPress automatically detects whether a URL you include in the Embed block is a video and wraps the correct HTML embed code around that URL to make sure that the video player appears in your post.

In this section, I am inserting a video of a puppy, but instead of a link to the video, I'm going to embed the video in my post so that readers can click the Play button to view the video right on my page. Follow these next steps to embed a video from a third-party service using the Embed block:

1. **Click the Add Block icon at the top left of the Edit Post screen, and select the YouTube block in the Embed section.**

   WordPress adds the YouTube block to your post in the Edit Post screen, as shown in Figure 6-24.

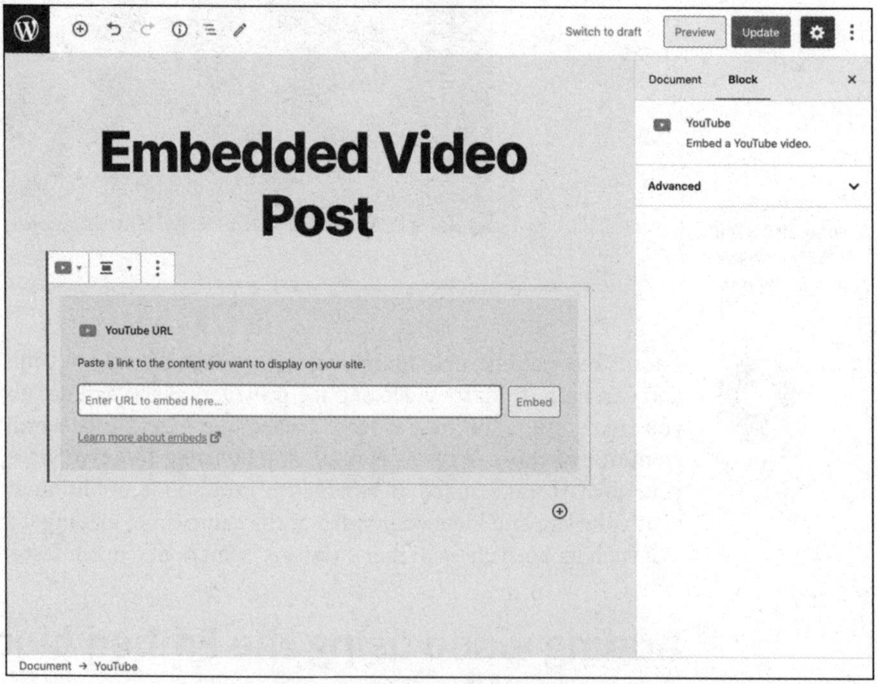

**FIGURE 6-24:**
Adding the
YouTube block to
a post.

2. **Enter the desired YouTube URL in the text field; press the Enter key.**

   WordPress embeds the video in your post, as shown in Figure 6-25.

3. **(Optional) Add a caption in the text field below the video player, if desired.**

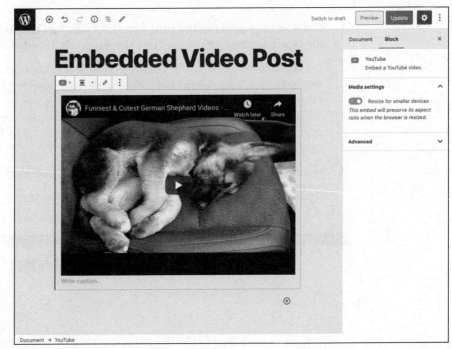

FIGURE 6-25:
A YouTube
video embedded
in a post.

4. **Use the YouTube block toolbar to set display options for the video.**

    This toolbar offers the following options:

    - *Align Left*: This block-formatting option positions the video within the YouTube block on the left side of the page.

    - *Align Center*: This block-formatting option positions the video within the YouTube block in the center of the page.

    - *Align Right*: This block-formatting option positions the video within the YouTube block on the right side of the page.

    - *Wide Width:* Click this option to set the width of the video to span the width of the content on the page.

    - *Full Width*: Click this option to increase the width of the video to span the full width of the screen on which you're viewing the content. The left and right edges of the video extend all the way to the left and right sides of the viewing screen.

    - *Edit Video:* Click this option if you want to change the YouTube URL.

    - *More Options*: The settings here are the same as the ones discussed in Chapter 5.

5. **In the settings panel on the right side of the Edit Post screen, configure the Media Settings for the YouTube video you embedded in step 2.**

    The option to resize for smaller devices is on by default; the video will shrink on smaller devices, but its aspect ratio will be preserved. Preserving the aspect ratio means that the height and width of the video will remain in proportion, not stretched or otherwise distorted. This feature is especially important on small devices such as mobile phones, so I would recommend keeping this setting toggled on.

6. **Publish your post, or continue editing content and publish the post later.**

    Figure 6-26 shows what the post looks like with the embedded YouTube video.

TECHNICAL
STUFF

I'm using the Embed block in this example to illustrate how you can embed a video from YouTube, but several Embed blocks allow you to embed all kinds of data from different services on the web. Chapter 5 covers the Embed block in greater detail.

# Inserting Audio Files into Your Blog Posts

Audio files can be music files or voice recordings, such as recordings of you speaking to your readers. These files add a nice personal touch to your site. You can easily share audio files on your blog by using the Audio block in the WordPress Block Editor. After you insert an audio file into a post, your readers can listen to it on their computers or download it to an MP3 player and listen to it while driving to work, if they want.

To insert an audio file to your site, follow these steps:

1. **Click the Add Block icon at the top left of the Edit Post screen, and select the Audio block in the Common Blocks section.**

   WordPress adds the Audio block to your post in the Edit Post screen, as shown in Figure 6-27.

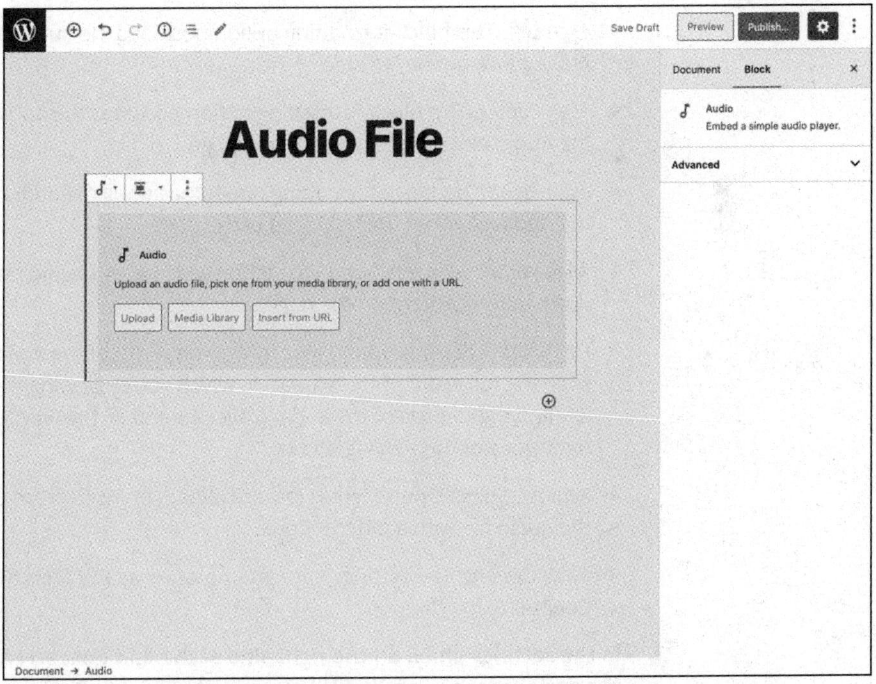

**FIGURE 6-27:**
Adding the Audio block to a post in the Edit Post screen.

2. **Click the Upload button in the Audio block.**

   A window opens, displaying the files that exist on your computer.

3. **Select the audio file you want to upload from your computer, and click Open.**

   The audio is uploaded from your computer to your web server and is inserted into the Audio block on the Edit Post screen.

4. **(Optional) Add a caption in the text field below the audio player, if desired.**

5. **Use the Audio block toolbar to set display options for the audio file.**

   The toolbar offers the following options:

   - *Change Block Type:* Clicking this option allows you to change the type of block you're using. If you want to change this Audio block to a File block, click the Change Block Type icon and then select the File block to swap it. For the Audio block, the only available block type you can change to is File.

   - *Align Left:* This block-formatting option positions the audio player within the Audio block on the left side of the page.

   - *Align Center:* This block-formatting option positions the audio player within the Audio block in the center of the page.

   - *Align Right:* This block-formatting option positions the audio player within the Audio block on the right side of the page.

   - *Wide Width:* Click this option to set the width of the audio player to span the width of the content on the page.

   - *Full Width:* Click this option to increase the width of the audio player to span the full width of the screen on which you're viewing the content. The left and right edges of the audio player extend all the way to the left and right sides of the viewing screen.

   - *Replace:* Select Open Media Library, Upload, or Insert from URL to replace the audio file with a different one.

   - *More Options:* The settings here are the same as the ones discussed in Chapter 5.

6. **In the settings panel on the right side of the Edit Post screen, configure the options for the audio file you uploaded in step 3:**

   - *Autoplay:* By default, this option is set to off. Turn it on to set the audio to start playing automatically when your visitors load this post in their browser.

- *Loop:* By default, this option is set to off. Turn it on to set the audio to play, on repeat, when your visitors load this post in their browser.

- *Preload:* The default attribute for the Audio block is None; to change it, make a different choice from the Preload drop-down menu. "Adding video from your computer" earlier in this chapter describes the Preload attribute in detail.

7. **Save and publish your post, or add more content and publish the post later.**

   Figure 6-28 displays a post with an embedded audio player for an uploaded audio file.

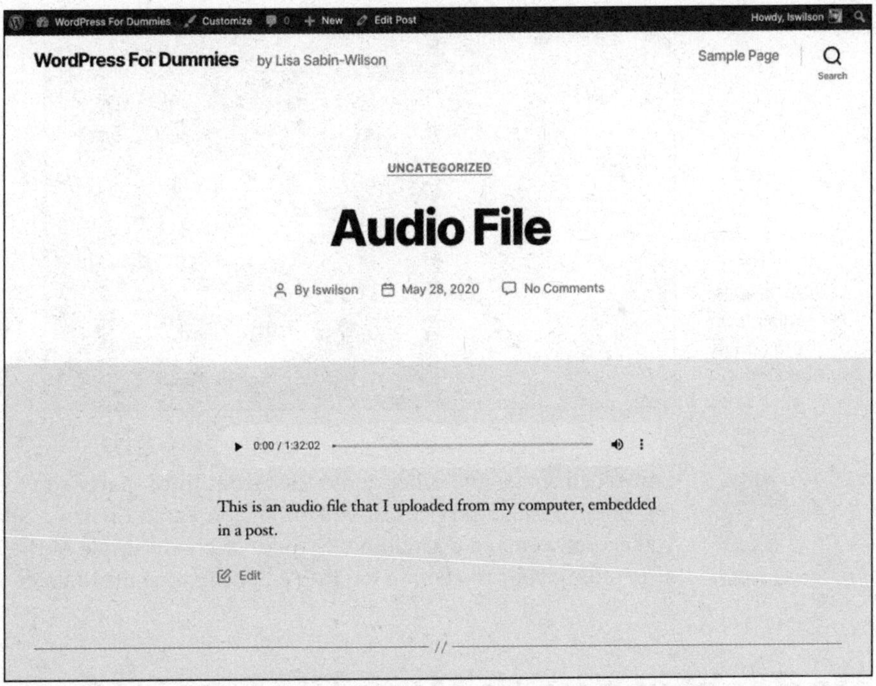

**FIGURE 6-28:**
An audio player embedded in a post.

# Inserting Audio Using the Embed Block

In this chapter and in Chapter 5, I mention the services that you can embed by using the Embed block in the WordPress Block Editor. Some of those services allow you to embed audio files from sources such as Spotify, Soundcloud, Mixcloud, and ReverbNation, to name a few. To embed audio from any of these sources, follow the same steps you took in the "Inserting Audio Files into Your

Blog Posts" section of this chapter to embed video. All you need is the direct URL for the audio you want to embed. Figure 6-29 shows a post with an embedded audio file from Spotify.

FIGURE 6-29:
An audio player
from Spotify
embedded in a
post.

**REMEMBER** Embedded songs and other audio files from third-party services obey the licensing rules of those services. If you embed an audio file from Spotify, for example, you actually embed a random 25- to 30-second sample of the song, with a message telling your readers to log into a Spotify account to listen to the full version.

# Podcasting with WordPress

When you provide regular episodes of an audio show that visitors can download to a computer and listen to on an audio player, you're *podcasting*. Think of a podcast as a weekly radio show that you tune into, except that it's hosted on the Internet rather than on a radio station.

Several plugins for podcasting are available in WordPress so that you can easily insert audio files in your WordPress posts and pages. The plugins that are dedicated to podcasting provide features that go beyond embedding audio files in a website. Some of the most important of these features include

>> **Archives:** You can create an archive of your audio podcast files so that your listeners can catch up on your show by listening to past episodes.

>> **RSS feed:** An RSS feed of your podcast show gives visitors the opportunity to subscribe to your syndicated content so that they can be notified when you publish future episodes.

>> **Promotion:** A podcast isn't successful without listeners, right? You can upload your podcast to services such as Apple Podcasts (https://apps.apple.com/us/app/apple-podcasts/id525463029) so that when people search for podcasts by subject, they find your podcast.

These plugins go beyond audio-file management. They're also dedicated to podcasting and all the features you need:

>> **Simple Podcasting** (https://wordpress.org/plugins/simple-podcasting): Simple Podcasting includes full Apple Podcasts support and allows you to upload audio files using the usual WordPress methods (Audio blocks), but the plugin includes a Podcast Block in the new WordPress block editor for easy publishing of podcasts on your website.

>> **Seriously Simple Podcasts** (https://wordpress.org/plugins/seriously-simple-podcasting): This plugin uses the native WordPress interface with minimal settings to make it as easy as possible to podcast with WordPress. You can run multiple podcasts; obtain stats on who's listening; do both audio- and videocasting; and publish to popular services such as iTunes, Google Play, and Stitcher.

TIP

I discuss web hosting requirements in Chapter 2. If you're a podcaster who intends to store audio files in your web hosting account, you may need to increase the storage and bandwidth for your account so that you don't run out of space or incur higher fees from your web hosting provider. Discuss these issues with your web hosting provider to find out up front what you have to pay for increased disk space and bandwidth.

Chapter **7**

# Making the Most of WordPress Plugins

H alf the fun of running a WordPress-powered website is playing with the thousands of plugins that you can install to extend your site's functions and options. WordPress plugins are like those really cool custom rims you put on your car: Although they don't come with the car, they're awesome accessories that make your car better than all the rest.

By itself, WordPress is a very powerful program for web publishing, but by customizing WordPress with *plugins* — add-on programs that give WordPress almost limitless ways to handle web content — you can make it even more powerful. You can choose any plugins you need to expand your online possibilities. Plugins can turn your WordPress installation into a full-featured gallery for posting images on the web, an online store to sell your products, a user forum, or a social networking site. WordPress plugins can be simple, adding a few minor features, or complex enough to change your entire WordPress site's functionality.

In this chapter, you find out what plugins are, how to find and install them, and how they enhance your site to make it unique. Using plugins can also greatly improve your readers' experiences by providing them various tools they can use to interact and participate — just the way you want them to!

I assume that you already have WordPress installed on your web server. If you're skipping around in the book and haven't yet installed WordPress on your web server, you can find instructions in Chapter 3.

REMEMBER

WordPress.com users can't install or configure plugins on their hosted blogs. I don't make the rules, so please don't kill the messenger.

# Finding Out What Plugins Are

A *plugin* is a small program that, when added to WordPress, interacts with the software to provide some extensibility to the software. Plugins aren't part of the core software; neither are they software programs themselves. They typically don't function as stand-alone software. They require the host program (Word-Press, in this case) to function.

Plugin developers are the people who write these gems and share them with the rest of us — usually for free. As is WordPress, many plugins are free to anyone who wants to further tailor and customize a site to meet specific needs.

Thousands of plugins are available for WordPress — certainly way too many for me to list in this chapter alone. I could, but then you'd need heavy machinery to lift this book off the shelf! Here are just a few examples of things that plugins let you add to your WordPress site:

>> **Email notification:** Your biggest fans can sign up to have an email notification sent to them every time you update your website.

>> **Social media integration:** Allow your readers to submit your content to some of the most popular social networking services, such as Twitter, Facebook, and Reddit.

>> **Stats program:** Keep track of where your traffic is coming from; which posts on your site are most popular; and how much traffic is coming through your website on a daily, monthly, and yearly basis.

Chapter 15 gives you a peek at some of the most popular plugins on the scene today. In the meantime, this chapter takes you through the process of finding plugins, installing them on your WordPress site, and managing and troubleshooting them.

# Extending WordPress with plugins

WordPress by itself is an amazing tool. The features built into WordPress are meant to be the ones that you'll benefit from most. All the desired site features that fall *outside* what is built into WordPress are considered to be the territory of plugins.

There's a popular saying among WordPress users: "There's a plugin for that." The idea is that if you want WordPress to do something new, you have a good chance of finding an existing plugin that can help you do what you want. Currently, more than 57,000 plugins are available on the Plugins page of the WordPress.org website (`https://wordpress.org/plugins`), and this number is growing each day. In addition, thousands of additional plugins are available outside the WordPress.org site for free or for a fee. So if you have an idea for a new feature for your site, you just may find a plugin for that feature.

Suppose that you want to add recipes to your site. An Internet search for *wordpress plugin recipes* results in links to the WP Recipe Maker plugin (`https://wordpress.org/plugins/wp-recipe-maker`) and the WP Ultimate Recipe plugin (`https://wordpress.org/plugins/wp-ultimate-recipe`). You can find even more recipe-related plugins with a more in-depth search.

# Distinguishing between plugins and themes

Because themes can contain large amounts of code and add new features or other modifications to WordPress, you may wonder how plugins differ from themes. In reality, only a few technical differences exist, but the ideas of what plugins and themes are supposed to be are quite different. (For more about themes, see Chapter 8.)

At the most basic level, the difference between plugins and themes is that they reside in different directories. Plugins are in the `wp-content/plugins` directory of your WordPress site. Themes are in the `wp-content/themes` directory.

**TECHNICAL STUFF**

The `wp-content/plugins` and `wp-content/themes` directories are set up this way by default. You can change both of these locations, but this change is very rarely made. The possibility that plugin and theme locations will change is something to be aware of if you're working on a WordPress site and are having a hard time locating a specific plugin or theme directory.

The most important difference between plugins and themes is that a WordPress site always has one, and only one, active theme, but it can have as many active plugins as you want — even none. This difference is important because it means that switching from one theme to another prevents you from using the features of the old theme. By contrast, activating a new plugin doesn't prevent you from making use of the other active plugins.

**REMEMBER**

Plugins are capable of changing nearly every aspect of WordPress. The Multiple Themes plugin, for example (available at `https://wordpress.org/plugins/jonradio-multiple-themes`), allows you to use different themes for specific parts of your WordPress site. Thus, you can overcome even the limitation of having only one active theme on a site by using a plugin.

Because WordPress can have only one theme but many plugins activated at one time, it's important that the features that modify WordPress be limited to plugins, whereas themes remain focused on the appearance of the site. For you, this separation of functionality and appearance is the most important difference between plugins and themes.

**TIP**

This separation of plugins for functionality and themes for appearance isn't enforced by WordPress, but it's a good practice to follow. You can build a theme that includes too much functionality, and you may start to rely on those functions to make your site work, which ultimately makes switching to another theme difficult.

## THE DIFFERENCES BETWEEN PLUGINS AND THEMES

Other technical differences separate plugins and themes. The differences matter mostly to developers, but it could be important for you to know these differences as a nondeveloper WordPress user:

- Plugins load before the theme, which gives plugins some special privileges over themes, and the result can be that one or more plugins prevent the theme from loading. The built-in WordPress functions in the `wp-includes/pluggable.php` file can be overridden with customized functions, and only plugins load early enough to override these functions.

- Themes support a series of structured template files and require a minimum set of files to be valid. By comparison, plugins have no structured set of files and require only a single .php file with a comment block at the top to tell WordPress that the file is a plugin.

- Themes support child themes, wherein one theme can require another theme to be present to function; no such feature is available to plugins.

The functionality role of plugins doesn't mean that control of the appearance of a WordPress site is limited to themes. Plugins are just as capable of modifying the site's appearance as a theme is. The WP Mobile X plugin, for example (available at `https://wordpress.org/plugins/wp-mobile-x`), can provide a completely different version of your site to mobile devices such as smartphones by replacing the functionality of the theme when the user visits the site on a mobile device.

# Exploring the Plugins Page

Before you start installing plugins on your site, it's important to explore the Plugins page of your WordPress Dashboard and understand how to manage the plugins after you install them. Click the Installed Plugins link on the Plugins menu of your WordPress Dashboard to view the Plugins screen, shown in Figure 7-1.

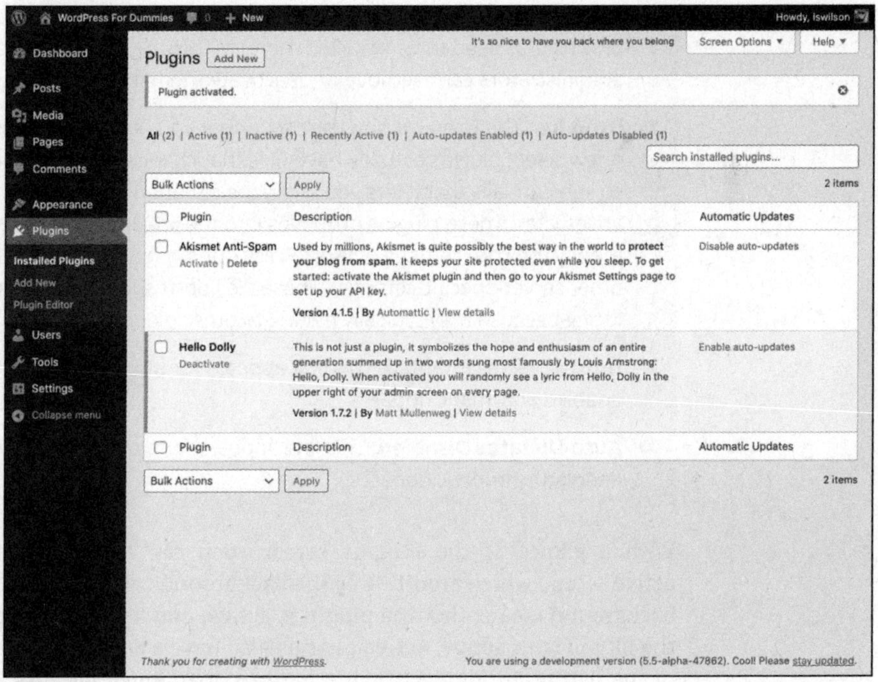

**FIGURE 7-1:**
Manage your plugins with the Plugins screen of the WordPress Dashboard.

The Plugins screen is where you manage all the plugins you install on your WordPress site. By default, the Plugins screen displays a full listing of all the WordPress plugins you currently have installed. You can filter the listing of plugins by using the links shown directly below the screen's title:

- >> **All:** This view is the default view for the Plugins screen, showing all plugins currently installed on your WordPress website, whether they're active or not.

- >> **Active:** Clicking this link shows a list of the plugins that are currently activated and in use on your WordPress website.

- >> **Inactive:** Clicking this link changes the display and shows the list of plugins that are installed but not currently active. (*Note:* This link displays on the Plugins screen only if you have inactive plugins.)

- >> **Recently Active:** This link appears only if you deactivated a plugin recently. It's helpful if you know that you've deactivated a plugin but can't remember which one.

- >> **Must Use:** This link appears only if you have plugins installed in the /wp-content/mu-plugins folder. Must Use plugins are standard WordPress plugins that need to be installed manually (usually by a developer); site administrators can't remove or deactivate them from the Dashboard.

- >> **Drop-Ins:** The Drop-Ins link appears only if you have drop-in plugins installed. A few select plugins actually have files that change the behavior of WordPress so substantially that it lets you know the situation, just in case a problem arises later. These plugins drop files into the wp-content directory that modify the core settings of WordPress (mostly having to do with caching or other server-specific settings). These files don't show up in the main plugin listing because they typically belong to other plugins.

- >> **Auto-Updates Enabled:** This link appears for plugins for which you've enabled automatic updates.

- >> **Auto-Updates Disabled:** This link appears for plugins for which you haven't enabled automatic updates.

With a glance at the Plugins screen, you can easily tell which plugins are active — and which aren't — by the background color of the plugins listed. A blue background means that the plugin is active, and a white background means that the plugin isn't active. Active plugins also have a bold blue border on the left side of the listing. In Figure 7-1, it's hard to distinguish because the images in this book are black and white, but you will see in your browser that the background of the Akismet plugin is white, and the background of the Hello Dolly plugin is blue.

Akismet isn't active, but Hello Dolly is. (The figures in this book are black and white, so you can't see the blue color, but you can see the darker and lighter shades in Figure 7-1.)

**TIP**

You can mass-manage your plugins on the Plugins screen. You can deactivate all your plugins simultaneously by selecting the box to the left of each plugin name, choosing Deactivate from the Bulk Actions drop-down menu at the top or bottom of the page (as shown in Figure 7-1), and clicking the Apply button. Likewise, you can activate, upgrade, or delete the plugins listed by choosing Activate, Update, or Delete from the Bulk Actions drop-down menu. To select all your plugins with one click, select the box to the left of the Plugins heading.

The Plugins screen displays plugins in three columns, which give details for each plugin:

>> **Plugin:** This column lists the plugin name so that you can find it easily when browsing the Plugins screen. Directly below the plugin name, you see a few links for easy plugin management:

- *Activate:* This link appears below the titles of inactive plugins. Click the link to activate a plugin.

- *Deactivate:* This link appears below the titles of active plugins. Click the link to deactivate a plugin.

- *Delete:* This link appears below the titles of inactive plugins. Click the link to delete the plugin from your site. (See more about this topic in "Uninstalling Plugins" later in this chapter.)

**TIP**

Sometimes, a plugin has a separate settings page; in that case, an additional link labeled Settings is displayed below the plugin name. Clicking that link takes you to the settings page for that plugin.

>> **Description:** This column lists a description for the plugin. Depending on the plugin, you may also see brief instructions on using the plugin. Directly below the description are the version number of the plugin, the plugin author's name, and a link to the website where you can read more information about the plugin.

>> **Automatic Updates:** This column, by default, displays an Enable Auto-Updates link, which means that the plugin requires you to manually update it when there is a new version available. When you click the Enable AutoUupdates link, it changes to Disable Auto-Updates, which enables you to disable automatic updates for the plugin. For more information about plugin updates, see "Discovering the one-click plugin update" later in this chapter.

# Identifying Core Plugins

Some plugins hold a very special place in WordPress, in that they ship with the WordPress software and are included by default in every WordPress installation.

For the past few years, two plugins have held this special position:

» **Akismet:** The Akismet plugin has the sole purpose of protecting your blog from comment spam. Although other plugins address the issue of comment spam, the fact that Akismet is packaged with WordPress and works quite well means that most WordPress users rely on Akismet for their needs.

» **Hello Dolly:** The Hello Dolly plugin helps you get your feet wet in plugin development, if you're interested. It was first released with WordPress version 1.2 and is considered to be the oldest WordPress plugin. When the plugin is active, the tops of your Dashboard pages show a random lyric from the song "Hello, Dolly!"

Figure 7-1 shows the core plugins in a new installation of WordPress.

**REMEMBER**

The idea of core plugins is to offer a base set of plugins to introduce you to the concept of plugins while also providing a benefit. The Akismet plugin is useful because comment spam is a big issue for WordPress websites. The Hello Dolly plugin is useful as a nice starting point for understanding what plugins are and how they're coded.

Although WordPress automatically includes these plugins, your site doesn't have to run them. Plugins are disabled by default; you must activate them manually. You can delete core plugins, just as you can delete any other plugins, and they won't be replaced when you upgrade WordPress

Future versions of WordPress may offer different sets of core plugins. It's possible that one or both of the current core plugins will cease being core plugins and that other plugins will be included. Although this topic has been much discussed in WordPress development circles over the past few years, at this writing, no definitive decisions have been made, so the current set of core plugins is likely to stay for a while longer.

## Incorporating Akismet

Akismet was created by the folks at Automattic — the same folks who bring you the Jetpack plugin (see Chapter 15). Akismet is the answer to comment and trackback spam.

Akismet is already included in every WordPress installation; you don't have to worry about downloading and installing it, because it's already there. Follow these steps to activate and begin using Akismet:

1. **Click the Plugins link on the left navigation menu of the Dashboard to load the Plugins screen.**

2. **Click the Activate link below the Akismet plugin name and description.**

   The Akismet screen loads (see Figure 7-2).

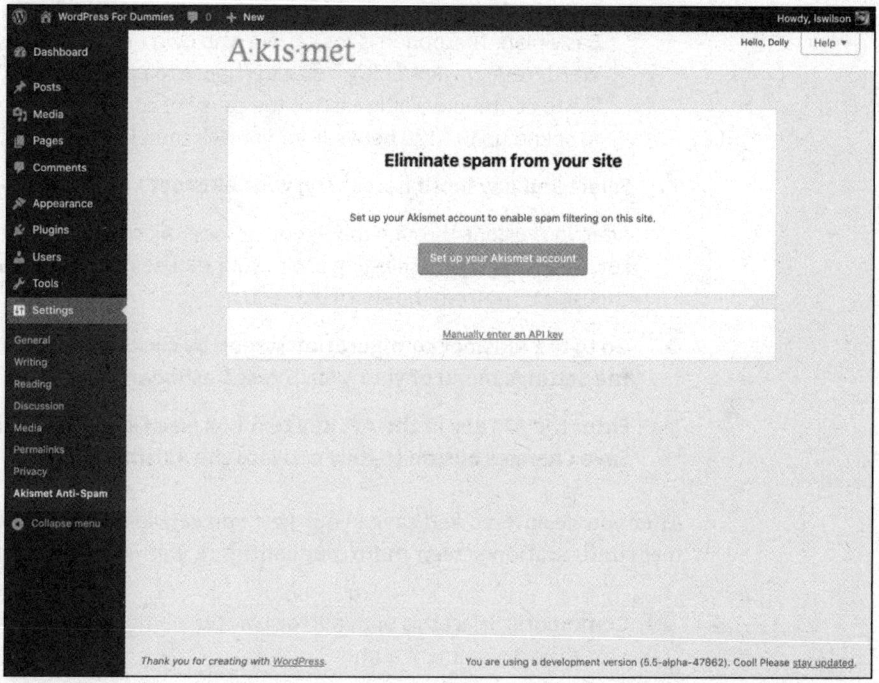

**FIGURE 7-2:**
The Akismet
screen.

3. **If you don't already have an Akismet account, click the Set Up Your Akismet Account button to navigate to the Akismet website, and skip to step 5.**

4. **If you already have an Akismet account and API key, enter it in the Manually Enter an API Key text field; then click the Use This Key button to save your changes.**

   You can stop here if you already have a key, but if you don't have an Akismet key, keep following the steps in this section.

   An *API key* is a string of numbers and letters that functions like a unique password given to you by Akismet; it's the key that allows your WordPress.org application to communicate with your Akismet account.

**5.** **Click the Set Up Your Akismet Account button.**

The sign-up page on the Akismet website opens.

**6.** **Choose among these options for obtaining an Akismet key:**

- *Enterprise:* $50 per month for people who own commercial or professional sites or blogs and want additional security screening and malware protection.

- *Plus:* $5 per month for people who own a small commercial or professional site or blog.

- *Personal:* Name your price. Type the amount you're willing to pay for the Basic plan. This option is for people who own one small, personal WordPress-powered blog. You can choose to pay nothing ($0), but if you'd like to contribute a little cash to the cause of combating spam, you can opt to spend up to $120 per year for your Akismet key subscription.

**7.** **Select and pay for (if necessary) your Akismet key.**

After you've gone through the signup process, Akismet provides you an API key. Copy that key by selecting it with your mouse pointer, right-clicking, and choosing Copy from the shortcut menu.

**8.** **Go to the Akismet configuration screen by clicking the Akismet link on the Settings menu of your WordPress Dashboard.**

**9.** **Enter the API key in the API Key text box (see Figure 7-3), and click the Save Changes button to fully activate the Akismet plugin.**

After you've entered and saved your key, you can select three options on the Akismet configuration screen to further configure your spam protection:

» **Comments:** Select this option to display the number of approved comments next to each comment author.

» **Strictness:** By default, Akismet puts spam in the Spam comment folder for you to review at your leisure. If you feel that this setting isn't strict enough, you can have Akismet silently delete the worst and most pervasive spam so that you never have to see it.

» **Privacy:** By default, Akismet isn't configured to display a privacy notice below the comments form on your website. Select the Display a Privacy Notice Under Your Comment Forms option to tell WordPress that you'd like a privacy notice to be displayed, as shown in Figure 7-4.

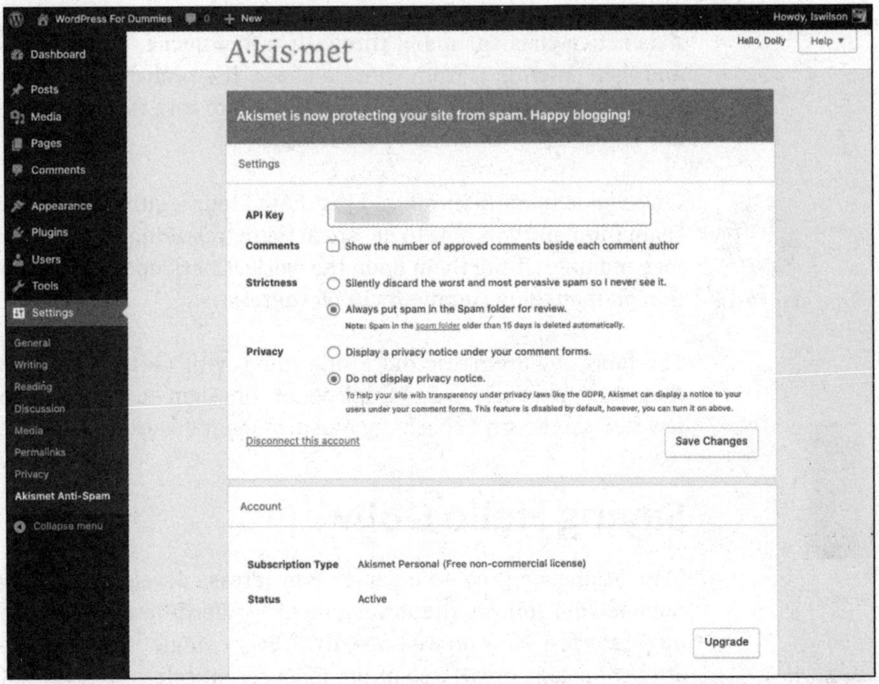

**FIGURE 7-3:**
Akismet
verification
confirmation
message on the
Akismet
configuration
screen.

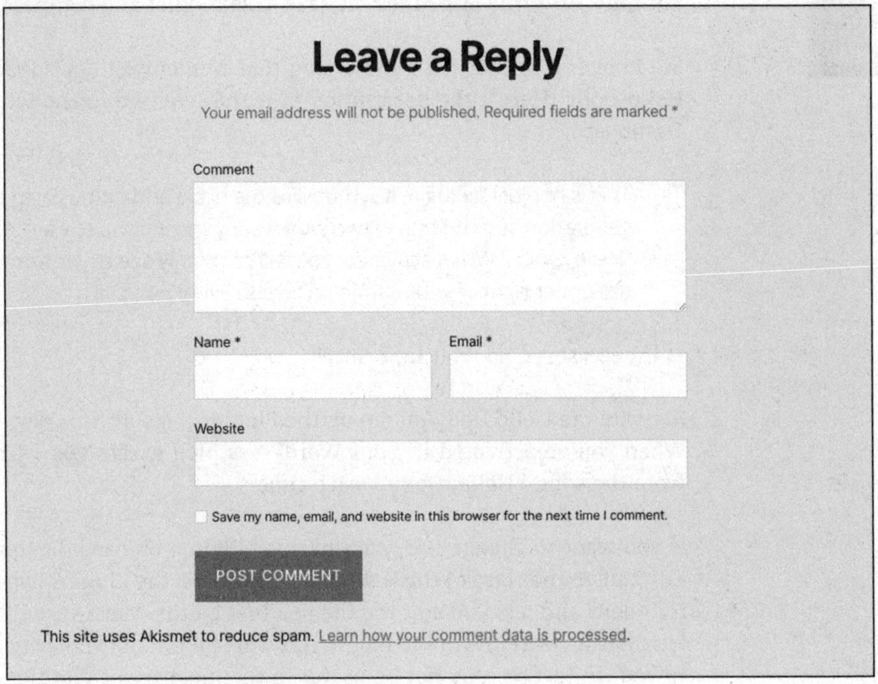

**FIGURE 7-4:**
The Akismet
privacy notice
displayed below
the comments
form.

Akismet catches spam and throws it into a queue, holding the spam for 15 days and then deleting it from your database. It's probably worth your while to check the Akismet spam page once a week to make sure that the plugin hasn't captured any legitimate comments or trackbacks.

**REMEMBER**

Check your spam filter often. I just found four legitimate comments caught in my spam filter and was able to de-spam them, releasing them from the binds of Akismet and unleashing them upon the world. Check out Chapter 5 for more information on managing comments in WordPress.

The folks at Automattic did a fine thing with Akismet. Since the emergence of Akismet, I've barely had to think about comment or trackback spam except for the few times a month I check my Akismet spam queue.

## Saying Hello Dolly

Matt Mullenweg, co-founder of WordPress, developed the Hello Dolly plugin. Anyone who follows the development of WordPress knows that Mullenweg is a huge jazz fan. How do we know this? Every single release of WordPress is named after some jazz great. One of the most recent releases of the software, for example, is named Parker, after jazz great Charlie Parker; another release was named Coltrane, after the late American jazz saxophonist and composer John Coltrane.

So, knowing this, it isn't surprising that Mullenweg developed a plugin named Hello Dolly. Here's the description of it that you see in the Plugins page of your Dashboard:

> This is not just a plugin, it symbolizes the hope and enthusiasm of an entire generation summed up in two words sung most famously by Louis Armstrong: "Hello, Dolly." When activated, you will randomly see a lyric from "Hello, Dolly" in the upper right of your admin screen on every page.

Is it necessary? No. Is it fun? Sure!

Activate the Hello Dolly plugin on the Plugins page of your WordPress Dashboard. When you've activated it, your WordPress blog greets you with a different lyric from the song "Hello, Dolly!" each time.

If you want to change the lyrics in this plugin, you can edit them by clicking the Edit link to the right of the Hello Dolly plugin on the Plugins page. The Plugin Editor opens and lets you edit the file in a text editor. Make sure that each line of the lyric has its own line in the plugin file. This plugin may not seem to be very useful to you (in fact, it may not be useful to the majority of WordPress users), but the real purpose of the plugin is to give WordPress plugin developers a simple example of how to write a plugin.

**TIP**

This book doesn't cover how to create your own plugins, but if you're interested in that topic, you may want to check out my other book, *WordPress All-in-One For Dummies*, 4th Edition (John Wiley & Sons, Inc.), which covers that topic in detail.

## Discovering the one-click plugin update

For a lot of reasons, mainly security reasons and feature updates, always use the most up-to-date versions of the plugins in your blog. With everything you have to do every day, how can you possibly keep up with knowing whether the plugins you're using have been updated?

You don't have to. WordPress does it for you.

WordPress notifies you when a new update is available for a plugin in four ways:

>> **Updates link:** The Updates link on the Dashboard menu displays a red circle with a white number indicating how many plugins have updates available. (In Figure 7-5, one plugin on my site has an update available.) Click the Updates link to see which plugins have updates available.

>> **Toolbar:** When a new update is available, a small icon appears to the right of your site title on the toolbar, as shown in Figure 7-5.

>> **Plugins menu's title:** The Plugins menu's title displays a red circle with a white number (refer to Figure 7-5). As with the Updates link, the number indicates how many plugins have updates available.

>> **Plugins page:** Figure 7-6 shows the Plugins page. Below the Akismet plugin is the message There is a new version of Akismet available. View version 4.15 details or update now.

WordPress gives you not only a message that a new version of the plugin is available, but also a link to a page where you can download the new version, or a link that you can click to update the plugin right there and then — WordPress's one-click plugin update.

Click the Update Now link, and WordPress grabs the new files off the WordPress. org server, uploads them to your plugins directory, deletes the old plugin, and activates the new one. (If a plugin is deactivated at the time it's updated, WordPress gives you the option to activate the plugin after your update process is completed.) Figure 7-7 shows the Updated message that you see on the Plugins page after the plugin has been upgraded.

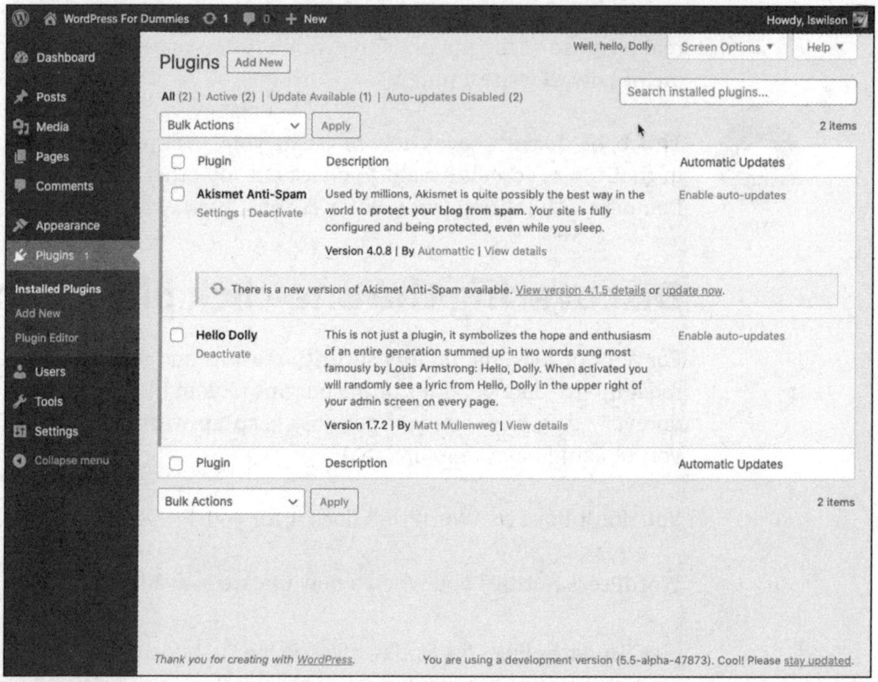

**FIGURE 7-5:**
WordPress tells you when a new plugin version is available.

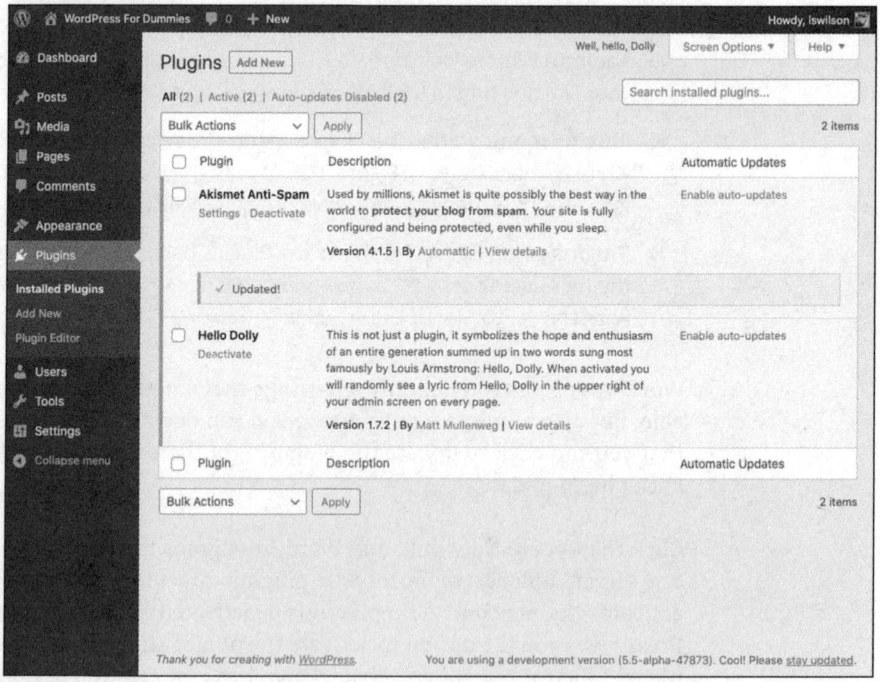

**FIGURE 7-6:**
The Akismet plugin with an available update.

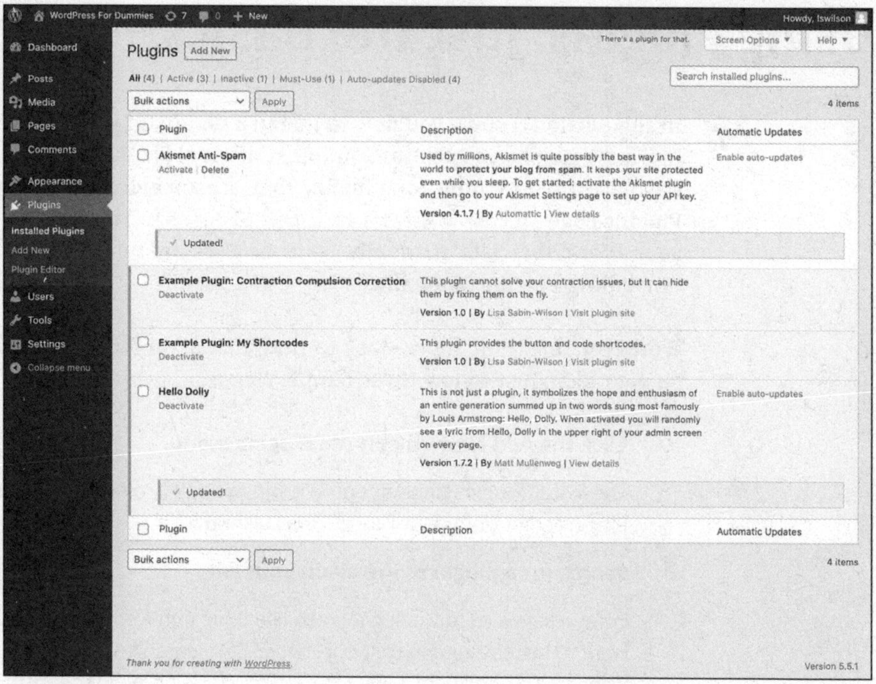

**FIGURE 7-7:**
An Updated message appears after you successfully update a plugin.

WordPress notifies you of an out-of-date plugin and provides the one-click-upgrade function only for plugins that are on the official WordPress Plugins page (`https://wordpress.org/plugins`). If a plugin you're using isn't listed in the directory, the notification and one-click upgrade function won't be available for that plugin.

Whatever you do, do *not* ignore the plugin update messages that WordPress gives you. Plugin developers usually release new versions because of security problems or vulnerabilities that require an upgrade. If you notice that an upgrade is available for a plugin you're using, stop what you're doing and upgrade it. An upgrade takes only a few seconds.

For the automatic plugin upgrade to work, your plugin directory (`/wp-content/plugins`) must be writable on your web server, which means that you should have set permissions of 755 or 777 (depending on your web server configuration). See Chapter 3 for information about changing file permissions on your web server, or contact your web-hosting provider for assistance.

# Using Plugins: Just the Basics

In this section, I show you how to install a plugin on your WordPress site by using the built-in plugins feature. Autoinstallation of plugins from your WordPress Dashboard works only for plugins that are included on the official WordPress Plugins page (https://wordpress.org/plugins). (You can also install plugins on your WordPress site manually — a process that I cover in the next section, "Installing Plugins Manually.")

WordPress makes it super-easy to find, install, and then activate plugins for use on your blog. Just follow these simple steps:

1. **Click the Add New link on the Plugins menu.**

   The Install Plugins page opens, allowing you to browse the official WordPress Plugins page from your WordPress Dashboard.

2. **Search for a plugin to install on your site.**

   Enter a keyword for a plugin you'd like to search for. For this example, you want to install a plugin that improves SEO (search engine optimization) for your website. To find it, enter **SEO** in the Search text box on the Install Plugins page; then click the Search button.

   Figure 7-8 shows the results page for the SEO search phrase. The first plugin listed, Yoast SEO, is the one you want to install.

   You can also discover new plugins by clicking any of the categories at the top of the Add Plugins screen, such as Featured, Popular, and Recommended.

   TIP

3. **Click the More Details link.**

   A Description window opens, displaying information about the Yoast SEO plugin (including a description, version number, and author name) and an Install Now button.

4. **Click the Install Now button.**

   You go to the Installing Plugins page of your WordPress Dashboard, where you find a confirmation message that the plugin has been downloaded, unpacked, and successfully installed.

5. **Specify whether to activate the plugin or proceed to the Plugins page.**

   Two links are shown below the confirmation message:

   - *Activate Plugin:* Click this link to activate the plugin you just installed on your site.

   - *Return to Plugin Installer:* Click this link to go to the Install Plugins page without activating the plugin.

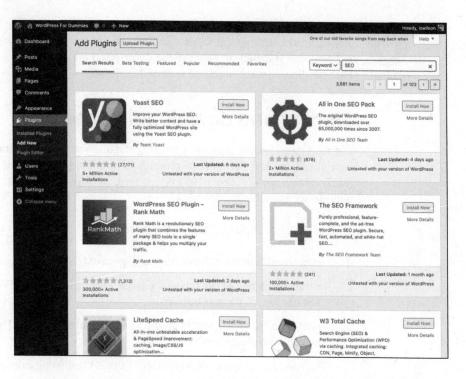

**FIGURE 7-8:**
The Add Plugins
screen's search
results for
SEO-related
plugins.

**WARNING**

Autoinstallation of plugins from your WordPress Dashboard works in most web-hosting configurations. Some web-hosting services, however, don't allow the kind of access that WordPress needs to complete autoinstallation. If you get any errors or find that you're unable to use the plugin autoinstallation feature, get in touch with your web-hosting provider to find out whether it can assist you.

**TIP**

If the Dashboard displays any kind of error message after you install the plugin, copy the message and paste it into a support ticket in the WordPress.org support forum (`https://wordpress.org/support`) to elicit help from other WordPress users about the source of the problem and the possible solution. When you post about the issue, provide as much information about the issue as possible, including a screen shot or pasted details.

# Installing Plugins Manually

Installing plugins from the Dashboard is so easy that you'll probably never need to know how to install a plugin manually via Secure File Transfer Protocol (SFTP). (Chapter 3 explains how to use SFTP.) But the technique is still helpful to know in case the WordPress Plugins page is down or unavailable.

**REMEMBER**

Installing the SEO plugin takes you through the process, but keep in mind that every plugin is different. Reading the description and installation instructions for each plugin you want to install is very important.

## Finding and downloading the files

The first step in using plugins is locating the one you want to install. The absolute best place to find WordPress plugins is the Plugins page of WordPress.org (https://wordpress.org/plugins), where, at this writing, more than 57,000 plugins are available for download.

To find the Twitter plugin, follow these steps:

1. **Go to the official WordPress Plugins page at** https://wordpress.org/plugins.

2. **In the search box at the top of the page, enter the keyword SEO; then click the Search Plugins button.**

3. **Locate the Yoast SEO plugin on the search results page (see Figure 7-9), and click the plugin's name.**

   The Yoast SEO plugin page opens (see Figure 7-10), offering a description and other information about the plugin.

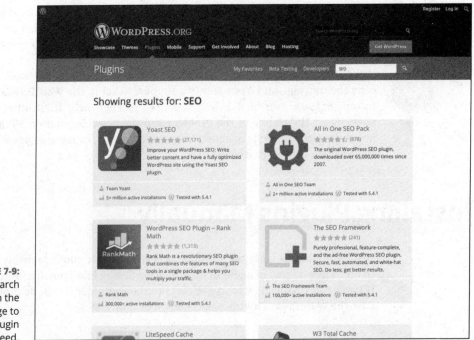

**FIGURE 7-9:**
Use the search feature on the Plugins page to find the plugin you need.

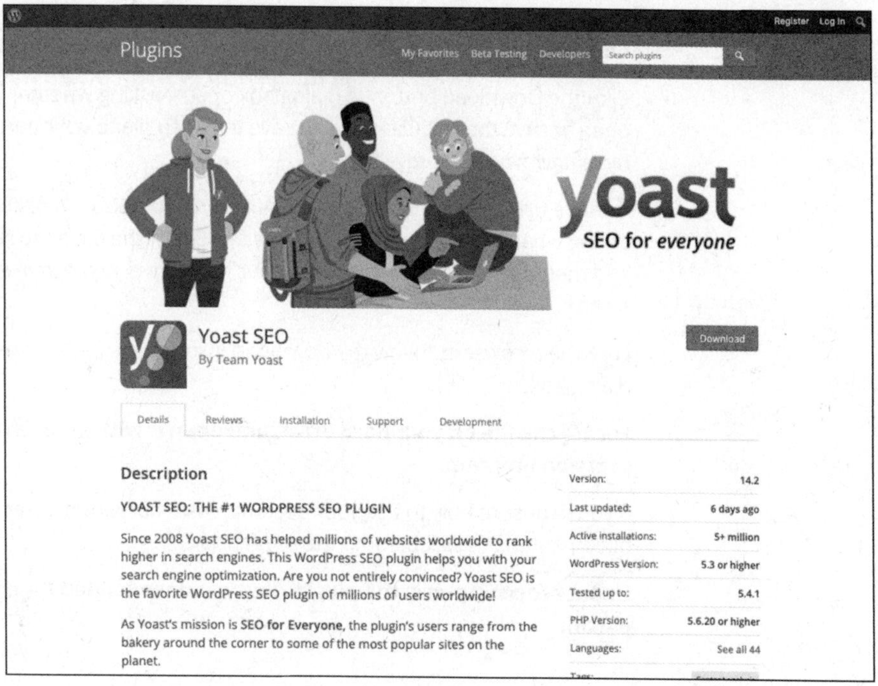

**FIGURE 7-10:**
The download
page for the
Yoast SEO plugin.

In Figure 7-10, take note of the important information on the right side of the page:

- *Version:* The number shown in this area is the most recent version of the plugin.

- *Last Updated:* This date is when the author last updated the plugin.

- *Active Installations:* This number tells you how many times this plugin has been downloaded and installed by other WordPress users.

- *Tested Up To:* This section tells you what version of WordPress this plugin is compatible up to. If it says that the plugin is compatible up to version 5.0, for example, you usually can't use the plugin with versions later than 5.0. I say *usually* because the plugin developer may not update the information in this section — especially if the plugin files themselves haven't changed. The best way to check is to download the plugin, install it, and see whether it works! (Figure 7-10 shows that the Yoast SEO plugin is compatible up to WordPress version 5.4.1.)

- *Ratings:* With a rating system of 1 to 5 stars (1 being the lowest and 5 being the highest), you can see how other WordPress users have rated this plugin.

4. **Click the Download button for the plugin version you want to download.**

   Click the Download button. A dialog box opens, asking whether you want to open or save the file. Click Save to save the `.zip` file to your hard drive, and *remember where you saved it.*

   If you're using Mozilla Firefox, click the Download button. A dialog box opens, asking what Firefox should do with the file. Select the Save File radio button and then click OK to save the file to your hard drive. Again, *remember where you saved it.*

   For other browsers, follow the download instructions in the corresponding dialog box.

5. **Locate the file on your hard drive, and open it with your favorite decompression program.**

   If you're unsure how to use your decompression program, refer to the documentation available with the program.

6. **Unpack (decompress) the plugin files you downloaded for the Yoast SEO plugin.**

## Reading the instructions

Frequently, the plugin developer includes a `readme.txt` file inside the `.zip` file. Do what the title of the file says: Read it. Often, it contains the same documentation and instructions that you'll find on the plugin developer's website.

Make sure that you read the instructions carefully and follow them correctly. Ninety-nine percent of WordPress plugins have great documentation and instructions from the developer. If you don't follow the instructions correctly, the best scenario is that the plugin just won't work on your site. At worst, the plugin will create all sorts of ugly errors, requiring you to start the plugin installation over.

You can open `readme.txt` files in any text-editor program, such as Notepad or WordPad on a PC or TextEdit on a Mac.

In the case of the Yoast SEO plugin, the `readme.txt` file contains a lot of information about the plugin, as well as handy information about SEO, in general.

Every plugin is different in terms of where the plugin files are uploaded and what configurations and setup are necessary to make the plugin work on your site. Read the installation instructions carefully, and follow those instructions to the letter to install the plugin correctly on your site.

# Uploading and Activating Plugins

In this section, I show you how to upload the plugin files to your web server. In earlier versions of WordPress, you needed to upload the unpacked plugin files to your web server via SFTP (see Chapter 3). Now all you need to do is upload the .zip file you downloaded from the WordPress Plugins page. Be sure that you're logged in to your WordPress Dashboard.

Unpacking the .zip file you've downloaded is helpful because it can contain files that give you insight into the use of the plugin itself. Locate the plugin files you unpacked on your hard drive. In the event that the plugin developer didn't include a readme.txt file with instructions, check the plugin developer's website for instructions on how to install the plugin on your WordPress site. Specifically, the documentation in the readme.txt file and/or on the plugin's website should address the following points:

>> What directory on your web server you upload the plugin files to.

>> What to do if you need to change permissions for any of the plugin files after you upload them to your web server. (See Chapter 3 if you need information on changing file permissions.)

>> What to do if you need to set specific configurations in the plugin file to make it work.

>> What to do if you need to modify your theme template files to include the plugin's functions on your site.

## Uploading a new plugin

To install the Yoast SEO plugin via the WordPress Dashboard, follow these easy steps:

1. **Click the Add New link on the Plugins menu.**

   The Install Plugins page of your Dashboard opens.

2. **Click the Upload Plugin link at the top of the Install Plugins page.**

   The resulting page gives you an interface for uploading a plugin in .zip format.

3. **Click the Choose File button.**

   In the resulting File Upload dialog box, you can locate the .zip file for the plugin you'd like to install. In this case, the file is wordpress-seo.14.2.zip (see Figure 7-11). Click the file to select it and then click the Open button to return to the Install Plugins page.

FIGURE 7-11:
Uploading a
plugin zip file via
the Dashboard.

4. **Click the Activate Plugin button.**

   WordPress uploads the plugin's zip file into the /wp-content/plugins/ folder
   on your web server, unpacks it, and installs it. Figure 7-12 shows the Installing
   Plugins page, with messages for you during and after the (ideally, successful)
   installation.

## Activating a plugin

All plugins listed on the Plugins screen are active or inactive except the Drop-In
plugins, which are active by default. When you want to activate an inactive plugin,
follow these easy steps:

1. **Click the Installed Plugins link on the Plugins menu.**

   The Plugins screen opens, listing all the plugins installed on your WordPress site.

2. **Locate the plugin you want to activate.**

   If you have a lot of plugins installed on your WordPress website, you can
   narrow your search by clicking the Inactive link, which lists the inactive plugins.

3. **Click the Activate Plugin button below the plugin name.**

   The Plugins screen refreshes, and the plugin you activated now appears as an
   active plugin on the page.

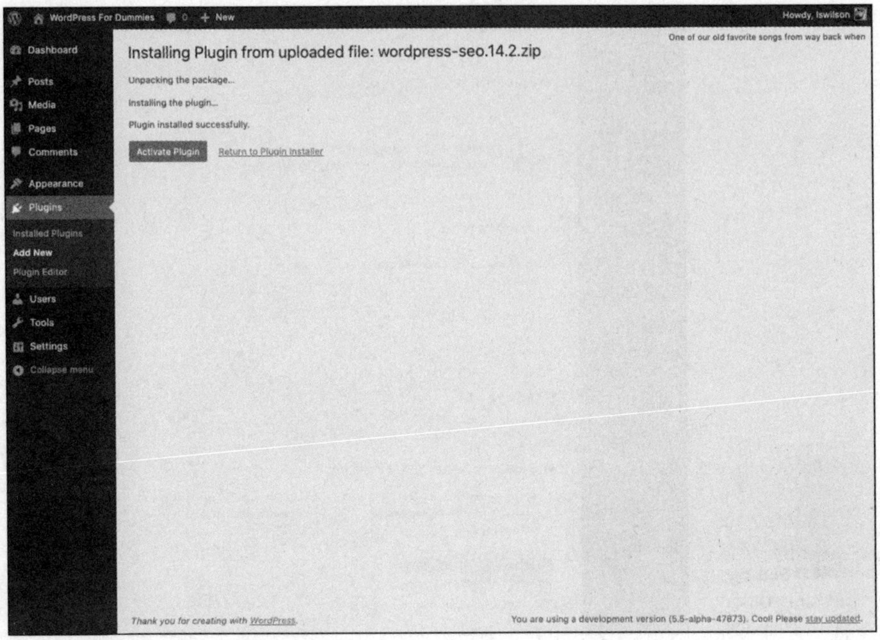

**FIGURE 7-12:**
Successful plugin upload via the Dashboard.

# Setting Plugin Options

Some, but not all, WordPress plugins have an administration page where you can set options specific to that plugin. You may find the plugin administration page in any of these places:

>> The Settings page (click the Settings menu)

>> The Tools menu (located on the navigation menu)

>> The Plugins menu (located on the navigation menu)

>> A Dashboard menu (located on the navigation menu)

>> The Admin toolbar (at the top of any Dashboard page)

You can open the Yoast SEO Settings page (see Figure 7-13), for example, by clicking the SEO link on the main Dashboard navigation menu.

**FIGURE 7-13:**
The Yoast
SEO Settings
administration
page.

# Uninstalling Plugins

After all this talk about installing and activating plugins, what happens if you install and activate a plugin and then decide that it just isn't what you want? Don't worry — you aren't stuck with a plugin that you don't want. WordPress lets you be fickle and finicky in your plugin choices!

To uninstall a plugin, follow these steps:

1. **Click the Installed Plugins link on the Plugins menu.**

   The Plugins screen opens.

2. **Locate the plugin you want to uninstall.**

3. **Click the Deactivate link below the plugin's title.**

   The Plugins screen refreshes, and the plugin appears as deactivated (or inactive).

4. **Click the Delete link below the plugin's title.**

   The Plugins screen opens, and a confirmation dialog box appears, asking you whether you're sure you want to delete this plugin (see Figure 7-14).

5. **Click the OK button.**

   The Plugins screen refreshes, and the plugin you just deleted is gone from the list of plugins, with a message at the top confirming the deletion.

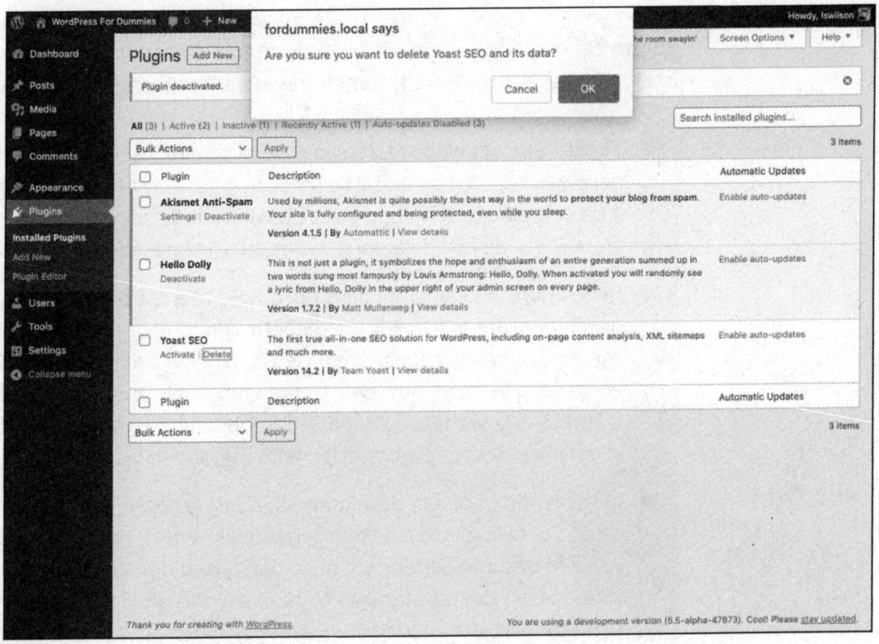

FIGURE 7-14:
Confirmation
dialog box on the
Plugins screen.

Bang! You're done. That's all it takes.

**REMEMBER**

Don't forget to remove any bits of code that you may have added to your theme templates for that particular plugin; otherwise, ugly error messages may appear on your site.

# Understanding the Open-Source Environment

The WordPress software was built on an existing platform called b2. Matt Mullenweg, co-founder of WordPress, was using b2 as a blogging platform when the developer of that program abandoned it. What did this change mean for its users? It meant no more development unless someone somewhere picked up the ball and continued with the platform. Enter Mullenweg and WordPress.

Apply the same concept to plugin development, and you'll understand that plugins sometimes fall by the wayside and drop off the face of the Earth. Unless someone takes over when the original developer loses interest, future development of that plugin ceases. It's important to understand that most plugins are developed in an open-source environment, which means a few things for you, the end user:

>> The developers who created your favorite plugins aren't obligated to continue development. If they find a new hobby or simply tire of the work, they can give it up completely. If no one picks up where they left off, you can kiss those plugins goodbye if they don't work with the latest WordPress release.

>> Developers of popular plugins don't hold to a specific timetable. Generally, developers are extremely good about updating their plugins when new versions of WordPress are released or when a security bug or flaw is discovered. Keep in mind, however, that no timetable exists for these developers to follow. Many of these folks have day jobs, classes, or families that can keep them from devoting as much time to the project as you want them to.

>> In the world of plugin development, it's easy come, easy go. Beware of the pitfalls of falling in love with any particular WordPress plugin. Don't let your website become dependent on a plugin, and don't be surprised if a plugin you love doesn't exist tomorrow. You can use the plugin as long as it continues to work for you, but when it stops working (such as with a new WordPress release or a security exploit that makes it unusable), you have a tough decision to make. You can

- Stop using the plugin and try to find a suitable alternative

- Hope that another developer takes over the project when the original developer discontinues his involvement

- Try to find someone to provide a fix for you (in which case, you'll more than likely have to pay that someone for her time)

I don't want to make the world of WordPress plugins sound like gloom and doom, but I do think it's very important for you to understand the dynamics in play. Consider this section to be food for thought.

# Finding Plugins Beyond WordPress.org

The exact number of plugins available beyond those offered through WordPress. org is unknown, but it's easily more than thousands, which means that a great variety of plugins are out there. These plugins can be difficult to discover, but I can point you to some good starting places.

Many of the plugins that aren't listed on the Plugins page are paid plugins, and the official WordPress Plugins page allows only free plugins. If a plugin is for sale (costs more than a penny), it can't be listed on the Plugins page, so the author needs to find other methods of listing and promoting their products.

Over the past few years, the market for commercial plugins has grown tremendously. It isn't possible to list all the companies that currently offer WordPress plugins in this chapter, so the following list is a sampling. It's a way of introducing you to the world of plugins other than those offered on WordPress.org.

Each of the following sites offers WordPress plugins:

>> **iThemes** (https://ithemes.com): Although it started as a theme developer, iThemes branched out into developing plugins. The most popular offering is BackupBuddy, a plugin for backing up and restoring your sites.

>> **Gravity Forms** (https://www.gravityforms.com): For many WordPress users, Gravity Forms is the plugin to pay for. Typically, it's the first and last recommendation people give someone who wants to create forms in WordPress.

>> **CodeCanyon** (https://codecanyon.net/category/wordpress): With thousands of plugins, this online marketplace is the paid plugin version of WordPress.org's Plugins page. CodeCanyon is a collection of plugins from various developers rather than plugins from a single company.

>> **WooCommerce** (https://woocommerce.com): WooCommerce is a popular plugin with a full-featured e-commerce solution that you can integrate into your WordPress site. Essentially, this plugin turns your WordPress site into an online store.

Although these sites give you a taste of what commercial plugin sites have to offer, having other sources of information about new, exciting plugins can be helpful. Many popular WordPress news sites talk about all things WordPress, including reviews and discussions of specific plugins. Check out the following sites if you want to know more about what plugins are being talked about:

>> **WPBeginner** (www.wpbeginner.com): This site is dedicated to helping new WordPress users get up and running quickly. It features an active blog on a variety of topics. The site often features posts about how to use plugins to create specific types of solutions for your site.

>> **Post Status** (https://poststatus.com): Post Status is an all-things-WordPress news site. If there's buzz on a topic in the WordPress world, you're likely to find discussions about it here.

TIP

One great thing about using a community or news site to discover new plugins is that you aren't alone in deciding whether to trust a plugin. You can get some other opinions before you take a chance on a plugin.

If you aren't finding what you want on the Plugins page, don't know anyone who offers the solution you're looking for, and aren't seeing anything on community sites, it's time to go to a trusty search engine (such as Google), and see what you can find.

A good way to start is to search for the words *wordpress* and *plugin* along with one word to a few words describing the feature you want. If you want a plugin that provides advanced image-gallery features, for example, search for *wordpress plug-in image gallery.* As long as your search isn't too specific, you're likely to get many results. The results often contain blog posts that review specific plugins or list recommended plugins.

WARNING

Some developers include malware, viruses, and other unwanted executable code in their plugin code. Your best bet is to use plugins from the official WordPress website or to purchase plugins from a reputable seller. Do your research first.

# Comparing Free and Commercial Plugins

Thousands of plugins are available for free, and thousands of plugins have a price. What are the benefits of free plugins versus paid plugins? This question is a tough one to answer.

It's tempting to think that some plugins are better than others, which is why they cost money. Unfortunately, things aren't that simple. Some amazing plugins that I'd gladly pay for are free, and some terrible plugins that I wouldn't pay for have a cost.

Often, a paid plugin includes support, which means that the company or person selling the plugin is offering assurance that if you have problems, you'll receive support and updates to address bugs and other issues.

Free plugins typically list places to make support requests or to ask questions, but nothing ensures that the developer will respond to your requests within a certain period — or at all. Even though developers have no obligation to help with support requests from their plugin's users, many developers work hard to help users with reported issues and other problems. Fortunately, because many free plugins are offered on the Plugins page, and because the Plugins page includes a built-in support forum and rating score, you can easily see how responsive a plugin author is to support issues.

I believe that the commercial plugin model works in an environment of tens of thousands of free plugins because many WordPress users want the assurance that when they have problems, they have a place to ask questions and get help.

So if people can get paid to produce plugins, why are so many plugins free? This question is another great one.

One reason why so many plugins are available for free is that many WordPress developers are generous people who believe in sharing their plugins with the community. Other developers feel that having their plugins available to the millions of WordPress users via the Plugins page is a great way to market their talents, which can lead to contract work and employment. Buzzwords on a résumé are far less valuable than a plugin you wrote that was downloaded thousands or millions of times.

Another reason to release a free plugin is to entice people to pay for upgrades — a model often referred to as *freemium*. Freemium plugins often have paid plugins that add features to the free plugin. Thus, the freemium model is a mix of the free and paid plugin models and gets the best of both worlds. The free plugin can be listed on the Plugins page, giving the plugin a large amount of exposure. You can get a feel for how the plugin functions, and if you want the additional features, you can purchase the paid plugin.

An example of the freemium model is the WooCommerce plugin (https://woocommerce.com). The main plugin is available for free on the Plugins page (https://wordpress.org/plugins/woocommerce), yet it supports a large number of paid plugins to add more features. By itself, the WooCommerce plugin turns the site into a shopping cart. To extend this functionality, paid plugins are available to add payment processing for specific credit-card processors, shipping, download managers, and many other features.

The biggest difference between free and paid plugins is that sometimes, you won't find what you need in a free plugin and have to go with a paid plugin. In the end, what you download is up to you. Many great free plugins are available, and so are many great paid plugins. If you want the features offered by a paid plugin and are willing to pay the price, paid plugins can be very good investments for your site.

**TIP**

Many free plugins have links to donation pages. If you find a free plugin to be valuable, please send a donation to the developer. Most developers of free plugins say that they rarely, if ever, receive donations. Even a few dollars can really encourage the developer to keep updating old plugins and releasing new free plugins.

## DEVELOPING PLUGINS — A COMMUNITY ACTIVITY

Although plugins are written and developed by people who have the skills required to do so, the WordPress user community is also largely responsible for the ongoing development of plugins. Ultimately, the end users are the ones who put those plugins to the true test on their own sites. Those same users are also the first to speak up and let the developers know when something isn't working right, helping the developers troubleshoot and fine-tune their plugins.

The most popular plugins are created by developers who encourage open communication with the user base. Overall, WordPress is one of those great open-source projects in which the relationship between developers and users fosters a creative environment that keeps the project fresh and exciting every step of the way.

IN THIS CHAPTER

» Finding free WordPress themes

» Downloading, installing, and activating themes

» Browsing and installing themes from your Dashboard

» Going with commercial themes

» Exploring the default theme

» Using widgets on your site

Chapter **8**

# Finding and Installing WordPress Themes

In previous chapters, I cover how to use the WordPress platform to publish your posts and pages. In those chapters, you discover how to categorize your posts, build your link lists, and set the publishing and profile options in the WordPress Dashboard. In this chapter, I focus on the visual look and format of your site — in other words, how other people see your site after you start publishing your content.

In Chapter 7, I introduce WordPress plugins and discuss some of the thousands of free plugins you can use to add functionality to your site. Similarly, thousands of free themes are available for you to download and use. This chapter shows you where to find them and takes you through the processes of downloading, installing, and using them.

# Getting Started with Free Themes

WordPress comes packaged with one very useful default theme called Twenty Twenty (named after the year 2020 and released in version 5.3 of WordPress). Most bloggers who use WordPress usually don't waste any time at all in finding a theme that they like better than the default theme. The Twenty Twenty theme is meant to get you started. Although you're not limited to the default theme, it's a very functional theme for a basic website. Feel free to use it to get you started.

Free WordPress themes, such as those I discuss in Chapter 16, are popular because of their appealing designs and their ease of installation and use. They're great tools to use when you launch your new site, and if you dabble a bit in graphic design and CSS (Cascading Style Sheets), you can customize one of the free WordPress themes to fit your own needs. (See Chapter 9 for some resources and tools for templates and template tags, as well as a few great CSS references.)

With thousands of free WordPress themes available and new ones appearing all the time, your challenge is to find the right one for your site. Here are a few things to remember while you explore (and also see the nearby sidebar "Are all WordPress themes free?" for information about free versus commercial themes):

» **Free themes are excellent starting places.** Find a couple of free themes, and use them as starting points for understanding how themes work and what you can do with them. Testing free themes, their layouts, and their options helps you identify what you want in a theme.

» **You'll switch themes frequently.** Typically, you'll find a WordPress theme that you adore, and then, a week or two later, you'll find another theme that fits you or your site better. Don't expect to stay with your initial choice. Something new will pop up on your radar screen. Eventually, you'll want to stick with a theme that fits your needs best and doesn't aggravate visitors because of continual changes.

» **You get what you pay for.** Although a plethora of free WordPress themes exist, you receive limited or no support for those themes. Free themes are often labors of love. The designers have full-time jobs and responsibilities; they release these free projects for fun, passion, and a desire to contribute to the WordPress community. Therefore, you shouldn't expect (or demand) support for these themes. Some designers maintain active and helpful forums to help users, but those forums are rare. Just be aware that with most free themes, you're on your own.

## ARE ALL WordPress THEMES FREE?

Not all WordPress themes are created equal, and it's important for you, the user, to know the difference between free and commercial themes:

- **Free:** These themes are free, period. You can download and use them on your website at absolutely no cost. It's a courtesy to include a link to the designer in the footer of your site, but you can even remove that link if you want.

- **Commercial:** These themes cost money. You usually find commercial themes available for download only after you've paid $10 to $500, or more. The designer feels that these themes are a cut above the rest and, therefore, worth the money you spend for them. Commercial themes also come with a mechanism to obtain support for the use of the theme. I provide information on where to find commercial themes in the "Exploring Premium Theme Options" section of this chapter.

>> **Download themes from reputable sources.** Themes are essentially pieces of software; therefore, they can contain things that could be scammy, spammy, or potentially harmful to your site or computer. It's vital that you do your homework by reading online reviews and downloading themes from credible, trusted sources. The best place to find free WordPress themes is the Themes page at https://wordpress.org/themes.

By using free themes, you can have your site up and running with a new design — without the help of a professional — pretty fast. And with thousands of themes available, you can change your theme as often as you want.

## Finding free themes

Finding the theme that fits you best may take some time, but with thousands of themes available, you'll eventually find one that suits you. Trying out several free themes is like trying on different outfits for your site. You can change outfits as needed until you find just the right theme.

You can visit the official WordPress Themes page at https://wordpress.org/themes (see Figure 8-1).

## Avoiding unsafe themes

Although free themes are great, you want to avoid some things when finding and using free themes. Like everything on the web, themes have the potential to be abused. Although free themes were conceived to allow people (namely, designers and developers) to contribute work to the WordPress community, they've also been used to wreak havoc for users. As a result, you need to understand what to watch out for and what to avoid.

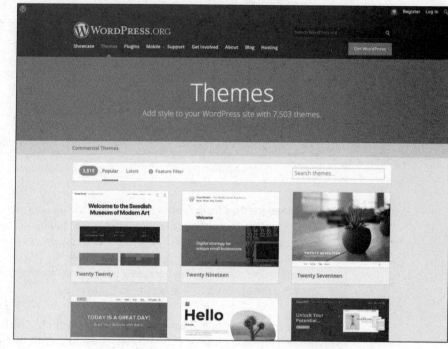

Here are some things to avoid when searching for free themes:

>> **Spam links:** Many free themes outside the WordPress Themes page include links in the footer or sidebars, and these links can be good or bad. The good uses of these links are designed to credit the original designer and possibly link to their website or portfolio. You should maintain these links as a show of appreciation to the creator because the links help increase the designer's traffic and clients. Spam links, however, aren't links to the designer's site; they're links to sites you may not ordinarily associate with or endorse on your site. The best examples are links in the footer that go to odd, off-topic, and uncharacteristic keywords or phrases, such as *weight loss supplement* or *best flower deals*. Mostly, this spam technique is used to increase the advertised site's search engine ranking for that particular keyword by adding another link from your site — or, worse, to take the site visitor who clicks it to a site that's unrelated to the linked phrase.

>> **Hidden and malicious code:** Unfortunately, the WordPress community has received reports of hidden, malicious code within a theme. This hidden code can produce spam links, security exploits, and abuses on your WordPress site. Hackers install code in various places that run this type of malware. Unscrupulous theme designers can, and do, place code in theme files to insert hidden malware, virus links, and spam. Sometimes, you see a line or two of encrypted code that looks like it's part of the theme code. Unless you have a great deal of knowledge of PHP or JavaScript, you may not know that the theme is infected with dangerous code.

>> **Themes that lack continued development:** WordPress software continues to improve with each new update. Two or three times a year, WordPress releases new software versions, adding new features, security patches, and numerous other updates. Sometimes, a code function is superseded or replaced, causing a theme to break because it hasn't been updated for the new WordPress version. Additionally, because software updates add new features, the theme needs to be updated accordingly. Because free themes typically come without any warranty or support, one thing you should look for — especially if a theme has many advanced back-end options — is whether the developer is actively maintaining the theme for current versions of WordPress. Active maintenance typically occurs more with plugins than with themes, but it's worth noting.

>> **Endlessly searching for free themes:** Avoid searching endlessly for the perfect theme. Trust me — you won't find it. You may find a great theme and then see another with a feature or design style that you like, but the new theme may lack certain other features. Infinite options can hinder you from making a final decision. Peruse the most popular themes on the WordPress Themes page, choose five that fit your criteria, and then move on. You always have the option to change a theme later, especially if you find the vast amount of choices in the directory to be overwhelming.

The results of these unsafe theme elements can range from simply annoying to downright dangerous, affecting the integrity and security of your computer or your hosting account. For this reason, the WordPress Themes page is considered to be a safe place from which to download free themes. WordPress designers develop these themes and upload them to the directory, and the folks behind the WordPress platform vet each theme on the official WordPress Themes page, themes that contain unsafe elements simply aren't allowed.

TIP

The WordPress Themes page isn't the only place to find free WordPress themes, but it's the place to find the most functional and *safest* themes available. Safe themes contain clean code and fundamental WordPress functions to ensure that your website functions with the minimum requirements. The WordPress.org website lists the basic requirements that theme designers have to meet before their themes are accepted into the directory; you can find that listing of requirements at https://wordpress.org/themes/getting-started. I highly recommend that you stick to the WordPress Themes page for themes to use on your site; you can be certain that those themes don't contain any unsafe elements or malicious code.

**TIP**

If you suspect or worry that you have malicious code on your site — either through a theme you're using or a plugin you've activated — the absolute best place to get your site checked is the Sucuri website (https://sitecheck.sucuri.net), which offers a free website malware scanner. Sucuri provides expertise in the field of web security, for WordPress users in particular, and it even has a free plugin you can install to check your WordPress site for malware or malicious code. You can find that plugin at https://wordpress.org/plugins/sucuri-scanner.

## Previewing themes on the Themes page

While you're visiting the WordPress Themes page, you can easily browse the various themes by using the following features:

>> **Search:** Type a keyword in the search box near the top of the page (refer to Figure 8-1), and press the Enter (Return on a Mac) key. A new page opens, displaying themes related to the keyword you searched for.

>> **Featured:** Click the Featured link to view the themes that WordPress has chosen to feature in the directory. WordPress changes the featured themes listing regularly.

>> **Popular:** Click the Popular link to view the themes that have been downloaded most often.

>> **Latest:** Click the Latest link to view themes recently added to the directory.

>> **Feature Filter:** Click the Feature Filter to view choices available to filter your theme search by, such as layout, features, and subject.

When you find a theme in the directory that you want to examine more closely, click the More Info button that appears when you hover your mouse over the theme, and do one of the following:

>> **Download:** Click this button to download the theme to your computer.

>> **Preview:** Click the Preview button on the Themes page to open a preview window (see Figure 8-2).

**Hueman** *By Nicolas*

Preview    Download

Version: 3.5.8
Last updated: May 15, 2020
Active Installations: 60,000+
PHP Version: 5.3 or higher
Theme Homepage →

## Ratings

★★★★★ *5 out of 5 stars.*    See all >

5 stars ▬▬▬▬▬ 618
4 stars ▬ 13
3 stars ▪ 7
2 stars ▪ 3
1 star ▪ 9

Add my review

The Hueman theme loads fast and is 100% mobile-friendly according to Google. One of the best rated theme for blogs and magazines on WordPress.org. Powering 70K+ websites around the world.

## Tags:

Custom Colors, Custom Menu, Featured Images, Flexible Header, Full Width Template, Left Sidebar, One Column, Post Formats, Right Sidebar, Sticky Post, Theme Options, Threaded Comments, Three Columns, Translation Ready, Two Columns

## Support

Got something to say? Need help?

View support forum

Downloads Per Day    Report

# Installing a Theme

After you find a WordPress theme, you can install the theme on your WordPress site via SFTP (Secure File Transfer Protocol) or the WordPress Dashboard.

To install a theme via SFTP, follow these steps:

1. **Download the theme file from the Themes page.**

   Typically, theme files are provided in compressed format (`.zip`).

   (I discuss how you can peruse the WordPress Themes page from your WordPress Dashboard in the next section, "Browsing the free themes.")

2. **Unzip or extract the theme's `.zip` file.**

   You see a new folder on your desktop, typically labeled with the corresponding theme name. (Refer to Chapter 3 for information on using SFTP.)

3. **Upload the theme folder to your web server.**

   Connect to your hosting server via SFTP, and upload the extracted theme folder to the `/wp-content/themes` folder on your server (see Figure 8-3).

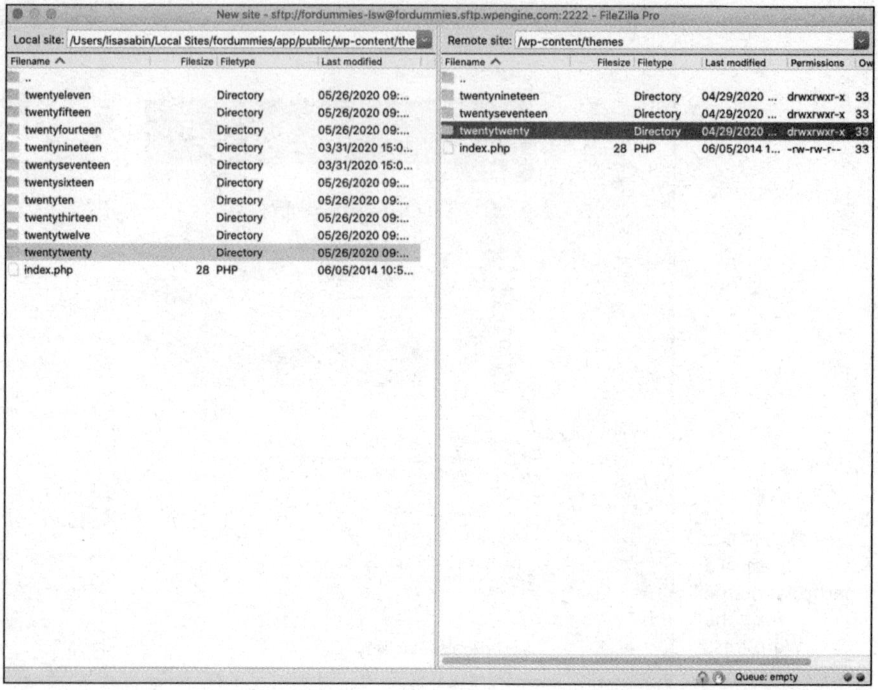

FIGURE 8-3:
Upload and
download panels
via SFTP.

To install a theme via the Dashboard's theme installer, follow these steps:

1. **Download the theme file from the Themes page to your desktop.**

   Typically, theme files are provided in compressed format (`.zip`). When you use this method, you don't extract the `.zip` file, because the theme installer does that for you.

2. **Log in to your WordPress Dashboard.**

3. **Click the Themes link on the Appearance menu.**

   The Themes screen appears.

4. **Click the Add New button.**

   The Add Themes screen appears, displaying a submenu of links.

5. **Click the Upload Theme button.**

   The panel displays a utility to upload a theme in `.zip` format.

6. **Upload the `.zip` file you downloaded in step 1.**

   Click the Choose File button and then locate and select the `.zip` file you stored on your computer.

**7.** **Click the Install Now button.**

WordPress unpacks and installs the theme in the appropriate directory for you. Figure 8-4 shows the result of installing a theme via this method.

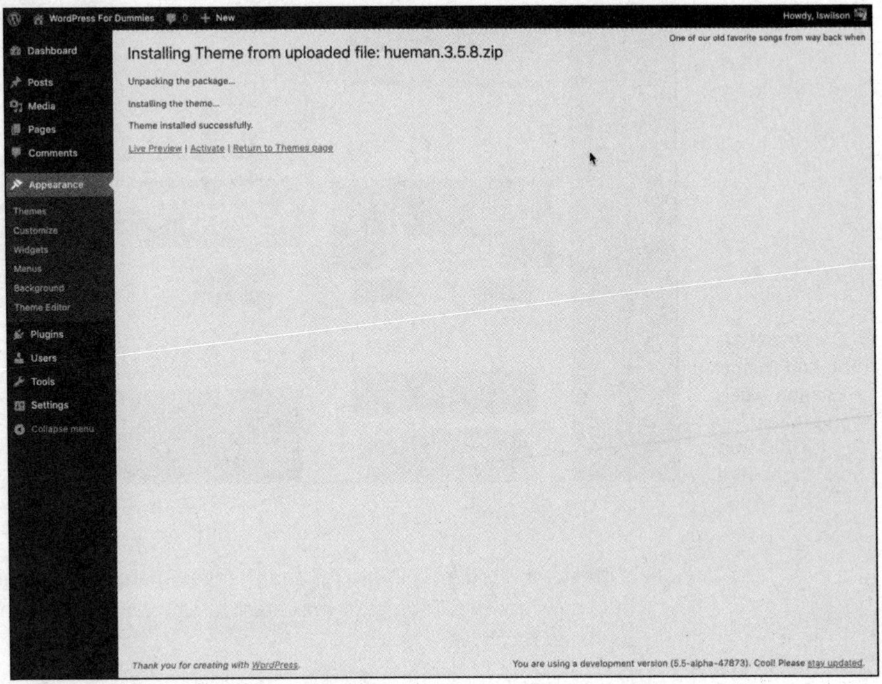

FIGURE 8-4:
Installing a
theme via the
Dashboard's
theme installer.

# Browsing the free themes

Finding free themes via the Add Themes screen is extremely convenient because it lets you search the Themes page from your WordPress site. Start by clicking the Themes link on the Appearance menu of the WordPress Dashboard and then click the Add New button to open the Add Themes screen (see Figure 8-5).

After you navigate to the Add Themes screen, you see the following items:

>> **Featured:** The Featured link takes you to themes that WordPress.org has selected. The themes in this section are favorites of the members of the Themes team.

>> **Popular:** If you don't have a theme in mind, the themes in this section are some of the most popular. I recommend that you install and test-drive one of these selections for your site's first theme.

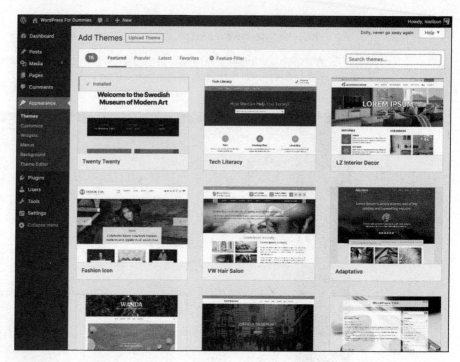

**FIGURE 8-5:**
The Add Themes screen, where you can find free themes from your Dashboard.

» **Latest:** As WordPress improves and changes, many themes need updating to add new features. Themes in the Latest category are themes that have been updated recently.

» **Favorites:** If you marked themes as favorites on the WordPress website, you can find them on the Add Themes screen. After you click the Favorites link, fill in your WordPress.org username in the text field, and click the Get Favorites button. (**Note:** This feature works only if you're logged in to your WordPress. org user account.)

» **Feature Filter:** This link gives you a variety of filters to choose among to find the theme you're looking for. You also can filter by Layout, Features, and Subject. After you select your desired filters, click the Apply Filters button to view the themes that match your set filters.

» **Search:** If you know the name of a free theme, you can easily search for it by keyword, author, or tag. You also can refine your search based on specific features of the theme, including color, layout, and subject (such as Holiday).

After you find the theme that you want, click the Install button that appears when you hover your mouse over the theme's thumbnail.

# Previewing and activating a theme

After you upload a theme via SFTP or install it via the theme installer, you can preview and activate your desired theme.

**TIP**

The WordPress Theme Preview option allows you to see how the theme would look on your site without actually activating it. If you have a site that's receiving traffic, it's best to preview any new theme before activating it to ensure that you'll be happy with its look and functionality. If you're trying to decide among several new theme options, you can preview them all before changing your live site.

To preview your new theme, follow these steps:

1. **Log in to your WordPress Dashboard.**

2. **Click the Themes link on the Appearance menu.**

   The Themes screen appears, displaying your current (activated) theme and any themes that are installed in the /wp-content/themes directory on your web server.

3. **Preview the theme you want to use.**

   Click the Live Preview button that appears when you hover your mouse over the theme thumbnail. A preview of your site with the theme appears in a pop-up window, as shown in Figure 8-6.

4. **(Optional) Configure theme features.**

   Some, but not all, themes provide customization features. Figure 8-6 shows the customization options for the Twenty Twenty theme:

   - Site Identity
   - Colors
   - Theme Options
   - Cover Template
   - Background Image
   - Menus
   - Widgets
   - Homepage Settings
   - Additional CSS

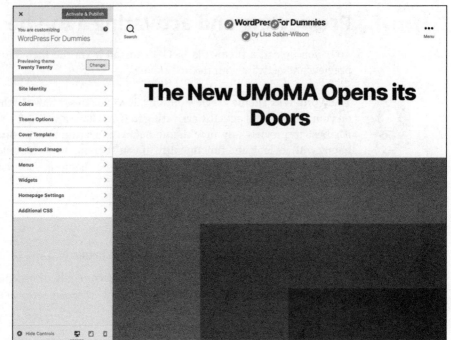

5. **Choose whether to activate the theme.**

   Click the Activate & Publish button in the top-right corner of the configuration panel to activate your new theme with the options you set in step 4, or close the preview by clicking the Cancel (X) button in the top-left corner of the panel.

To activate a new theme without previewing it, follow these steps:

1. **Log in to your WordPress Dashboard.**

2. **Click the Themes link on the Appearance menu.**

   The Themes screen appears, displaying your current (activated) theme and any themes that are installed in the /wp–content/themes directory on your web server.

3. **Find the theme you want to use.**

4. **Click the Activate button that appears when you hover your mouse over the theme thumbnail.**

   The theme immediately becomes live on your site.

# Exploring Premium Theme Options

Thousands of free WordPress themes are available, but you may also want to consider premium (for purchase) themes for your site. Remember the adage "You get what you pay for" when considering free services or products, including WordPress and free themes.

Typically, when you download and use something that's free, you get no assistance with the product or service. Requests for help generally go unanswered. Therefore, your expectations should be lower because you aren't paying anything. When you pay for something, you usually assume that you'll get support or service for your purchase and that the product is of high (or acceptable) quality.

WordPress, for example, is available free. But you have no guarantee of support while using the software except for the active WordPress support forum. Moreover, you have no right to demand service.

Here are some things to consider when contemplating a premium theme. (I selected the commercial companies listed later in this chapter based on these criteria.)

>> **Selection:** Many theme developers offer a rich, diverse selection of themes, including themes designed for specific industries, topics, or uses (such as videos, blogs, real estate, or magazines). Generally, you can find a good, solid theme to use for your site from one source.

>> **Innovation:** To differentiate them from their free counterparts, premium themes include innovative features, such as theme settings or advanced options that extend WordPress to help you do more.

>> **Great design with solid code:** Although many beautiful themes are available for free, premium themes are professionally coded and nicely designed, cost thousands of dollars, and require dozens of hours to build, which simply isn't feasible for many developers of free themes.

>> **Support:** Most commercial companies have full-time support staff to answer questions, troubleshoot issues, and point you to resources beyond their support. Often, premium theme developers spend more time helping customers troubleshoot issues outside the theme products. Therefore, purchasing a premium theme often gives you access to a dedicated support community, which you can ask about advanced issues and upcoming WordPress features; otherwise, you're on your own.

>> **Stability:** No doubt you've purchased a product or service from a company only to find later that the company has gone out of business. If you choose to use a premium theme, purchase a theme from an established company with a solid business model, a record of accomplishment, and a dedicated team devoted to building and supporting quality products.

Although some free themes have some or all of the features in the preceding list, for the most part, they don't. Keep in mind that just because a designer calls a theme *premium* doesn't mean that the theme has passed through any kind of quality review. The view of what constitutes a premium theme can, and will, differ from one designer to the next.

Fully investigate any theme before you spend your money on it. Here are some things to check out before you pay:

>> Email the designer who's selling the premium theme, and ask about a support policy.

>> Find people who've purchased the theme, and contact them to ask about their experiences with the theme and the designer.

>> Carefully read any terms that the designer has published on their site to find any licensing restrictions that exist.

>> If the premium theme designer has a support forum, ask whether you can browse the forum to find out how actively the designer answers questions and provides support. Are users waiting weeks to get their questions answered, for example, or does the designer seem to be on top of support requests?

>> Search online for the theme and the designer. Often, users of premium themes post about their experiences with the theme and the designer. You can find both positive and negative information about the theme and the designer before you buy.

The developers in the following list are doing some amazingly innovative things with WordPress themes, and I highly recommend that you explore their offerings:

>> **WP Astra** (https://wpastra.com): WP Astra (see Figure 8-7) emphasizes an experience using popular site-builder plugins such as Beaver Builder (https://www.wpbeaverbuilder.com) and Elementor (https://elementor.com).

>> **Organic Themes** (https://organicthemes.com/themes): Organic Themes (see Figure 8-8) has a great team, paid support moderators, and WordPress themes that are as solid (from a code standpoint) as they are beautiful.

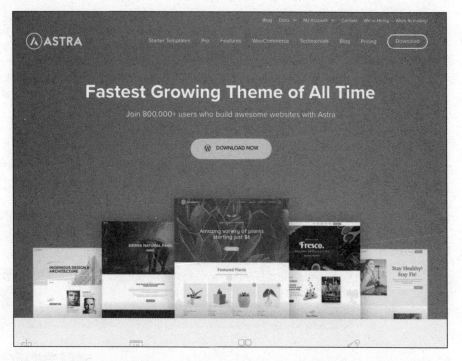

**FIGURE 8-7:**
WP Astra,
provider of
commercial
WordPress
themes.

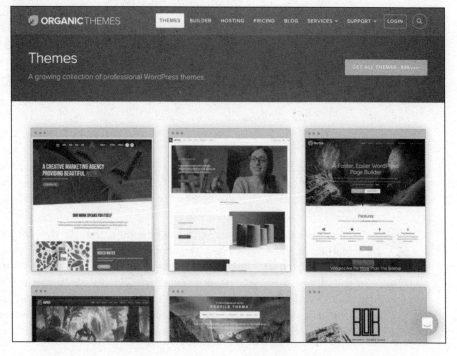

**FIGURE 8-8:**
Organic Themes,
another provider
of commercial
WordPress
themes.

>> **Woocommerce Theme Store** (https://woocommerce.com/product-category/themes): WooThemes (see Figure 8-9) has a wide selection of high-quality themes with excellent options and support for ecommerce websites that use the WooCommerce plugin. The most popular theme is Canvas, a highly customizable theme with more than 100 options you can use to personalize your site via a theme options panel.

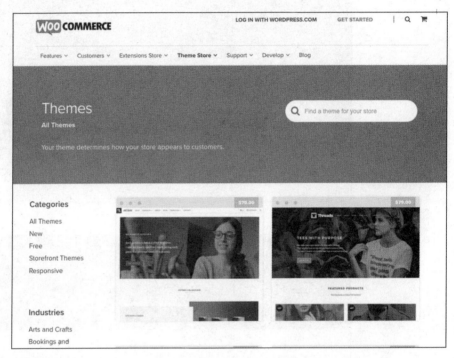

**FIGURE 8-9:**
WooThemes has premium themes, community, and support.

>> **Press75** (https://press75.com): Press75 (see Figure 8-10) offers niche themes for photography, portfolios, and video. Check out the On Demand theme for a great example (https://press75.com/view/on-demand).

TIP

You can't find, preview, or install premium themes by using the Add Themes feature of your WordPress Dashboard. You can find, purchase, and download premium themes only from third-party websites. After you find a premium theme you like, you need to install it via the SFTP method or by using the Dashboard upload feature. (See the earlier "Installing a Theme" section.) You can find a very nice selection of premium themes on the WordPress website at https://wordpress.org/themes/commercial.

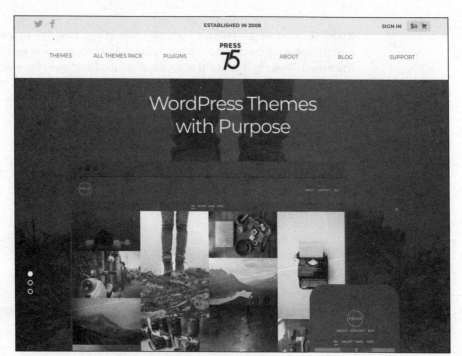

**FIGURE 8-10:**
Press75 offers premium themes, demos, and theme packages.

# Examining the Default Theme: Twenty Twenty

The Twenty Twenty theme is a powerful theme that takes full advantage of the block editor, allowing you to create pages with endless layout options and custom background colors. The theme also offers multiple-page templates, widget-ready areas, and built-in mobile and tablet support.

The members of the WordPress team who worked on Twenty Twenty (`https://wordpress.org/themes/twentytwenty/`) describe it as follows:

> Our default theme for 2020 is designed to take full advantage of the flexibility of the block editor. Organizations and businesses have the ability to create dynamic landing pages with endless layouts using the group and column blocks. The centered content column and fine-tuned typography also makes it perfect for traditional blogs. Complete editor styles give you a good idea of what your content will look like, even before you publish. You can give your site a personal touch by changing the background colors and the accent color in the Customizer. The colors of all elements on your site are automatically calculated based on the colors you pick, ensuring a high, accessible color contrast for your visitors.

These features make Twenty Twenty an excellent base for many of your theme customization projects. This chapter shows you how to manage all the features of the default Twenty Twenty theme, such as handling layouts, editing background colors, using custom navigation menus, and using widgets on your site to add some great features.

## Exploring the layout and structure

If you just want a simple look for your site, look no further than Twenty Twenty. This theme offers a clean design style that's highly customizable. As such, the font treatments are sharp and easy to read. Many of the new built-in features allow you to make simple yet elegant tweaks to the theme, including uploading new feature images, using the block editor for layouts, and adjusting the background colors. Figure 8-11 shows the Twenty Twenty WordPress default theme, out of the box — that is to say, without any customizations.

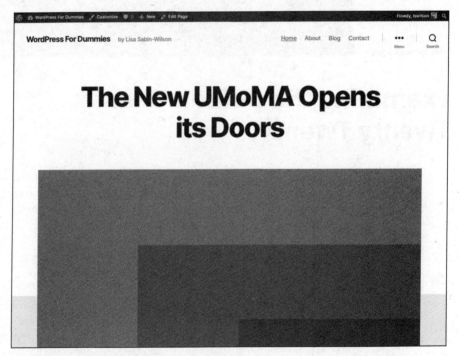

**FIGURE 8-11:** The default theme for WordPress: Twenty Twenty.

Some of the Twenty Twenty theme's distinctive layout features include

» **Site Identity:** Twenty Twenty gives you the ability to upload and display your logo in the header of your website.

>> **Customizer panel for colors:** The Twenty Twenty theme allows you to configure colors for the background of your website, as well as the background colors for the header and footer of your website. By default, the color scheme for the Twenty Twenty theme is a tan background with a red accent color and black text and content. You can easily change the colors of the header and site background by using the Customizer's color picker. Figure 8-12 shows the Twenty Twenty theme with the Customizer open to the Customizing Colors settings.

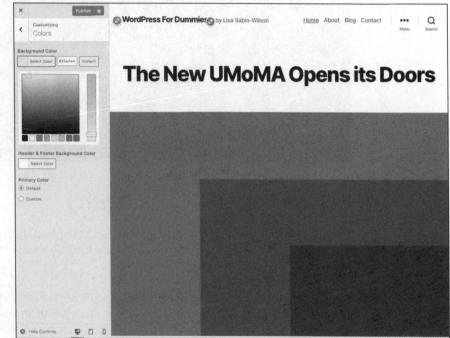

**FIGURE 8-12:**
The Twenty Twenty theme with custom colors.

>> **Theme Options:** The Theme Options section enables you to add a search form in the header of your website and show the author's bio at the bottom of posts, and it allows you to show the full text or excerpts of your posts on archive pages.

>> **Cover Page:** Twenty Twenty's Cover Page template for WordPress, shown in Figure 8-13, comes in handy for a home page or a specific landing page for a product or service because it provides a different layout and experience from the rest of the default templates in Twenty Twenty. You can customize the color of the overlay of the featured image and the color of the text that's displayed on top of that image.

FIGURE 8-13:
The Twenty
Twenty Cover
Page layout.

## Customizing the site identity

Twenty Twenty's Site Identity option allows you to upload a custom logo image for your WordPress site. The option is called Site Identity because the logo you use will be displayed on every page of your website.

The recommended dimensions for your logo media are 120 pixels wide by 90 pixels high. If your logo is larger, you can crop it after you've uploaded it to WordPress, although cropping with a graphics program (such as Adobe Photoshop) is the best way to get exact results.

Twenty Twenty displays a text version of the title of your website by default. Uploading a custom logo replaces the text version of your site title with an image of your choosing. To install a custom logo image, follow these steps with the Twenty Twenty theme activated:

**1.** **On the WordPress Dashboard, click the Customize link on the Appearance menu.**

The Customizer panel appears on the left side of the screen.

**2.** **Click the Site Identity link.**

The Customizing Site Identity section opens in the Customizer panel.

**3.** **In the Logo section, click the Select Logo button.**

The Select Logo screen opens (see Figure 8-14).

**FIGURE 8-14:**
Selecting an
image to use as a
logo image.

**4.** **Select an image from the Media Library, or upload an image from your computer.**

**5.** **Click the Select and Crop button.**

This step opens the Crop Image screen, shown in Figure 8-15.

**6.** **(Optional) Crop the image to your liking.**

To resize and crop your image, drag one of the eight tiny handles located around the image. You can also click within the image and move the entire image up or down to get the placement and cropping effect that you want.

**7.** **Click the Crop Image button to crop your header image.**

**8.** **Click the Publish button in the top-right corner of the Customizer panel to save your changes.**

Figure 8-16 shows the Twenty Twenty theme on my website with a custom logo image.

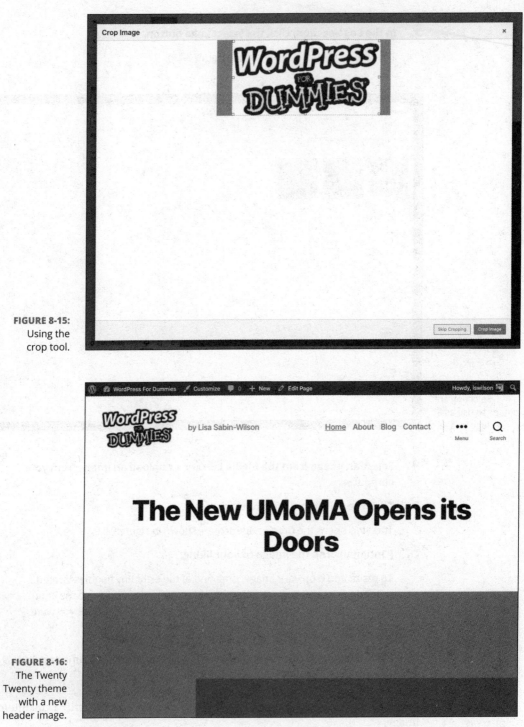

FIGURE 8-15:
Using the
crop tool.

FIGURE 8-16:
The Twenty
Twenty theme
with a new
header image.

## Customizing colors

After you explore the Site Identity settings, you may want to pick a background color or change the color of the header text. The default background color in the Twenty Twenty theme is tan, but you can change the color scheme. Here's how:

1. **On the WordPress Dashboard, click the Customize link on the Appearance menu.**

   The Customizer panel opens on the left side of your screen.

2. **Click the Colors link.**

   The Customizing Colors screen opens.

3. **Click Select Color below the Background Color heading.**

   The Select Color button changes to a color picker.

4. **Choose a color.**

   You can click anywhere on the color picker (see Figure 8-17) or type a six-digit hexadecimal code (*hex code,* for short) if you know your preferred color.

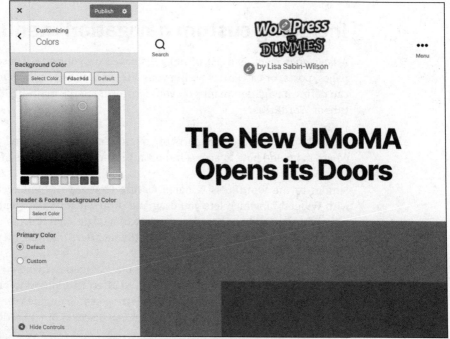

**FIGURE 8-17:**
The Twenty Twenty theme's color options.

**TIP**

Color values are defined in HTML and CSS by six-digit hexadecimal codes starting with the # sign, such as #000000 for black or #FFFFFF for white. (As noted in Chapter 10, adjusting hexadecimal colors is one of the easiest ways to tweak the colors in your theme for a new look.)

5. **Click Select Color below the Header & Footer Background Color heading.**

6. **Repeat the instructions from step 4 to change the background colors of the header and footer of your website.**

7. **Select the Custom option below the Primary Color heading.**

   A color picker opens.

8. **Choose the colors for links, buttons, and featured images on your website.**

9. **When you've finalized your selections, click the Publish button.**

   WordPress saves and publishes the color changes you made.

**TIP**

The WordPress Customizer appears on the left side of your computer screen, and a preview of your site displays on the right side. As you're making changes in the Customizer, you see a preview of the changes those options make on your site.

## Including custom navigation menus

A *navigation menu* is a list of links displayed on your site. These links can be to pages, posts, or categories within your site, or they can be links to other sites. You can define navigation menus on your site by using the built-in Custom Menu feature of WordPress.

Navigation menus are vital parts of your site's design. They tell your site visitors where to go and how to access important information or areas of your site.

Similar to the WordPress Widgets feature (covered in "Enhancing Your Theme with Widgets"), which lets you drag and drop widgets, the Menus feature offers an easy way to add and reorder a variety of navigational links to your site, as well as create secondary menu bars (if your theme offers multiple menu areas).

It's to your advantage to provide at least one navigation menu on your site so that readers can see everything your site has to offer. In addition to providing a visual navigation aide to your visitors, you're letting search engines see the links you've added to a menu; then the search engines can discover other sections and pages of your site to add to their listings.

TIP

The Menus feature also improves WordPress by making it easy to use websites that need multiple diverse navigational features compared with those that a typical website uses.

Twenty Twenty comes with the appropriate code in the navigation menus for this robust feature. By default, Twenty Twenty offers five menu locations to include custom menus: Social Media, Desktop Horizontal Menu, Desktop Extended Menu, Mobile Menu, and Footer Menu.

TIP

Chapter 12 gives you all the information you need to work with the custom menu features in WordPress.

# Enhancing Your Theme with Widgets

WordPress *widgets* are helpful tools built into the WordPress application that allow you to arrange the display of content on your blog sidebar, such as recent posts and archive lists. With widgets, you can arrange and display the content of the sidebar of your site without having to know a single bit of PHP or HTML.

*Widget areas* are the regions of your theme where you can insert and arrange content, such as a list of your recent blog posts or links to your favorite site. You can also add custom menus by dragging and dropping (and editing) available widgets (shown on the Dashboard's Widget page) into the corresponding areas.

Many widgets offered by WordPress (and those added by some WordPress themes and plugins) provide drag-and-drop installation of advanced functions that normally are available only if you write code directly in your theme files.

Click the Widgets link on the Appearance menu of the Dashboard to see the available widgets on the Widgets screen. This feature is a big draw because it lets you control what features you use and where you place them without having to know a lick of code.

The Twenty Twenty theme's widget-ready sections include Footer #1 and Footer #2, shown in Figure 8-18.

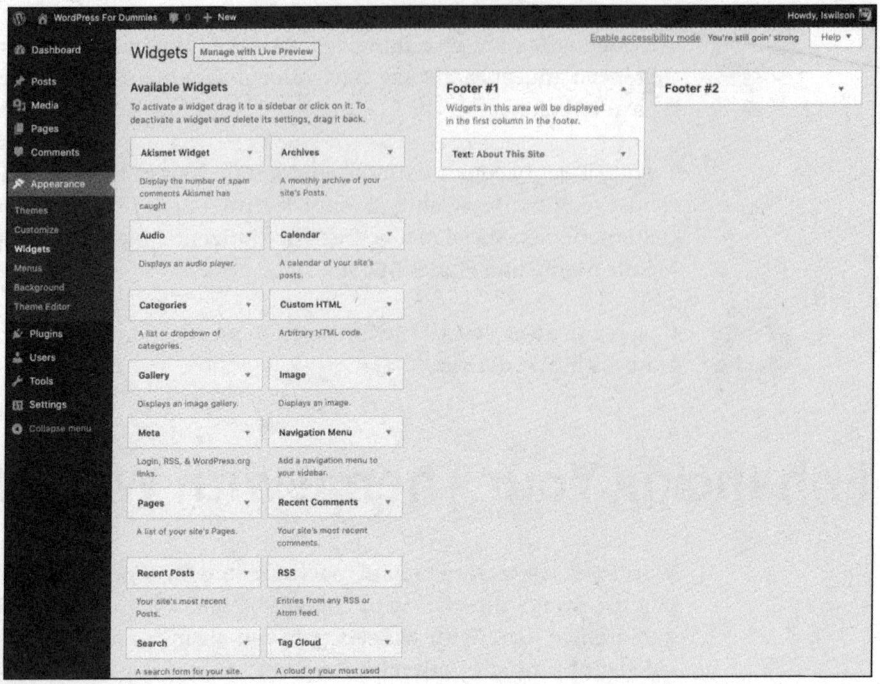

**FIGURE 8-18:**
This page
displays available
widgets and
widget-ready
areas.

## Adding widgets to your website

The Widgets screen lists all the widgets that are available for your WordPress site. On the right side of the Widgets screen is the Footer #1 and Footer #2 areas designated in the Twenty Twenty theme (refer to Figure 8-18). You drag your selected widget from the Available Widgets section to your chosen widget area on the right. To add a Search box to the right side of the footer area of the default layout, for example, drag the Search widget from the Available Widgets section to the Footer #2 widget area.

To add a new widget to your sidebar, follow these steps:

1. **Find the widget you want to use.**

   The widgets are listed in the Available Widgets section. For the purpose of these steps, choose the Recent Posts widget.

2. **Drag and drop the Recent Posts widget to the Footer #1 widget section on the right side of the page.**

   The widget is now located in the Footer #1 widget section, and the content of the widget appears on your site's sidebar.

**3.** **Click the arrow to the right of the widget title.**

Options for the widget appear. Each widget has different options that you can configure. The Recent Posts widget, for example, lets you configure the title, the number of recent posts you want to display (the default is 5; the maximum is 15), and the date (see Figure 8-19).

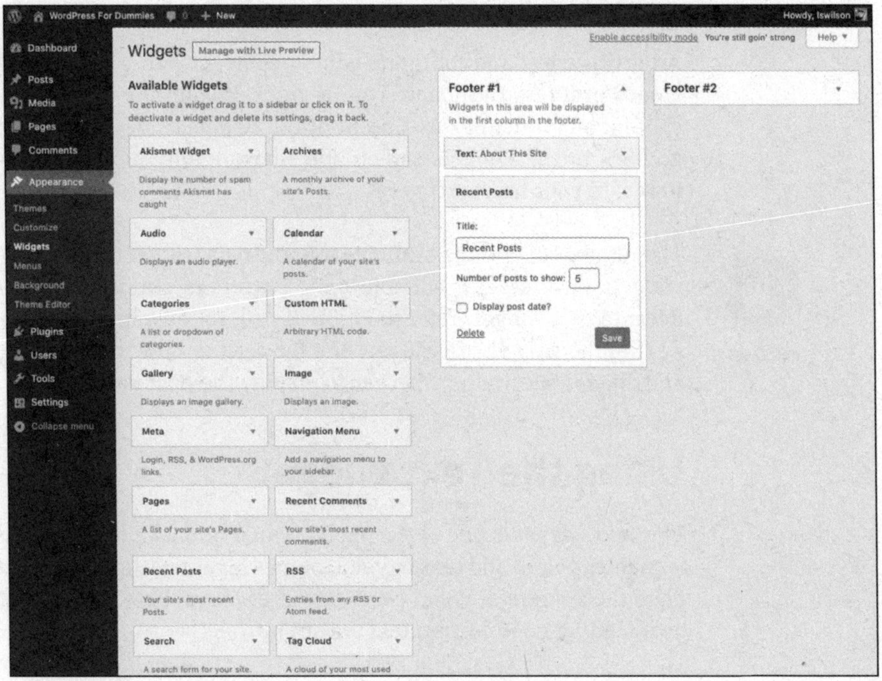

**FIGURE 8-19:**
Editing the Recent Posts widget.

**4.** **Select your options, and click the Save button.**

The options you've set are saved.

**5.** **Arrange your widgets in the order in which you want them to appear on your site by dragging and dropping them in the list.**

Repeat this step until your widgets are arranged the way you want them.

**TIP**

To remove a widget from your sidebar, click the arrow to the right of the widget title to open the widget options; then click the Delete link. WordPress removes the widget from the right side of the page and places it back in the Available Widgets list. If you want to remove a widget but want WordPress to remember the settings that you configured for it, instead of clicking the Delete link, simply drag the widget into the Inactive Widgets area (shown in Figure 8-20) at the bottom of the Widgets screen. The widget and all your settings are stored for future use.

FIGURE 8-20:
The Inactive
Widgets area of
the Widgets
screen.

After you select and configure your widgets, click the Visit Site button at the top of your WordPress Dashboard (to the right of your site name). Your site reflects the content and order of the content of the Widgets screen. How cool is that? You can go back to the Widgets screen and rearrange items, as well as add and remove items, to your heart's content.

**REMEMBER**

The number of options available for editing a widget depends on the widget. Some widgets have several editable options; others simply let you write a title. The Recent Posts widget (refer to Figure 8-19), for example, has three options: one for editing the title of the widget, one for setting how many recent posts to display, and one for specifying whether to display the post date.

## Using the Text widget

The Text widget is one of the most popular and useful WordPress widgets because it enables you to add text to your themes layout without editing the theme's template files. Therefore, you can designate several types of information on your site by including your desired text within it.

Here are some examples of how you can use the Text widget:

» **About Us.** Add a paragraph about you, or about your business, that appears in the widget areas of the theme.

» **Display business hours of operation.** Display the days and hours of your business operation where everyone can easily see them.

» **Announce special events and notices.** If your organization has a special sale, an announcement about a staff member, or an important notice about weather closings, for example, you can use the Text widget to post this information to your site in a few seconds.

To add the Text widget, follow these steps:

1. **On the WordPress Dashboard, click the Widgets link on the Appearance menu.**

2. **Find the Text widget in the Available Widgets section.**

3. **Drag the Text widget to the desired widget area.**

   The Text widget opens, as shown in Figure 8-21.

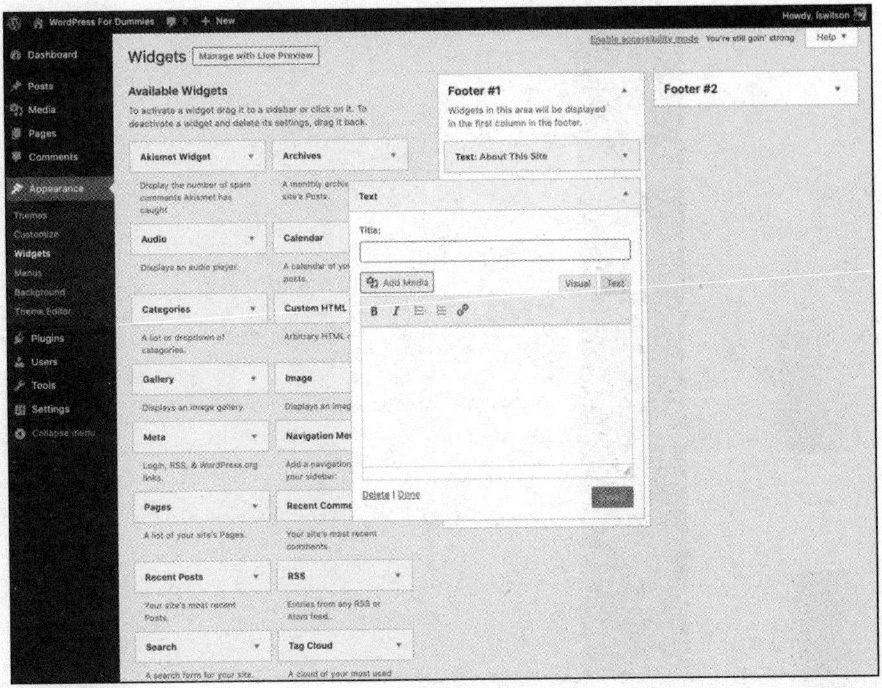

**FIGURE 8-21:**
The Text widget.

4. **Add a widget headline in the Title field and any desired text in the Content box.**

5. **Click the Save button.**

6. **Click the Close link at the bottom of the Text widget.**

## Using the RSS widget

The RSS widget allows you to pull headlines from almost any RSS (Really Simple Syndication) feed, including recent headlines from your other WordPress sites. You can also use it to pull in headlines from news sites or other sources that offer RSS feeds. This practice is commonly referred to as *aggregation*, which means that you're gathering information from a syndicated RSS feed source to display on your site.

After you drag the RSS widget to the appropriate widget area, the widget opens, and you can enter the RSS Feed URL you want to display. Additionally, you can easily tweak other settings, as shown in Figure 8-22, to add information to the widget area for your readers.

Follow these steps to add the RSS widget to your blog:

1. **Add the RSS widget to the Footer #1 area of the Widgets screen.**

   Follow the steps in "Adding widgets to your website" earlier in this chapter.

2. **Click the arrow to the right of the RSS widget's name.**

   The widget opens, displaying options you can configure.

3. **In the Enter the RSS Feed URL Here text box, type the RSS URL of the site you want to add.**

4. **Type the title of the RSS widget.**

   This title appears in your site above the links from this site. If I wanted to add the RSS feed from my business site, for example, I'd type **WebDevStudios Feed**.

5. **Select the number of items from the RSS feed to display on your site.**

   The drop-down menu gives you a choice of 1 to 20.

6. **(Optional) Select the Display the Item Content? check box.**

   Selecting this check box tells WordPress that you also want to display the content of the feed (usually, the content of the post from the feed URL). If you want to display only the title, leave the check box deselected.

7. **(Optional) Select the Display Item Author If Available? check box.**

   Select this option if you want to display the author's name with the item's title.

8. **(Optional) Select the Display Item Date? check box.**

   Select this option if you want to display the date when the item was published along with the item's title.

9. **Click the Save button.**

   WordPress saves all the options and reloads the Widgets page with your RSS widget intact. Figure 8-23 shows the footer of my website displaying the feed from WebDevStudios, which shows five posts from my blog.

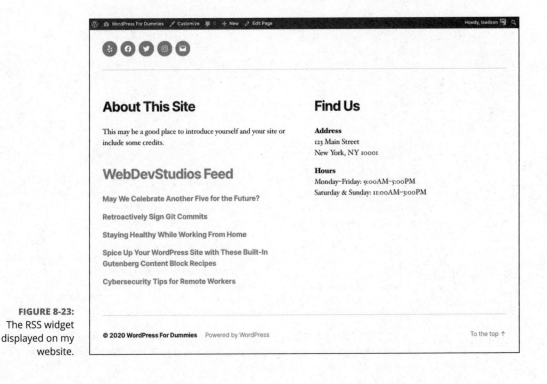

**FIGURE 8-23:** The RSS widget displayed on my website.

# 4

# Customizing WordPress

**IN THIS PART . . .**

Learn all about the basic templates used to make up a WordPress theme, including basic template tags.

Utilize basic HTML markup and CSS to tweak the look and feel of your existing theme.

Understand the basic concept behind Parent/Child themes in WordPress and learn how to create your own custom theme based on an existing theme.

Discover how to use WordPress as a content management system and add support for features such as custom post types, featured images, and post formats.

Find out what it takes to migrate your existing website to WordPress (if it's currently using a different publishing platform) or move it to a new hosting provider.

IN THIS CHAPTER

» Looking at the basic website
structure

» Exploring the required templates

» Understanding The Loop and Main
Index template

» Putting together a basic theme

» Using common template tags

Chapter **9**

# Understanding Themes and Templates

I f, like me, you like to get your hands dirty, you need to read this chapter. WordPress users who create their own themes do so in the interest of

» **Individuality:** You can have a theme that no one else has. (If you use one of the free themes, you can pretty much count on the fact that at *least* a dozen other WordPress websites will have the same look as yours.)

» **Creativity:** You can display your own personal flair and style.

» **Control:** You can have full control of how the site looks, acts, and delivers your content.

Many of you aren't at all interested in creating your own theme for your WordPress website, however. Sometimes, it's just easier to leave matters to the professionals and hire an experienced WordPress theme developer to create a custom look for your WordPress website or to use one of the thousands of free themes provided by WordPress designers (see Chapter 8). Chapter 16 also tells you where you can get ten free WordPress themes.

Creating themes requires you to step into the code of the templates, which can be a scary place sometimes — especially if you don't really know what you're looking at. A good place to start is to understand the structure of a WordPress website. Separately, the parts won't do you any good, but when you put them together, the real magic begins! This chapter covers the basics of doing just that, and near the end of the chapter, you find specific steps for putting your own theme together.

**TIP**

You don't need to know HTML to use WordPress. If you plan to create and design WordPress themes, however, you need some basic knowledge of HTML and Cascading Style Sheets (CSS). For assistance with HTML, check out *HTML5 and CSS3 All-in-One For Dummies,* 3rd Edition, by Andy Harris (published by John Wiley & Sons, Inc.).

# Using WordPress Themes: The Basics

A WordPress *theme* is a collection of WordPress templates made up of WordPress template tags. When I refer to a WordPress *theme,* I'm talking about the group of templates that makes up the theme. When I talk about a WordPress *template,* I'm referring to only one of the template files that contain WordPress template tags and functions. WordPress template tags make all the templates work together as a theme. These files include

>> **The theme's stylesheet** (style.css): The stylesheet provides the theme's name, as well as the CSS rules that apply to the theme.

>> **The main index template** (index.php): The index file is the first file that will be loaded when a visitor comes to your site. It contains the HTML as well as any WordPress functions and template tags needed on your home page.

>> **An optional functions file** (functions.php): This optional file is a place where you can add additional functionality to your site via PHP and/or WordPress-specific functions.

Template and functions files end with the .php extension. *PHP* is the scripting language used in WordPress, which your web server recognizes and interprets as such. These files contain more than just scripts, though. The PHP files also contain HTML markup, which is the basic markup language of web pages.

Within this set of PHP files is all the information your browser and web server need to make your website. Everything from the color of the background to the layout of the content is contained in this set of files.

**REMEMBER**

The difference between a template and a theme can cause confusion. *Templates* are individual files. Each template file provides the structure in which your content will display. A *theme* is a set of templates. The theme uses the templates to make the whole site.

## Understanding theme structure

Understanding where the WordPress theme files are located on your web server gives you the ability to find and edit them as needed. You can use two methods to view and edit WordPress theme files by following these steps:

1. **Connect to your web server via SFTP, and have a look at the existing WordPress themes on your server.**

   The correct location is /wp-content/themes/. When you open this folder, you find the /twentytwenty theme folder (left side of Figure 9-1).

   If a theme is uploaded to any folder other than /wp-content/themes, it won't work.

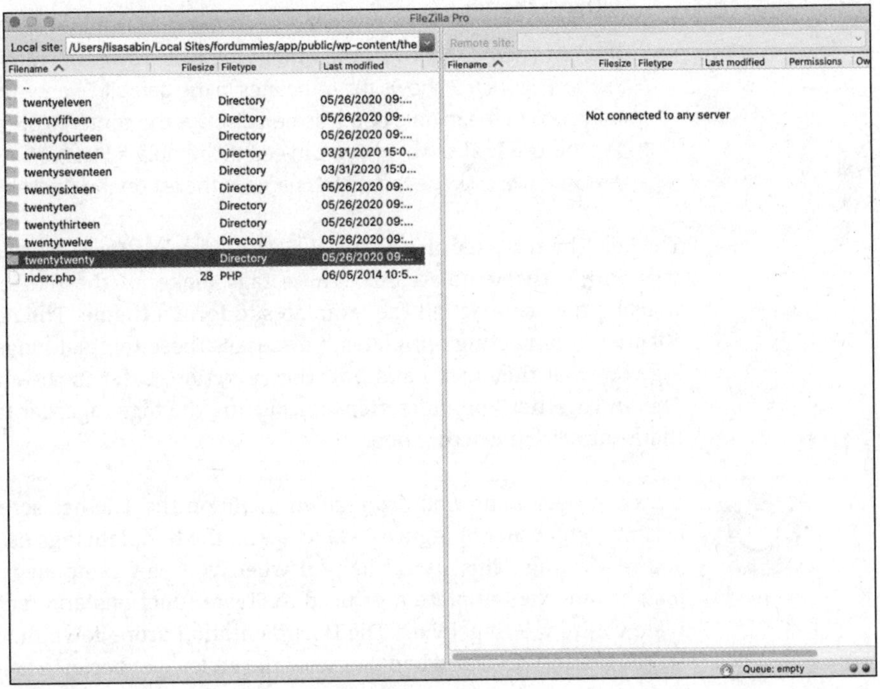

**REMEMBER**

FIGURE 9-1:
WordPress themes in the /wp-content/ themes folder in SFTP.

2. **Open the folder for the Twenty Twenty theme** (`/wp-content/themes/twentytwenty`), **and look at the template files inside.**

   When you open the Twenty Twenty theme's folder, you see several files. At minimum, you find these five templates in the default theme:

   - *Stylesheet* (`style.css`)

   - *Functions file* (`functions.php`)

   - *Header template* (`header.php`)

   - *Main index* (`index.php`)

   - *Footer template* (`footer.php`)

   These files are the main WordPress template files, and I discuss them in more detail in this chapter. There are several template files, however, and you should try to explore all of them if you can. Take a peek inside to see the different template functions they contain. Every WordPress theme is unique and contains its own number and type of files, but the five files I mention here usually exist in WordPress themes.

3. **Click the Editor link on the Appearance menu to look at the template files within a theme.**

   This Edit Themes page lists the various templates available within the active theme. (Figure 9-2 shows the templates in the default Twenty Twenty theme.) A text box in the middle of the screen displays the contents of each template, and this box is also where you can edit the template file(s). To view and edit a template file, click the template's name in the list on the right side of the page.

The Edit Themes page also shows the HTML markup (see Chapter 10) and template tags within the template file. These tags make all the magic happen on your website; they connect all the templates to form a theme. The next section of this chapter, "Connecting templates," discusses these template tags in detail, showing you what they mean and how they function. Later in the chapter, "Putting a Theme Together" provides steps for putting the tags together to create your own theme or edit an existing one.

**TIP**

Click the Documentation drop-down menu on the Themes screen (shown in the bottom-left corner of Figure 9-2) to see all the template tags used in the template you're viewing. This list is helpful when you edit templates, and it gives you insight into the template tags used to create functions and features in the template you're viewing. (*Note:* The Documentation drop-down menu on the Themes screen doesn't appear when you view the stylesheet because no template tags are used in the `style.css` template — only CSS.)

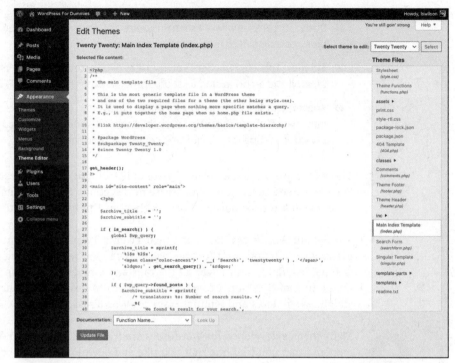

## Connecting templates

The template files don't work alone; for the theme to function, the files need one another. To tie these files together as one working entity, you use template tags to pull the information from each template — Header, Sidebar, and Footer — into the Main Index. I refer to this procedure as *calling* one template into another. (You can find more information in the "Getting Familiar with the Four Main Templates" section later in this chapter.)

# Contemplating the Structure of a WordPress Website

A WordPress blog, in its basic form, has four main areas. Template files for these four areas appear in the default theme that comes in every version of WordPress:

>> **Header:** This area usually contains the name of the site along with the site tagline or slogan. Sometimes, the header also contains a graphic or image.

- **Body:** This area is where the main content of your website appears, such as blog posts displayed in chronological order.

- **Sidebar:** This area is where you find lists of blog-related elements, such as the blogroll, the archives, and a list of recent posts.

- **Footer:** This area, at the bottom of the page, often contains links to further information about the website, such as who designed it, which company provides hosting for the site, and copyright information.

These four areas are the absolute bare bones of a *basic* WordPress blog template. You can extend these areas and create new sections that carry more information, of course, but for the purposes of this chapter, I'm focusing on the basics.

The default WordPress theme is called Twenty Twenty, and in my opinion, it's a pretty good starting point for you, especially if you're just getting your feet wet in web publishing. I don't cover all the tags and templates that the Twenty Twenty theme includes; rather, I touch on the basics to get you on your way to understanding templates and template tags for WordPress.

TIP

Many themes developed for WordPress are free for public use, and I strongly recommend that you find one that you like and download it. Use the free themes as places to get started in theme development. Really, why reinvent the wheel? With the free themes available today, most of the work has already been completed for you, and you may find it easier to use one of these themes than to start a theme from scratch.

Each free theme available for download is different, depending on what the developer included (such as CSS styling, display options, format, and layout). Experimenting with a few themes is a fun and useful way to find out more about the development of WordPress themes. A great place to find free WordPress themes is the official WordPress Themes page at https://wordpress.org/themes.

To build a WordPress theme that covers the basic areas of a website, you need four templates and a fifth file called a stylesheet:

- header.php

- index.php

- sidebar.php

- footer.php

- style.css

Each WordPress theme comes with a stylesheet (style.css), which drives the formatting and layout of your blog template in terms of where the elements are positioned on the page, what the font looks like, what colors your hyperlinks will be, and so on. As you may have already figured out, you don't use CSS to put content on your site; rather, you use CSS to style the content that's already there.

Chapter 10 provides information on tweaking the design of your theme by combining the template tags presented in this chapter with some CSS adjustments in your theme files.

Right now, I'm covering only the basics. At the end of this chapter, however, I provide some ideas on how you can use various templates to further extend your website functionality, using templates for categories, archives, static pages, multiple sidebars, and so on. After you build the basics, you can spread your wings and step into more advanced themes.

# Examining the Anatomy of a Template Tag

Some people are intimidated when they look at template tags. Really, they're just simple bits of PHP code that you can use inside a template file to display information dynamically. Before you start to play around with template tags in your WordPress templates, however, it's important to understand what makes up a template tag — and why.

WordPress is based in PHP (a scripting language for creating web pages) and uses PHP commands to pull information from the MySQL database. Every tag begins with a function to start PHP and ends with a function to stop PHP. In the middle of those two commands lives the request to the database that tells WordPress to grab the data and display it.

A typical template tag looks like this:

```
<?php bloginfo(); ?>
```

This entire example tells WordPress to do three things:

>> Start PHP (<?php).

>> Use PHP to get information from the MySQL database and deliver it to your blog (bloginfo();).

>> Stop PHP (?>).

In this case, `bloginfo();` represents the function of this template tag, which grabs information from the database to deliver it to your site. What information is retrieved depends on what tag function appears between the two PHP commands. As you may notice, a lot of starting and stopping of PHP happens throughout the WordPress templates. The process seems as though it would be resource-intensive, if not exhaustive — but it really isn't.

For every PHP command you start, you need a stop command. Every time a command begins with `<?php`, somewhere later in the code is the closing `?>` command. PHP commands that aren't structured properly cause really ugly errors on your site, and they've been known to send programmers, developers, and hosting providers into loud screaming fits.

# Getting Familiar with the Four Main Templates

In the following sections, I cover some of the template tags that pull in the information you want to include on your website. To keep this chapter shorter than 1,000 pages, I focus on the four main templates that get you going with creating your own theme or with editing the template tags in the theme you're currently using. Here are those four main templates:

>> Header

>> Main Index

>> Sidebar

>> Footer

*Templates* are individual files. A *theme* is a set of templates.

## The Header template

The Header template is the starting point for every WordPress theme because it tells web browsers the following:

>> The title of your site

>> The location of the CSS

>> The RSS-feed URL

» The site URL

» The tagline (or description) of the website

In many themes, the first elements in the header are a main image and the navigation. These two elements are usually in the `header.php` file because they load on every page and rarely change. The following statement is the built-in WordPress function to call the header template:

```
<?php get_header(); ?>
```

TIP

Every page on the web has to start with a few pieces of code. In every `header.php` file in any WordPress theme, you find these bits of code at the top:

» DOCTYPE (which stands for *document type declaration*) tells the browser which type of XHTML standards you're using. The Twenty Twenty theme uses `<!DOCTYPE html>`, which is a declaration for W3C (World Wide Web Consortium `https://www.w3.org/`) standards compliance mode and covers all major browsers.

» The `<html>` tag (*HTML* stands for *Hypertext Markup Language*) tells the browser which language you're using to write your web pages.

» The `<head>` tag tells the browser that the information contained within the tag shouldn't be displayed on the site; rather, that information is *about* the document.

In the Header template of the Twenty Twenty theme, these bits of code look like the following example, and you should leave them intact:

```
<!DOCTYPE html>
<html class="no-js" <?php language_attributes(); ?>
<head>
```

TIP

On the Edit Themes page, click the Header template link to display the template code in the text box. Look closely, and you see that the `<!DOCTYPE html>` declaration, `<html>` tag, and `<head>` tag show up in the template.

The `<head>` tag needs to be closed at the end of the Header template, which looks like this: `</head>`. You also need to include a fourth tag, `<body>`, which tells the browser where the information you want to display begins. Both the `<body>` and `<html>` tags need to be closed at the end of the template files (in `footer.php`), like this: `</body></html>`.

## Using bloginfo parameters

The Header template makes much use of one WordPress template tag in particular: `bloginfo();`. This tag is commonly used in WordPress themes.

What differentiates the type of information that a tag pulls in is a parameter. Parameters are placed inside the parentheses of the tag, enclosed in single quotes. For the most part, these parameters pull information from the settings on your WordPress Dashboard. The template tag to get your site title, for example, looks like this:

```
<?php bloginfo( 'name' ); ?>
```

Table 9-1 lists the parameters that are available for the `bloginfo();` template tag and shows you what the template tag looks like.

**TABLE 9-1**    ## Tag Values for bloginfo()

| Parameter | Information | Tag |
|-----------|-------------|-----|
| charset | Character settings, set on the General Settings screen | `<?php bloginfo( 'charset' ); ?>` |
| name | Site title, set by choosing Settings ⇨ General | `<?php bloginfo( 'name' ); ?>` |
| description | Tagline for your site, set by choosing Settings ⇨ General | `<?php bloginfo( 'description' ); ?>` |
| url | Your site's web address, set by choosing Settings ⇨ General | `<?php bloginfo( 'url' ); ?>` |
| stylesheet_url | URL of primary CSS file | `<?php bloginfo( 'stylesheet url' ); ?>` |
| pingback_url | Displays the trackback URL for your site on single post pages | `<?php bloginfo( 'pingback_url' ); ?>` |

## Creating <title> tags

Here's a useful tip about your blog's <title> tag: Search engines pick up the words used in the <title> tag as keywords to categorize your site in their directories.

The <title></title> tags are HTML tags that tell the browser to display the title of your website on the title bar of a visitor's browser. Figure 9-3 shows how the title of my business website sits on the title bar of the browser window.

Title bar

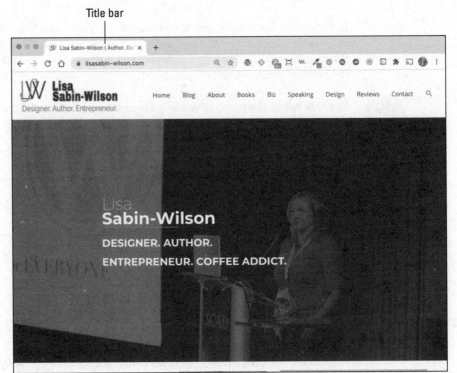

**FIGURE 9-3:**
The title bar of a browser.

**REMEMBER**

Search engines love the title bar. The more you can tweak that title to provide detailed descriptions of your site (otherwise known as *search engine optimization*, or *SEO*), the more the search engines love your blog site. Browsers show that love by giving your site higher rankings in their results.

The site `<title>` tag is the code that lives in the Header template between these two tag markers: `<title></title>`. In the default Twenty Twenty theme, that bit of code (located in the theme's `functions.php` template file, on line 92 looks like this:

```
add_theme_support( 'title-tag' );
```

The `add_theme_support( 'title-tag' );` in the `functions.php` template tells WordPress to place the title tag in the `<head>` section of the website. By adding theme support for the title tag, you're saying that this theme doesn't use a hard-coded `<title>` tag in the document head and that you expect WordPress to provide it.

It may help for me to put this example in plain English. The way that the `add_theme_support( 'title-tag' );` function displays the title is based on the type of page that's being displayed, and it shrewdly uses SEO to help you with the browser powers that be.

**REMEMBER**

The title bar of the browser window always displays your site name unless you're on a single post page. In that case, it shows your site title plus the title of the post on that page.

## Displaying your site name and tagline

Most WordPress themes show your site name and tagline in the header of the site, which means that those items are displayed in easy, readable text for all visitors (not just search engines) to see. My site name and tagline, for example, are

>> **Site name:** Lisa Sabin-Wilson

>> **Site tagline:** Designer. Author. Entrepreneur.

Refer to Figure 9-3 to see these two elements in the header of the site.

You can use the `bloginfo();` tag plus a little HTML code to display your site name and tagline. Most sites have a clickable title that takes you back to the main page. No matter where your visitors are on your site, they can always go back home by clicking the title of your site on the header.

To create a clickable title, use the following HTML markup and WordPress template tags:

```
<a href="<?php bloginfo( 'url' ); ?>"><?php bloginfo( 'name' ); ?></a>
```

The `bloginfo( 'url' );` tag is your main site Internet address, and the `bloginfo( 'name' );` tag is the name of your site (refer to Table 9-1). Therefore, the code creates a link that looks something like this:

```
<a href="http://yourdomain.com">Your Site Name</a>
```

The tagline generally isn't linked back home. You can display it by using the following tag:

```
<?php bloginfo( 'description' ); ?>
```

This tag pulls the tagline directly from the one that you've set up on the General Settings page of your WordPress Dashboard.

This example shows that WordPress is intuitive and user-friendly; you can do things such as change the site name and tagline with a few keystrokes in the Dashboard. Changing your options in the Dashboard creates the change on every page of your site; no coding experience is required. Beautiful, isn't it?

Often, these tags are surrounded by tags that look like these: `<h1></h1>` and `<p></p>`. These tags are HTML markup tags, which define the look and layout of the blog name and tagline in the CSS of your theme. I cover CSS further in Chapter 10.

The `header.php` template file also can include the `sidebar.php` template file, which means that it tells WordPress to execute and display all the template functions included in the Sidebar template (`sidebar.php`). The line of code from the Header template that performs this task looks like this:

```
get_sidebar();
```

In the following sections, I provide more information about including other templates by calling them in through template functions.

## The Main Index template

The Main Index template drags your blog posts out of the MySQL database and inserts them into your site. This template is to your blog what the dance floor is to a nightclub: where all the action happens.

The filename of the Main Index template is `index.php`. You can find it in the `/wp-content/themes/twentytwenty/` folder.

The first template tag in the Main Index template calls in the Header template, meaning that it pulls the information from the Header template into the Main Index template, as follows:

```
<?php get_header(); ?>
```

Your theme can work without calling in the Header template, but it'll be missing several essential pieces — the CSS and the blog name and tagline, for starters. Without the call to the Header template, your blog resembles the image shown in Figure 9-4.

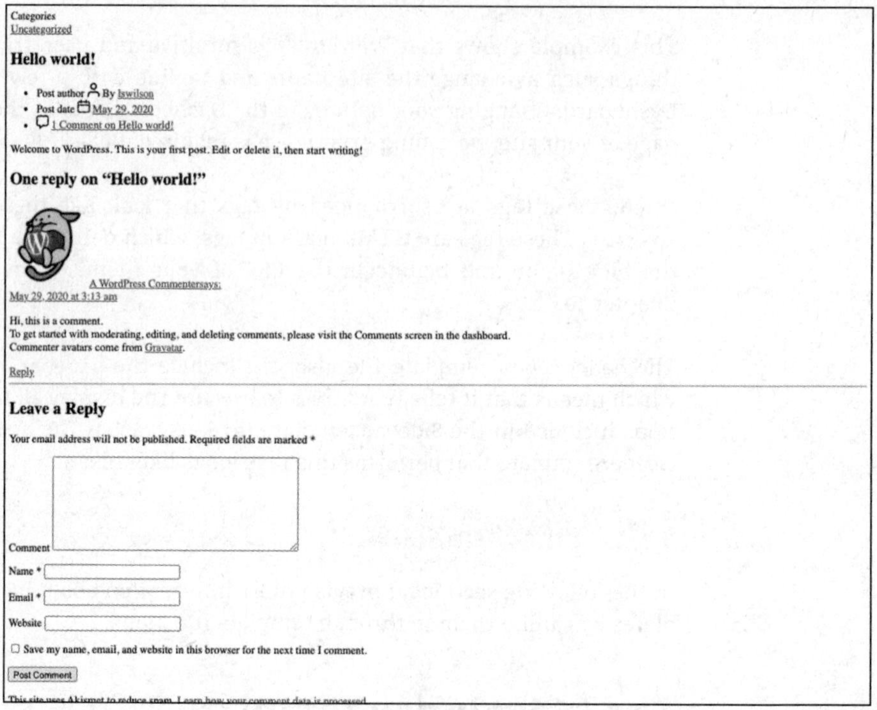

Figure content:

Categories
Uncategorized

**Hello world!**

- Post author   By jxwilson
- Post date  May 29, 2020
- 1 Comment on Hello world!

Welcome to WordPress. This is your first post. Edit or delete it, then start writing!

**One reply on "Hello world!"**

A WordPress Commentersays:
May 29, 2020 at 3:13 am

Hi, this is a comment.
To get started with moderating, editing, and deleting comments, please visit the Comments screen in the dashboard.
Commenter avatars come from Gravatar.

Reply

**Leave a Reply**

Your email address will not be published. Required fields are marked *

Comment

Name *

Email *

Website

☐ Save my name, email, and website in this browser for the next time I comment.

Post Comment

This site uses Akismet to reduce spam. Learn how your comment data is processed.

**FIGURE 9-4:**
A WordPress
blog missing
the call to
the header.
It's naked!

The Main Index template in the default theme, Twenty Twenty, calls in three other files in a similar fashion:

» `get_template_part( 'template-parts/content', get_post_type() );`

This function includes the Post Type–specific template for the content.

» `get_template_part( 'template-parts/pagination' );`

This function calls in the template file named `pagination.php`, which handles the output of pagination for the website.

» `get_template_part( 'template-parts/footer-menus-widgets' );`

This function calls in the template file named `footer-menus-widgets.php`, which handles the display of the menus and widgets in the footer of the website.

REMEMBER

Earlier in this chapter, in the "Connecting templates" section, I explain the concept of calling in a template file by using a function or template tag. That's exactly what the Main Index template does with the four functions for the Header, Loop, Sidebar, and Footer templates explained in this section.

## The Loop

I'm not talking about America's second-largest downtown business district, originating at the corner of State and Madison streets in Chicago. I could write about some interesting experiences I've had there . . . but that would be a different book.

*The Loop* in this case is a function that WordPress uses to display content on your site, such as blog posts and page content. The Loop has a starting point and an ending point; anything placed between those points is used to display each post, including any HTML, PHP, or CSS tags and codes.

Here's a look at what the WordPress Codex calls the world's simplest index:

```php
<?php
  get_header();
    if (have_posts()) :
    while (have_posts()) :
      the_post();
        the_content();
    endwhile;
    endif;
  get_sidebar();
  get_footer();
?>
```

Here's how the code works:

1. The template opens the php tag.

2. The Loop includes the header, meaning that it retrieves anything contained in the header.php file and displays it.

3. The Loop begins with the while (have_posts()) : bit.

4. Anything between the while and the endwhile repeats for each post that displays.

   The number of posts that displays is determined in the settings section of the WordPress Dashboard.

5. If your site has posts (and most sites do, even when you first install them), WordPress proceeds with The Loop, starting with the piece of code that looks like this:

   ```php
   if (have_posts()) :
   while (have_posts()) :
   ```

This code tells WordPress to grab the posts from the MySQL database and display them on your page.

6. The Loop closes with this tag:

```
endwhile;
endif;
```

Near the beginning of The Loop template is a template tag that looks like this:

```
if (have_posts()) :
```

In plain English, that tag says If [this blog] has posts.

7. If your site meets that condition (that is, if it has posts), WordPress proceeds with The Loop and displays your posts. If your site doesn't meet that condition (that is, doesn't have posts), WordPress displays a message that no posts exist.

8. When The Loop ends (at endwhile), the index template goes on to execute the files for the sidebar and footer.

Although it's simple, The Loop is one of the core functions of WordPress.

**WARNING**

Misplacement of the while or endwhile statements causes The Loop to break. If you're having trouble with The Loop in an existing template, check your version against the original to see whether the while statements are misplaced.

**REMEMBER**

In your travels as a WordPress user, you may run across plugins or scripts with instructions that say something like this: "This must be placed within The Loop." That Loop is the one I discuss in this section, so pay particular attention. Understanding The Loop arms you with the knowledge you need to understand and tackle your WordPress themes.

The Loop is no different from any other template tag; it must begin with a function to start PHP, and it must end with a function to stop PHP. The Loop begins with PHP and then makes a request: "While there are posts in my website, display them on this page." This PHP function tells WordPress to grab the blog-post information from the database and return it to the web page. The end of The Loop is like a traffic cop with a big red stop sign telling WordPress to stop the function.

**REMEMBER**

You can set the number of posts displayed per page in the Reading Settings page of the WordPress Dashboard. The Loop abides by this rule and displays only the number of posts per page that you've set.

## The big if

PHP functions in a pretty simple, logical manner. It functions by doing what you and I do on a daily basis: make decisions based on questions and answers. PHP deals with three conditional statements:

» if

» then

» else

The basic idea is this: IF this, THEN that, or ELSE this.

# The Sidebar template

The Sidebar template in WordPress has the filename `sidebar.php`. The sidebar usually appears on the left or right side of the main content area of your Word-Press theme. (The default Twenty Twenty theme doesn't have a sidebar by design.) It's a good place to put useful information about your site, such as a site summary, advertisements, or testimonials.

Many themes use widget areas in the sidebar template to display content on WordPress pages and posts. The following line of code is the built-in WordPress function that calls the Sidebar template:

```php
<?php get_sidebar(); ?>
```

This code calls the Sidebar template and all the information it contains into your page.

In the "Using Tags with Parameters for Sidebars" section later in this chapter, you find information on template tags to use in the sidebar to display the usual sidebar elements, such as a list of the most recent posts or a list of categories.

## The Footer template

The Footer template in WordPress has the filename `footer.php`. The footer is generally at the bottom of the page and contains brief reference information about the site, such as copyright information, template design credits, and a mention of WordPress. Similar to the Header and Sidebar templates, the Footer template gets called into the Main Index template through this bit of code:

```php
<?php get_footer(); ?>
```

**TIP**

You can write calls for the Sidebar and Footer templates that are written as part of a larger call:

```php
<?php
  get_sidebar();
  get_footer();
?>
```

It's possible, and common, to wrap two template tags in one PHP function that way they appear directly after one another. The examples I give in this chapter separate them into single functions to make sure that you're clear about what the actual function is.

This code calls the Footer and all the information it contains into your website page.

## Other templates

To make your website work properly, WordPress uses all the theme files together. Some files, such as the header and footer, are used on every page; others, such as the Comments template (`comments.php`), are used only at specific times to pull in specific functions.

When someone visits your site, WordPress uses a series of queries to determine which templates to use.

You can include many more theme templates in your theme. Here are some of the other template files you may want to use:

>> **Comments template** (`comments.php`): The Comments template is required if you plan to host comments on your site; it provides all the template tags you need to display those comments. The template tag used to call the comments into the template is `<?php comments_template(); ?>`.

>> **Single Post template** (`single.php`): When your visitors click the title or permalink of a post you've published to your site, they're taken to that post's individual page. There, they can read the entire post, and if you have comments enabled, they see the comments form and can leave comments.

>> **Page template** (`page.php`): You can use a Page template for static pages of your WordPress site.

>> **Search Results** (`search.php`): You can use this template to create a custom display of search results on your site. When someone uses the search feature to search your site for specific keywords, this template formats the return of those results.

>> **404 template** (`404.php`): Use this template to create a custom 404 page, which is the page visitors get when the browser can't find the page requested and returns that ugly 404 Page Cannot Be Found error.

**REMEMBER**

The templates in the preceding list are optional. If these templates don't exist in your WordPress `themes` folder, nothing breaks. The Main Index template handles the default display of these items (the single post page, the search results page, and so on). The only exception is the Comments template. If you want to display comments on your site, you must include that template in your theme.

# Putting a Theme Together

In this section, you put together a basic theme by using the information on templates and tags I've provided so far in this chapter. Template files can't do a whole lot by themselves. The real power comes when they're put together.

## Connecting the templates

WordPress has built-in functions to include the main template files — such as `header.php`, `sidebar.php`, and `footer.php` — in other templates. An `include` function is a custom PHP function built into WordPress, allowing you to retrieve the content of one template file and display it along with the content of another template file. Table 9-2 shows the templates and the functions that include them.

TABLE 9-2

## Template Files and include Functions

| Template | include Function |
|----------|------------------|
| header.php | `<?php get_header(); ?>` |
| sidebar.php | `<?php get_sidebar(); ?>` |
| footer.php | `<?php get_footer(); ?>` |
| search.php | `<?php get_search_form(); ?>` |
| comments.php | `<?php comments_template(); ?>` |

If you want to include a file that doesn't have a built-in include function, you need a different piece of code. To add a unique sidebar (different from the default sidebar.php file within your existing theme) to a certain page template, name the sidebar file sidebar-page.php. To include it in another template, use the following code:

```php
<?php get_template_part('sidebar', 'page'); ?>
```

In this statement, the get_template_part('sidebar', 'page'); function looks through the main theme folder for the sidebar-page.php file and displays the sidebar. The beautiful part about the get template part() template tag is that WordPress looks for the sidebar-page.php template first, but if it doesn't find that template, it defaults to using the sidebar.php template.

In this section, you put together the guts of a basic Main Index template by using the information on templates and tags provided so far in this chapter. There seem to be endless lines of code when you view the index.php template file in the Twenty Twenty theme, so I've simplified the file for you in a list of steps. These steps should give you a basic understanding of the WordPress Loop and common template tags and functions that you can use to create your own.

**WARNING**

The theme you're creating in this chapter won't win you any awards. In fact, it's pretty ugly, as you'll see when you activate it on your website. It doesn't have many of the features that you expect from a WordPress theme, it's not responsive on mobile devices, and so on. The purpose of the example theme is to get you comfortable with the basic concepts and mechanics of themes, including using the template tags that make themes work and understanding how to put different templates together. I didn't want to complicate this theme for beginners. Take the concepts you learn here and apply them as you move forward with more advanced theme development — and please don't use this theme anywhere on the web.

You create a new WordPress theme by using some of the basic WordPress templates. The first steps in pulling everything together are as follows:

**1.** **Connect to your web server via SFTP, click the `wp-content` folder, and then click the `themes` folder.**

This folder contains the themes that are currently installed in your WordPress site. (See Chapter 3 for more information on SFTP.)

**2.** **Create a new folder and call it `mytheme`.**

In most SFTP programs, you can right-click the window and choose New Folder from the shortcut menu. (If you aren't sure how to create a folder, refer to your SFTP program's help files.)

**3.** **In your favored text editor (such as Notepad for the PC or TextEdit for the Mac), create and save the following files with the lines of code I've provided for each:**

- *Header template:* Create the file with the following lines of code and then save the file with the name `header.php`:

```
<!DOCTYPE html>
<html <?php language_attributes(); ?>>

<head>
<meta http-equiv="Content-Type" content="<?php bloginfo( 'html_type' );
    ?>; charset=<?php  bloginfo( 'charset' ); ?>"/>

<title><?php bloginfo( 'name' ); ?> <?php wp_title(); ?></title>

<link rel="stylesheet" href="<?php bloginfo( 'stylesheet_url' ); ?>"
    type="text/css" media="screen"/>

<?php wp_head(); ?>
</head>
<body <?php body_class(); ?>>

<div id="header">
<h1 class="site-name"><a href="<?php bloginfo( 'url' ); ?>"><?php
    bloginfo( 'name' ); ?></a></h1>
<p class="site-description"><?php bloginfo( 'description' ); ?></p>
</div>
        <div id="container">
```

- *Theme Functions:* Create the file with the following lines of code and then save it, using the filename `functions.php`:

```php
<?php
add_theme_support( 'title-tag' );

register_sidebar(
  array(
    'name' => __ ('Blog Sidebar'),
    'id' => 'sidebar',
    'before_widget' => '<div id="%1$s" class="widget %2$s">',
    'after_widget' => '</div>',
    'before_title' => '<h2 class="widget-title">',
    'after_title' => '</h2>'
  )
);

function register_my_menus() {
  register_nav_menus(
    array(
      'header-menu' => __( 'Primary' ),
    )
  );
}
add_action( 'init', 'register_my_menus' );

?>
```

The Theme Functions file registers the widget area for your site so that you're able to add widgets to your sidebar by using the WordPress widgets available on the Widget page of the Dashboard. This file also registers a navigation menu for you to use; you can manage it on the Menus page in the Dashboard.

- *Sidebar template:* Create the file with the following lines of code and then save it, using the filename `sidebar.php`:

```php
<div id="sidebar">
<?php
  if ( !dynamic_sidebar( 'sidebar' ) ) : ?>
    <p>Add Sidebar Widgets</p>
    <?php endif; // end dynamic sidebar ?>
</div>
```

The code here tells WordPress where you want the WordPress widgets to display in your theme; in this case, widgets are displayed on the sidebar of your site.

- *Footer template:* Create the file with the following lines of code and then save it with the filename footer.php:

```
</div>
<div id="footer">
  <p>
   <?php bloginfo( 'name' ) ; ?> is proudly powered by <a href="http://
   wordpress.org/">WordPress</a>.
</p>
</div>

<?php wp_footer(); ?>
</body>
</html>
```

- *Stylesheet:* Create the file with the following lines of code and then save it with the filename style.css:

```
/*
Theme Name: My Theme
Description: Basic Theme from WordPress For Dummies example
Author: Lisa Sabin-Wilson
Author URI: http://lisasabin-wilson.com
*/
body {
  color:#333;
  font:14px/1.4 "Lucida Grande", Calibri, Verdana, Arial, sans-serif;
  margin:40px auto;
  width:44em;
}
ol.commentlist {
  list-style:none;
  margin-left:0;
  padding-left:0;
}

ol.commentlist .comment-body {
  margin:1em 0;
}

ol.commentlist .avatar {
```

```
    float:right;
    margin:0 0 1em 1em;
}


/*
 * These are some default styles that WordPress core hooks into for
   images.
 * You can safely remove them, but WordPress.org recommends all themes
   have them.
 */
.alignleft {
  float:left;
  margin-right:1em;
  margin-bottom:1em;
}

.alignright {
  float:right;
  margin-left:1em;
  margin-bottom:1em;
}

.aligncenter {
  display: block;
  margin-left: auto;
  margin-right: auto;
}

.wp-caption {
  border: 1px solid #ddd;
  text-align: center;
  background-color: #f3f3f3;
  padding-top: 4px;
  margin: 10px;
}

.wp-caption img {
  margin: 0;
  padding: 0;
  border: 0 none;
}
```

```
.wp-caption p.wp-caption-text {
  font-size: 11px;
  line-height: 17px;
  padding: 0 4px 5px;
  margin: 0;
}
```

I provide more details on CSS in Chapter 10. This example just gives you some *very* basic styling to use in your sample theme. Table 9-3 gives you some common template tags used for posts in WordPress.

**TABLE 9-3**  **Template Tags for Posts**

| Tag | Function |
|-----|----------|
| get_the_date(); | Displays the date of the post. |
| get_the_time(); | Displays the time of the post. |
| the_title(); | Displays the title of the post. |
| get_permalink(); | Displays the permalink (URL) of the post. |
| get_the_author(); | Displays the post author's name. |
| the_author_posts_url(); | Displays the URL of the post author's site. |
| the_content( 'Read More...' ); | Displays the content of the post. (If you use an excerpt [next item], the words *Read More* appear and are linked to the individual post page.) |
| the_excerpt(); | Displays an excerpt (snippet) of the post. |
| the_category(); | Displays the category (or categories) assigned to the post. If the post is assigned to multiple categories, the categories are separated by commas. |
| comments_popup_link( 'No Comments', 'Comment (1)', 'Comments(%)' ); | Displays a link to the comments, along with the comment count for the post in parentheses. If no comments exist, WordPress displays a No Comments message. |
| next_posts_link( '&laquo; Previous Entries' ) | Displays the words Previous Entries linked to the previous page of blog entries. |
| previous_posts_link( 'Next Entries &raquo;' ) | Displays the words Next Entries linked to the next page of blog entries. |

**WARNING**

When you're typing templates, use a text editor such as Notepad or TextEdit. Using a word-processing program such as Microsoft Word opens a slew of problems in your code. Word-processing programs insert hidden characters and format quotation marks in a way that WordPress can't read.

Now that you have the basic theme foundation, the last template file you need to create is the Main Index template. To create a Main Index template to work with the other templates in your WordPress theme, open a new window in a text editor and then complete the following steps. (Type the text in each of these steps on its own line, and press the Enter key after typing each line so that each tag starts on a new line.)

Using the tags provided in Table 9-3, along with the information on The Loop and the calls to the Header, Sidebar, and Footer templates provided in earlier sections, you can follow the next steps for a bare-bones example of what the Main Index template looks like when you put the tags together.

**1.** **Type** `<?php get_header(); ?>`.

This template tag pulls the information in the Header template of your WordPress theme.

**2.** **Type** `<div id="content">`

This HTML markup tells the browser to look for the div id. `content` in the CSS file.

**3.** **Type** `if (have_posts()) : while (have_posts()) : the_post(); ?>`

This tag is an `if` statement that asks, "Does this blog have posts?" If the answer is yes, the tag grabs the post content information from your MySQL database and displays the posts in your blog.

**4.** **Type** `<div <?php post_class() ?> id="post-<?php the_ID(); ?>">`

This template tag starts The Loop.

**5.** **Type** `<h2 class="entry-title"><a href="<?php the_permalink() ?>" title="Permanent Link to <?php the_title_attribute(); ?>" rel="bookmark"><?php the_title(); ?></a></h2>`.

This tag tells your blog to display the title of a post that's clickable (linked) to the URL of the post, surrounded by HTML Header tags.

**6.** **Type** `<div class="entry-content">`.

This HTML markup tells the browser to look for the `div class: entry-content` in the CSS file.

**7. Type** `<?php the_content( 'Continued&hellip;' ); ?>`.

This template tag displays the actual content of the blog post. The `'Continued...'` portion of this tag tells WordPress to display the word `Continued`, which is hyperlinked to the post's permalink and takes the reader to the page where he or she can read the rest of the post. This tag applies when you're displaying a post excerpt, as determined by the actual post configuration on the Dashboard.

**8. Type** `</div>`.

This tag closes the div class HTML tag that was opened in Step 6.

**9. Type** `</div>`.

This closes the div class HTML tag that was opened in Step 4.

**10. Type** `<?php endwhile; ?>`.

This template tag ends The Loop and tells WordPress to stop displaying blog posts here. WordPress knows exactly how many times The Loop needs to work, based on the setting on the WordPress Dashboard. That setting is exactly how many times WordPress will execute The Loop.

**11. Type** `<p class="pagination"><?php posts_nav_link(); ?></p>`.

This tag provides pagination that allows the reader to click a link to navigate to the next or previous post on the site.

**12. Type** `<?php endif; ?>`.

This template tag ends the `if` statement from Step 3.

**13. Type** `</div>`.

This closes the div id HTML tag that was opened in Step 2.

**14. Type** `<?php get_sidebar(); ?>`.

This template tag calls in the Sidebar template and pulls that information into the Main Index template.

**15. Type** `<?php get_footer(); ?>`.

This template tag calls in the Footer template and pulls that information into the Main Index template. **Note:** The code in the `footer.php` template ends the `<body>` and `<html>` tags that were started in the Header template (`header.php`).

When you're done, the display of the Main Index template code looks like this:

```php
<?php get_header(); ?>
<div id="content">
  <?php
    if (have_posts()) :
    while (have_posts()) : the_post(); ?>

      <div <?php post_class() ?> id="post-<?php the_ID(); ?>">
        <h2 class="entry-title"><a href="<?php the_permalink() ?>"
title="Permanent Link
        to <?php the_title_attribute(); ?>" rel="bookmark"><?php the_
title(); ?></a></h2>

        <div class="entry-content">
          <?php the_content( 'Continued...' ); ?>
        </div>
      </div>

  <?php endwhile; ?>
    <p class="pagination"><?php posts_nav_link(); ?></p>
  <?php endif; ?>
</div>

<?php get_sidebar(); ?>
<?php get_footer(); ?>
```

**16.** Save this file as `index.php`, **and upload it to the** `mythemes` **folder.**

In Notepad or TextEdit, you can save the PHP file by choosing File ⇨ Save As. Type the name of the file in the File Name text box, and click Save.

**17.** **Activate the theme on the WordPress Dashboard, and view your site to see your handiwork in action!**

**TIP**

My Main Index template code has one template tag that I explain in Chapter 12: `<article <?php post_class() ?> id="post-<?php the_ID(); ?>">`. This tag helps you create some interesting styles in your template by using CSS, so check out Chapter 12 to find out all about it.

This simple, basic Main Index template doesn't have the standard HTML markup in it, so you'll find that the visual display of your site differs from the default Twenty Twenty theme. I use this example to give you bare-bones basics about the Main Index template and The Loop in action. Chapter 10 goes into detail about using HTML and CSS to create nice styling and formatting for your posts and pages.

**TIP**

If you're having a hard time typing the code provided in this section, I've made this sample theme available for download on my website. The .zip file contains the files discussed in this chapter so that you can compare your efforts with mine, electronically. You can download the file here: http://lisasabin-wilson.com/wpfd/my-theme.zip.

## Using additional stylesheets

Often, a theme uses multiple stylesheets for browser compatibility or consistent organization. If you use multiple stylesheets, the process of including them in the template is the same as for any other stylesheet.

To add a new stylesheet, create a directory in the root theme folder called css. Next, create a new file called mystyle.css within the css folder. To include the file, you must edit the header.php file. The following example shows the code you need to include in the new CSS file:

```
<link rel="stylesheet" href="<?php bloginfo( 'stylesheet_directory' );
?>/css/mystyle.css" type="text/css" media="screen"/>
```

Additional stylesheets come in handy when you're working with a concept called parent and child themes, which is the practice of creating a child theme that depends on a separate parent theme for features and functions. I write more about parent and child themes, as well as provide additional information about HTML and CSS, in Chapter 11.

## Customizing Your Posts with Template Tags

This section covers the template tags that you use to display the body of each blog post you publish. The body of a blog post includes information such as the post date and time, title, author name, category, and content. Table 9-3 earlier in this chapter lists the common template tags you can use for posts, which are available for you to use in any WordPress theme template. The tags in Table 9-3 work only if you place them within The Loop (covered in "Connecting the templates" earlier in this chapter and located in the index.php template file).

# Using Tags with Parameters for Sidebars

If you followed along in this chapter as I covered the Header and Main Index templates and tags, you have a functional WordPress site with blog posts and various metadata displayed in each post.

In this section, I give you the template tags for the items that are commonly placed on the sidebar of a site. I say *commonly placed* because it's possible to get creative with these template tags and place them in other locations (the Footer template, for example). To keep this introduction to Sidebar template tags simple, I stick with the most common use, leaving the creative and uncommon uses for you to try when you're comfortable with the basics.

This section also introduces *tag parameters,* which are additional options you can include in the tag to control some of its display properties. Not all template tags have parameters. You place tag parameters inside the parentheses of the tag. Many of the parameters discussed in this section were obtained from the Word-Press software documentation in the WordPress Code Reference page at `https://developer.wordpress.org/reference`.

Table 9-4 helps you understand the three variations of parameters used in WordPress.

**TABLE 9-4**     **Three Variations of Template Parameters**

| Variation | Description | Example |
|---|---|---|
| Tags without parameters | These tags have no additional options. Tags without parameters have nothing within the parentheses. | `the_tag();` |
| Tags with PHP function–style parameters | These tags have a comma-separated list of values placed within the tag parentheses. | `the_tag( '1,2,3' );` |
| Tags with query-string parameters | These tags generally have several parameters. This tag style enables you to change the value for each parameter without being required to provide values for all available parameters for the tag. | `the_tag( 'parameter=true ' );` |

You need to know these three types of parameters:

>> **String:** A line of text that can be anything from a single letter to a long list of words. A string is placed between single quotation marks and sets an option for the parameter or is displayed as text.

>> **Integer:** A positive or negative number. Integers are placed within the parentheses and either inside or outside single quotation marks. Either way, WordPress processes them correctly.

>> **Boolean:** A parameter that sets the options to `true` or `false`. This parameter can be numeric (0=`false` and 1=`true`) or textual. Boolean parameters aren't placed within quotation marks.

**REMEMBER**

The WordPress Code Reference page, located at `https://developer.wordpress.org/reference`, has every conceivable template tag and possible parameter known to the WordPress software. The tags and parameters that I share with you in this chapter are the ones used most often.

## The calendar

The `calendar` tag displays a calendar that highlights each day of the week on which you've posted a blog. Those days are also hyperlinked to the original post. Here's the tag to use to display the calendar:

```
<?php get_calendar(); ?>
```

The `calendar` tag has two parameters: $initial and $echo. Set the $initial parameter to `true` to display the day of the week with one letter (Friday = F, for example). Set this parameter to `false` to display the day of the week as a three-letter abbreviation (Friday = Fri, for example).

Set the $echo parameter to true to display the calendar output on your website, or set it to false if you do not wish to have the calendar displayed at all.

Here is an example of the template tag with parameters that display the calendar with single letters for the days of the week:

```
<?php get_calendar( true, true ); ?>
```

## List pages

The `<?php wp_list_pages(); ?>` tag displays a list of the static pages you can create on your WordPress site (such as About Me or Contact pages). Displaying a link to the static pages makes them available so that readers can click the links and read the content you've provided.

WordPress has a handy navigation menu–building tool (covered in Chapter 10) that allows you to build custom navigation menus. If you like the navigation tool, you may never need to use the wp_list_pages(); template tag. Still, I include it here because you may want to use it if you want to have complete control of how the list of pages appears on your website.

The ‹list› tag parameters use the string style. Table 9-5 lists the most common parameters used for the wp_list_pages template tag.

**TABLE 9-5**   **Most Common Parameters (Query-String) for wp_list_pages();**

| Parameter | Type | Description and Values |
|---|---|---|
| child_of | integer | Displays only the subpages of the page; uses the numeric ID for a page as the value and defaults to 0 (display all pages). |
| sort_column | string | Sorts pages with one of the following options:<br><br>'post_title' — Sorts alphabetically by page title (default).<br><br>'menu_order' — Sorts by page order (the order in which the pages appear on the Manage tab and Pages subtab of the Dashboard). |
| | | 'post_date' — Sorts by the date on which pages were created.<br><br>'post_modified' — Sorts by the time when the page was last modified.<br><br>'post_author' — Sorts the page listing by author.<br><br>author ID. #.'post_name' — Sorts alphabetically by the post slug.<br><br>'ID' — Sorts by numeric page ID. |
| exclude | string | Lists the numeric page ID numbers, separated by commas, that you want to exclude from the page list display (such as 'exclude=10, 20, 30'). No default value exists. |
| depth | integer | Uses a numeric value for the levels of pages displayed in the list of pages. Possible options are<br><br>0 — Displays all pages, including main and subpages (default).<br><br>–1 — Shows subpages but doesn't indent them in the list display.<br><br>1 — Shows only main pages (no subpages).<br><br>2, 3 . . . — Displays pages to the given depth. |

| Parameter | Type | Description and Values |
|---|---|---|
| show_date | string | Displays the date when the page was created or last modified. Possible options are<br><br>`' '` — Displays no date (default).<br><br>`'modified'` — Displays the date when the page was last modified.<br><br>`'created'` — Displays the date when the page was created. |
| date_format | string | Sets the format of the date to be displayed; defaults to the date format in the Options tab and General subtab of the Dashboard. |
| title_li | string | Types text for the heading of the page list; defaults to display the text Pages. If the value is empty (''), no heading is displayed. (`'title_li=My Pages"` displays the heading My Pages above the page list, for example.) |

TIP

An alternative to using the `wp_list_pages()` template tag to create a navigation system is to use the built-in WordPress Menus, which enables you to build custom menus that don't completely depend on your WordPress pages but can include links to posts, categories, and custom links that you define. Chapter 10 contains information about the Menus feature, along with the `wp_nav_menu()` template tag that you use to display menus.

Page lists are displayed in an *unordered list* (also called *bulleted list*). Whichever term you use, it's a list with a bullet before every page link.

The following tag and query string display a list of pages without the text heading "Pages" (in other words, display no title at the top of the page's link list):

```php
<?php wp_list_pages( 'title_li=' ); ?>
```

## QUERY STRING PARAMETERS

Take a look at the way query-string parameters are written:

`'parameter1=value&parameter2=value&parameter3=value'`

The entire string is surrounded by single quotation marks, and no white space is within the query string. Each parameter is joined to its value by the = character. When you use multiple parameters/values, you separate them with the & character. You can think of the string like this: parameter1=value**AND**parameter2=value**AND**parameter3=value. Keep this convention in mind for the remaining template tags and parameters in this chapter.

The next tag and query string display the list of pages sorted by the date when they were created, displaying the date along with the page name:

```
<?php wp_list_pages( 'sort_column=post_date&show_date=created' ); ?>
```

# Post archives

The `<?php wp_get_archives(); ?>` template tag displays the blog post archives in several ways, using the parameters and values shown in Table 9-6. Values that appear in bold are the default values set by WordPress. Here are just a few examples of what you can produce with this template tag:

» Display the titles of the last 15 posts you've made on your site

» Display the titles of the posts you've made in the past ten days

» Display a monthly list of archives

**TABLE 9-6** **Most Common Parameters (Query-String) for wp_get_archives();**

| Parameter and Type | Possible Values | Example |
|---|---|---|
| type (string)<br><br>Determines the type of archive to display. | monthly<br><br>daily<br><br>weekly<br><br>postbypost | `<?php wp_get_archives( 'type=postbypost' ); ?>`<br><br>Displays the titles of the most recent blog posts. |
| format (string)<br><br>Formats the display of the links in the archive list. | html — Surrounds the links with `<li> </li>` tags<br><br>option — Places archive list in drop-down-menu format<br><br>link — Surrounds the links with `<link> </link>` tags<br><br>custom — Uses your own HTML tags with the before and after parameters | `<?php wp_get_archives( 'format=html' ); ?>`<br><br>Displays the list of archive links where each link is surrounded by the `<li> </li>` HTML tags. |
| limit (integer)<br><br>Limits the number of archives to display. | If no value, all posts are displayed | `<?php wp_get_archives( 'limit=10' ); ?>`<br><br>Displays the last ten archives in a list. |

| Parameter and Type | Possible Values | Example |
|---|---|---|
| before (string)<br><br>Places text or formatting before the link in the archive list when the custom parameter is used. | No default | `<?php wp_get_archives( 'before=<strong>' ); ?>`<br><br>Inserts the `<strong>` HTML tag before each link in the archive link list |
| after (string)<br><br>Inserts text or formatting after the link in the archive list when the custom parameter is used. | No default | `<?php wp_get_archives( 'after=</strong>' ); ?>`<br><br>Inserts the `</strong>` HTML tag after each link in the archive link list |
| show_post_count (Boolean)<br><br>Displays the number of posts in the archive. You'd use this parameter if you use the type of monthly. | true or 1<br><br>false or 0 | `<? wp_get_archives( 'show_post_count=1' ); ?>`<br><br>Displays the number of posts in each archive after each archive link |

Here are a couple of examples of tags used to display blog-post archives.

This tag displays a linked list of monthly archives (November 2017, December 2017, and so on):

```
<?php wp_get_archives( 'type=monthly' ); ?>
```

This next tag displays a linked list of the 15 most recent blog posts:

```
<?php wp_get_archives( 'type=postbypost&limit=15' ); ?>
```

## Categories

WordPress lets you create categories and assign posts to a specific category (or multiple categories). Categories provide an organized navigation system that helps you and your readers find posts you've made on certain topics.

The `<?php wp_list_categories(); ?>` template tag lets you display a list of your categories by using the available parameters and values. (Table 9-7 shows some of the most popular parameters.) Each category is linked to the appropriate category page that lists all the posts you've assigned to it. The values that appear in bold are the default values set by WordPress.

**TABLE 9-7** **Most Common Parameters (Query-String) for wp_list_categories();**

| Parameter and Type | Possible Values | Example |
|---|---|---|
| orderby (string)<br><br>Determines how the category list will be ordered. | ID<br><br>name | `<?php wp_list_categories( 'orderby=name' ); ?>`<br><br>Displays the list of categories by name, alphabetically, as they appear on the Dashboard. |
| style (string)<br><br>Determines the format of the category list display. | list<br><br>none | `<?php wp_list_categories ( 'style=list' ); ?>`<br><br>Displays the list of category links where each link is surrounded by the `<li>` `</li>` HTML tags.<br><br>`<?php wp_list_categories( 'style=none' ); ?>`<br><br>Displays the list of category links with a simple line break after each link. |
| show_count (Boolean)<br><br>Determines whether to display the post count for each listed category. | true or 1<br><br>false or 0 | `<?php wp_list_categories( 'show_count=1' ); ?>`<br><br>Displays the post count, in parentheses, after each category list. Espresso (10), for example, means that the Espresso category has ten posts. |
| hide_empty (Boolean)<br><br>Determines whether empty categories should be displayed in the list (meaning a category with zero posts assigned to it). | true or 1<br><br>false or 0 | `<?php wp_list_categories( 'hide_empty=0' ); ?>`<br><br>Displays only those categories that currently have posts assigned to them. |
| feed (string)<br><br>Determines whether the RSS feed should be displayed for each category in the list. | rss<br><br>Default is no feeds displayed | `<?php wp_list_categories( 'feed=rss' ); ?>`<br><br>Displays category titles with an RSS link next to each one. |
| feed_image (string)<br><br>Provides the path/filename for an image for the feed. | No default | `<?php wp_list_categories( 'feed_image=/wp-content/images/feed.gif' ); ?>`<br><br>Displays the feed.gif image for each category title. This image is linked to the RSS feed for that category. |
| hierarchical (Boolean)<br><br>Determines whether the child categories should be displayed after each parent category in the category link list. | true or 1<br><br>false or 0 | `<?php wp_list_categories( 'hierarchical=0' ); ?>`<br><br>Doesn't display the child categories after each parent category in the category list. |

Here are a couple of examples of tags used to display a list of your categories.

This example, with its parameters, displays a list of categories sorted by name without showing the number of posts made in each category, as well as the RSS feed for each category title:

```
<?php wp_list_categories( 'orderby=name&show_count=0&feed=RSS' ); ?>
```

This example, with its parameters, displays a list of categories sorted by name with the post count showing as well as the subcategories of every parent category:

```
<?php wp_list_categories( 'orderby=name&show_count=1&hierarchical=1' ); '>
```

# Getting widgetized

Many themes are *widget-ready*, meaning that you can insert widgets into them easily. Widgets allow you to add functionality to your sidebar without having to use code. Some common widget functionalities include displaying recent posts, displaying recent comments, adding a search box for searching content on a site, and adding static text. Even widget-ready themes have their limitations, however. You may find that the theme you chose doesn't have widget-ready areas in all the places you want them. You can make your own, however.

In a WordPress theme, the Theme Functions template (functions.php) and the Sidebar template (sidebar.php) create the functionality and the possibility for widgets to exist within your theme. You're not limited as to where you place and use widgets. I'm using the Sidebar template (sidebar.php) in this example.

To add a widget-ready area to the WordPress Dashboard Widget interface, first register the widget in your theme's functions.php file as follows:

```
function my_widgets_init() {
  register_sidebar( array (
    'name' => __( 'Widget Name' ),
    'id' => 'widget-name',
    'description' => __( 'The primary widget area' ),
    'before_widget' => '<li id="%1$s" class="widget-container %2$s">',
    'after_widget' => "</li>",
    'before_title' => '<h3 class="widget-title">',
    'after_title' => '</h3>',
  ) );
}
add_action( 'widgets_init', 'my_widgets_init' );
```

You can insert this code directly below the first opening PHP tag (`<?php`).

Within that code, you see seven *arrays* (sets of values that tell WordPress how you want your widgets to be handled and displayed):

>> `name`: This name is unique to the widget and appears on the Widgets page of the Dashboard. It's helpful to register several widgetized areas on your site.

>> `id`: This array is the unique ID given to the widget.

>> `description` **(optional):** This array is a text description of the widget. The text placed here displays on the Widgets page of the Dashboard.

>> `before_widget`: This array is the HTML markup that gets inserted directly before the widget and is helpful for CSS styling purposes.

>> `after_widget`: This array is the HTML markup that gets inserted directly after the widget.

>> `before_title`: This array is the HTML markup that gets inserted directly before the widget title.

>> `after_title`: This array is the HTML markup that gets inserted directly after the widget title.

**REMEMBER**

Even though you use `register_sidebar` to register a widget, widgets don't have to appear on a sidebar; they can appear anywhere you want them to. The example code snippet earlier in this section registers a widget named Widget Name on the WordPress Dashboard. Additionally, it places the widget's content in an element that has the CSS class `widget` and puts `<h3>` tags around the widget's title.

Widgets that have been registered on the WordPress Dashboard are ready to be populated with content. On the Appearance menu of your site's Dashboard, you see a link titled Widgets. When you click the Widgets link, you see the new widget area you registered.

**TIP**

You can register an unlimited number of widgets for your theme. This flexibility allows you to create several different widgetized areas and widget features in different areas of your site. Chapter 12 goes into more detail about using Sidebar templates to create widgetized areas and features on your site.

IN THIS CHAPTER

» Exploring basic CSS

» Setting new background and theme colors

» Creating a header

» Customizing navigation menus

» Changing fonts

» Knowing HTML essentials

Chapter **10**

# Tweaking WordPress Themes

C hapter 8 shows how you can use free WordPress themes on your website. Many people are quite happy to use these themes without making any adjustments to them at all. I can't tell you, however, how many times people have asked me whether they can customize a theme that they've found. The answer to their question is always, "Of course you can make changes on your own."

The practice of changing a few elements of an existing WordPress theme is known as *tweaking.* Thousands of WordPress site owners tweak their existing themes on a regular basis. This chapter provides information on some of the most common tweaks you can make to your theme, such as changing the header image, changing the color of the background or the text links, and changing font styles. These changes are pretty easy to make, and you'll be tweaking your own themes in no time flat!

Using a theme exactly as a theme author released it is great. If a new version is released that fixes a browser compatibility issue or adds features offered by a new version of WordPress, a quick theme upgrade is very easy to do.

Chances are good, however, that you'll want to tinker with the design, add new features, or modify the theme structure. If you modify the theme, you won't be able to upgrade to a newly released version without modifying the theme again.

If only you could upgrade customized versions of themes with new features when they're released! Fortunately, child themes give you this best-of-both-worlds theme solution. Chapter 11 explores what child themes are, how to create a parent theme that's child-theme ready, and how to get the most out of using child themes.

**REMEMBER**

Before you go too wild with tweaking templates, make a backup of your theme so that you have the original files from which to restore it if necessary. You can back up your theme files by connecting to your web server via SFTP (see Chapter 3) and downloading your theme folder to your computer. When you have the original theme files safe and secure on your hard drive, feel free to tweak away, comfortable in the knowledge that you have a backup.

# Styling with CSS: The Basics

*Cascading Style Sheets (CSS)* are cascading sheets of style markup that control the appearance of content on a website. Every single WordPress theme you use in your site uses CSS. CSS provides style and design flair to the template tags in your templates. (See Chapter 9 for information about WordPress template tags.) The CSS for your WordPress theme is pulled in through the Header template (header.php) and is named style.css.

In the Functions template (functions.php) of most WordPress themes, you find the following line of code, which pulls the CSS (style.css) into the page to provide the formatting of the elements of your blog:

```
function mytheme_register_styles() {
    wp_enqueue_style( 'mytheme-style', get_stylesheet_uri(), array(), $theme_
version );
}

add_action( 'wp_enqueue_scripts', 'mytheme_register_styles' );
```

**WARNING**

Don't tweak the line of code that pulls in the style.css file; if you do, the CSS won't work for your site.

Chapter 9 covers the commonly used parameters for the bloginfo(); template tag used in WordPress themes.

# CSS selectors

With CSS, you can provide style (such as size, color, and placement) to the display of elements on your website (such as text links, header images, font size and colors, paragraph margins, and line spacing). *CSS selectors* contain names, properties, and values to define which HTML elements in the templates you'll style with CSS. You use CSS selectors to declare (or select) which part of the markup the style applies to. For example, if you were to assign a style to the h1 selector, that style would affect all <h1> tags in your HTML. Sometimes you want to do this; at other times, you want to affect only a subset of elements.

Table 10-1 provides some examples of CSS selectors and their use.

**TABLE 10-1**  **Basic CSS Selectors**

| CSS Selector | Description | HTML | CSS |
|---|---|---|---|
| body | Sets the style for the overall body of the site, such as background color and default fonts | `<body>` | `body {background-color: white;}` <br><br> The background color on all pages is white. |
| p | Defines how paragraphs are formatted | `<p>This is a paragraph</p>` | `p {color: black;}` <br><br> The color of the fonts used in all paragraphs is black. |
| h1, h2, h3, h4, h5, h6 | Provides bold headers for different sections of your site | `<h1>This is a site title</h1>` | `h1 {font-weight: bold;}` <br><br> A font surrounded by the `<h1>...</h1>` HTML tags is bold. |
| a | Defines how text links display on your site | `<a href= "http://wiley.com">Wiley Publishing</a>` | `a {color: red;}` <br><br> All text links appear in red. |

# Classes and IDs

With CSS IDs and classes, you can define more elements to style. Generally, you use IDs to style one broad element (such as your header section) on your page. Classes style, define, and categorize more specifically grouped items (such as images and text alignment, widgets, or links to posts). The differences between CSS IDs and classes are

>> **CSS IDs** are identified with the hash mark (#). #header, for example, indicates the header ID. Only one element can be identified with an ID per page.

>> **CSS classes** are identified with a period (.). .alignleft, for example, indicates aligning an element to the left.

Figure 10-1 shows the stylesheet from the default Twenty Twenty theme.

**FIGURE 10-1:**
A WordPress
theme stylesheet
(style.css).

IDs and classes define styling properties for different sections of your WordPress theme. Table 10-2 shows some examples of IDs and classes from the header.php template in the Twenty Twenty WordPress theme. Armed with this information, you'll know where to look in the stylesheet when you want to change the styling for a particular area of your theme.

**REMEMBER** If you find an element in the template code that says id (such as div id= or p id=), look for the hash symbol in the stylesheet. If you find an element in the template code that says class (such as div class= or p class=), look for the period in the stylesheet followed by the selector name.

**TABLE 10-2** **Connecting HTML with CSS Selectors**

| HTML | CSS Selector | Description |
|---|---|---|
| `<header id="site-header">` | `#site-header` | Styles the elements for the site-header ID in your template(s) |
| `<button class="search-toggle">` | `.search-toggle` | Styles the elements for the search-toggle class in your template(s) |
| `<div class= "header-titles">` | `.header-titles` | Styles the elements for your header-titles class in your template(s) |
| `<ul class= "primary-menu ">` | `.primary-menu` | Styles the elements for your primary-menu class in your template(s) |

## CSS properties and values

CSS properties are assigned to the CSS selector name. You also need to provide values for the CSS properties to define the style elements for the particular CSS selector you're working with.

The body selector that follows defines the overall look of your web page. background is a property, and #DDDDDD (a light gray) is the value; color is a property, and #222222 (very dark gray, almost black) is the value:

```
body {
   background: #DDDDDD;
   color: #222222;
}
```

**REMEMBER**

Every CSS property needs to be followed by a colon ( : ), and each CSS value needs to be followed by a semicolon ( ; ).

The fact that properties are assigned to selectors, as well as your options for the values, makes CSS a fun playground for personalizing your site. You can experiment with colors, fonts, font sizes, and more to tweak the appearance of your theme.

# Changing the Background Graphics with CSS

Using background graphics is an easy way to set your site apart from others that use the same theme. Finding a background graphic for your site is much like finding just the right desktop background for your computer. You can choose among a variety of background graphics for your site, such as photography, abstract art, and repeatable patterns.

You can find ideas for new, different background graphics by checking out some of the CSS galleries on the web, such as CSS Drive (`www.cssdrive.com`).

**REMEMBER**

Sites like these should be used only for inspiration, not theft. Be careful when using images from outside sources.

**WARNING**

You want to use only graphics and images that you have been given the right (through express permission or reuse licenses) to use on your site. For this reason, always purchase graphics from reputable sources, such as these online-graphics sites:

» **iStock** (`https://www.istockphoto.com`): iStock has an extensive library of stock photography, vector illustrations, video and audio clips, and Adobe Flash media. You can sign up for an account and search libraries of image files to find the one that suits you or your client best. The files that you use from iStock aren't free; you do have to pay for them. Be sure that you read the license for each image you use.

The site has several licenses. The cheapest one is the Standard license, which has some limitations. You can use an illustration from iStock in one website design, for example, but you can't use that same illustration in a theme design that you intend to sell multiple times (say, in a premium theme marketplace). Be sure to read the fine print!

» **Dreamstime** (`https://www.dreamstime.com`): Dreamstime is a major supplier of stock photography and digital images. Sign up for an account, and search the huge library of digital image offerings. Dreamstime does offer free images at times, so keep your eyes out for those! Also, Dreamstime has different licenses for its image files, and you need to pay close attention to them.

One nice feature is the royalty-free licensing option, which allows you to pay for the image one time and then use the image as many times as you like. You can't redistribute the image in the same website theme repeatedly, however, such as in a template that you sell to the public.

» **GraphicRiver** (`https://graphicriver.net`): GraphicRiver offers stock graphic files such as Adobe Photoshop images, design templates, textures, vector graphics, and icons, to name just a few. The selection is vast, and the cost to download and use the graphic files is minimal.

As with all graphic and image libraries, be sure to read the terms of use or any licensing attached to each of the files to make sure you're abiding by the legal terms.

» **Unsplash** (`https://unsplash.com`): Unsplash offers stock imagery that is licensed for both personal and commercial use. Best of all, the images are free.

To make the best use of background graphics, answer a few simple questions:

>> **What type of background graphic do you want to use?** You may want a repeatable pattern or texture, for example, or a black-and-white photograph of something in your business.

>> **How do you want the background graphic to display in your browser?** You may want to tile or repeat your background image in the browser window, or pin it to a certain position no matter what size your guest's browser is.

The answers to those questions determine how you install a background graphic in your theme design.

**REMEMBER**

When working with graphics on the web, use GIF, JPG, or PNG image formats. For images with a small number of colors (such as charts, line art, and logos), GIF format works best. For other image types (screen shots with text and images, blended transparency, and so on), use JPG (also called JPEG) or PNG.

In web design, the characteristics of each image file format can help you decide which file format you need to use for your site. The most common image file formats and characteristics include the following:

>> **JPG:** Suited for use with photographs and smaller images used in your web design projects. Although the JPG format compresses with lossy compression, you can adjust compression when you save a file in JPG format. That is, you can choose the degree (or amount) of compression that occurs, from 1 to 100. Usually, you won't see a great deal of image-quality loss with compression levels 1 through 20.

>> **PNG:** Suited for larger graphics used in web design, such as the logo or main header graphic that helps brand the overall look of the website. A .png file uses lossless image compression; therefore, no data loss occurs during compression, so you get a cleaner, sharper image. You can also create and save a .png file on a transparent canvas. .jpg files must have a white canvas or some other color that you designate.

>> **GIF:** Suited for displaying simple images with only a few colors. Compression of a .gif file is lossless; therefore, the image renders exactly the way you design it, without loss of quality. These files compress with lossless quality when the image uses 256 colors or fewer, however. For images that use more colors (higher quality), GIF isn't the greatest format to use. For images with a lot of colors, go with PNG format instead.

# Uploading an image for background use

If you want to change the background graphic in your theme, follow these steps:

1. **Upload your new background graphic via SFTP to the images folder in your theme directory.**

   Typically, the images folder is at wp-content/themes/themename/images.

2. **On the WordPress Dashboard, click the Editor link on the Appearance menu.**

   The Edit Themes screen displays.

3. **Click the Stylesheet (style.css) link on the right side of the page.**

   The style.css template opens in the text-editor box on the left side of the Edit Themes screen.

4. **Scroll down to find the body CSS selector.**

   The following code segment is a sample CSS snippet you can use to define the background color of your site (light gray in this example):

   ```
   body {
      background: #f1f1f1;
   }
   ```

5. **(Optional) Modify the background property values from the code in Step 4.**

   Change

   ```
   background: #f1f1f1;
   ```

   to

   ```
   background #FFFFFF url('images/newbackground.gif');
   ```

   With this example, you add a new background image (newbackground.gif) to the existing code and change the color code to white (#FFFFFF).

6. **Click the Update File button to save the stylesheet changes you made.**

   Your changes are saved and applied to your theme.

# Positioning, repeating, and attaching images

After you upload a background graphic, you can use CSS background properties to position it the way you want it. The main CSS properties —

background-position, background-repeat, and background-attachment —
help you achieve the desired effect.

Table 10-3 describes the CSS background properties and the available values for changing them in your theme stylesheet. If you're a visual person, you'll enjoy testing and tweaking values to see the effects on your site.

**TABLE 10-3    CSS Background Properties**

| Property | Description | Values | Example |
|---|---|---|---|
| background-position | Determines the starting point of your background image on your web page | bottom center<br><br>bottom right<br><br>left center<br><br>right center<br><br>center center | background-position: bottom center; |
| background-repeat | Determines whether your background image will repeat or tile | repeat (repeats infinitely)<br><br>repeat-y (repeats vertically)<br><br>repeat-x (repeats horizontally)<br><br>no-repeat (doesn't repeat) | background-repeat: repeat-y; |
| background-attachment | Determines whether your background image is fixed or scrolls with the browser window | fixed<br><br>scroll | background-attachment: scroll; |
| background-origin | Specifies the positioning area of the background images | padding-box<br><br>border-box<br><br>content-box<br><br>initial<br><br>inherit | background-origin: content-box; |
| background-clip | Specifies the painting area of the background images | border-box<br><br>padding-box<br><br>content-box<br><br>initial<br><br>inherit | background-clip: padding-box; |

Suppose that your goal is to *tile*, or repeat, the background image so that it scales with the width of the browser on any computer. To achieve this goal, open the stylesheet again, and change

```
background: #f1f1f1;
```

to

```
background: #FFFFFF;
background-image: url(images/newbackground.gif);
background-repeat: repeat;
```

If your goal is to display a fixed image that doesn't scroll or move when your site visitor scrolls through your website, you can use the background-position, background-repeat, and background-attachment properties to display it exactly as you want it to appear.

To achieve this look, add background-attachment: fixed, and change the background-repeat to no-repeat in your stylesheet to

```
background: #FFFFFF;
background-image: url(images/newbackground.gif);
background-repeat: no-repeat;
background-attachment: fixed;
```

**TIP**

As you become more comfortable with CSS properties, you can start using short-ening methods to make your CSS coding practice more efficient. The preceding block of code, for example, looks like this with shortened CSS practice:

```
background: #ffffff url(images/newbackground.gif) no-repeat fixed;
```

As you can see from these examples, changing the background graphic by using CSS involves setting options that depend on your creativity and design style more than anything else. When you use these options properly, CSS can take your design to the next level.

# Using Your Own Header Image

Most themes have a header image that appears at the top of the page. This image is generated by a graphic defined in the CSS value for the property that represents the header area or through the use of a feature in WordPress called a custom header. To install a custom logo in the header of your site, follow these steps with the Twenty Twenty theme activated:

1. **On the WordPress Dashboard, click the Customize link on the Appearance menu.**

   The Customizer screen opens on the left side of your screen.

2. **Click the Site Identity link.**

   The Customizing Site Identity screen opens within the Customizer.

3. **Click the Select Logo button below the Logo header.**

   The Select Logo window opens, allowing you to select an image from your Media Library or upload a new image from your computer.

4. **Click the Select button.**

   This step opens the Crop Image screen, shown in Figure 10-2.

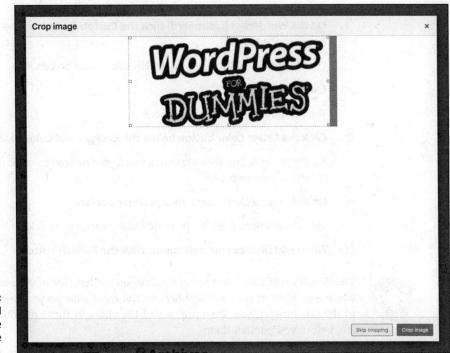

**FIGURE 10-2:**
Adjust the dotted lines to choose the area of the image to display.

5. **(Optional) Crop the image to your liking.**

   To resize and crop your image, drag one of the eight tiny boxes located at the corners and in the middle of the image. You can also click within the image, and move the entire image up or down to get the placement and cropping effect that you want.

6. **Click the Crop Image button to crop your header image.**

7. **Click the Publish button in the top-right corner of the Customizing Site Identity screen to save your changes.**

# Customizing Colors in Twenty Twenty

After you explore the Site Identity settings, you may want to pick a background color or change the color of the header text. The default background color in the Twenty Twenty theme is white, but you can change the color scheme. Here's how:

1. **On the WordPress Dashboard, click the Customize link on the Appearance menu.**

   The Customizer panel opens on the left side of your screen.

2. **Click the Colors link.**

   The Customizing Colors screen opens.

3. **Click the Select Color button below the Background Color heading.**

   A color picker opens, allowing you to select your desired color for the background of your website.

4. **On the color picker, select your preferred color.**

   Click anywhere on the color picker to set the background color.

5. **When you finalize your selections, click the Publish button.**

TIP

The WordPress Customizer panel displays on the left side of your computer screen while a preview of your site displays on the right side. As you're making changes in the Customizer, you see a preview of the changes those options make on your site before you publish them.

# Creating Custom Navigation Menus

A *navigation menu* is a list of links displayed on your site. These links can be to pages, posts, or categories within your site, or they can be links to other sites. You can define navigation menus on your site by using the built-in Custom Menu feature of WordPress.

It's to your advantage to provide at least one navigation menu on your site so that readers can see everything your site has to offer. Providing visitors a link — or several — to click maintains the point-and-click spirit of the web.

## Building custom navigation menus

After you add the menu feature to your theme (or when you're using a theme that has the menu feature), building a custom navigation menu is easy. Just follow these steps:

1. **Click the Menus link on the Appearance menu of your Dashboard.**

   The Menus page opens.

2. **Click the Create a New Menu link.**

   The Menu Structure panel opens.

3. **Type a name in the Menu Name box, and click the Create Menu button.**

   The Menus page reloads with your new menu displayed, ready for you to add links to it.

4. **Add links to your new menu.**

   WordPress gives you four ways to add links to your new menu (see Figure 10-3):

   - *Pages:* Click the View All link to display a list of all the page(s) you've published on your site. Select the box next to the page names you want to add to your menu. Then click the Add to Menu button.

   - *Posts:* Click the View All link to display a list of all the posts you've published on your site. Select the box next to the post name you'd like to add to your menu; then click the Add to Menu button.

   - *Custom Links:* In the URL field, type the URL of the website that you want to add (such as https://www.google.com). Next, in the Label text field, type the name of the link that you want to display on your menu (such as Google). Then click the Add to Menu button.

   - *Categories:* Click the View All link to display a list of all the categories you've created on your site. Select the box next to the category names you want to add to the menu. Then click the Add to Menu button.

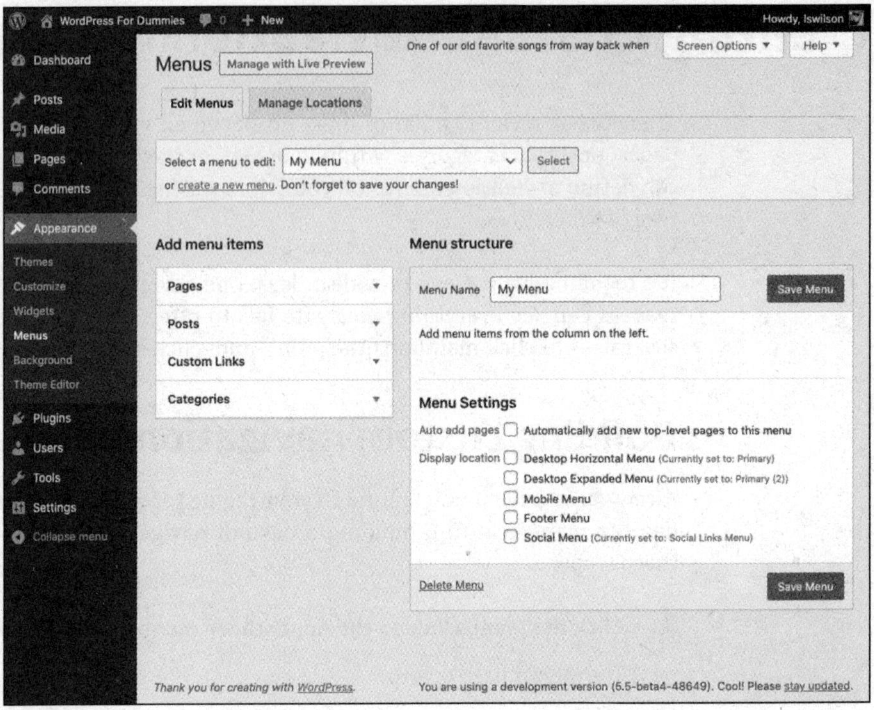

**FIGURE 10-3:**
The Custom
Menus options
on the Menus
page of the
Dashboard.

5. **Review your menu choices on the right side of the page.**

   When you add menu items, the column on the right side of the Menus page populates with your menu choices.

6. **Edit your menu choices, if needed.**

   Click the Edit link to the right of the menu link name to edit the information of each link on your new menu.

7. **Save your menu before leaving the Menus page.**

   Be sure to click the Save Menu button (refer to Figure 10-3). A message appears, confirming that the new menu has been saved.

**TIP** You can create as many menus as you need for your website. Just follow the parameters for the menu template tag to make sure that you're pulling in the correct menu in the correct spot on your theme. Pay attention to the menu ID or menu name in the template tag. You find more options for your navigation menus by clicking the Screen Options tab in the top-right corner of your Dashboard. On that tab, you can add things such as Posts and Custom Post Types to your menu options, as well as descriptions of menu items.

By default, the HTML markup for the menu is generated as an unordered list, and the main container for the menu has a CSS class that contains the name you gave the menu in Step 3 in the previous list. I named my menu Main, and the HTML markup in the source code for that menu looks like this:

```
<div class="menu-main-container">
  <ul id="menu-main" class="menu">

    <li id="menu-item-56" class="menu-item menu-item-type-custom menu-item-
          object-custom current-menu-item current_page_item menu-item-home
          menu-item-56"><a href="http://fordummies.local/" aria-
          current="page">Home</a>
    </li>

    <li id="menu-item-57" class="menu-item menu-item-type-post_type menu-item-
          object-page menu-item-57"><a href="http://fordummies.local/
          about/">About</a>
    </li>

    <li id="menu-item-58" class="menu-item menu-item-type-post_type menu-item-
          object-page menu-item-has-children menu-item-58"><a href="http://
          fordummies.local/blog">Blog</a>

      <ul class="sub-menu">
        <li id="menu-item-60" class="menu-item menu-item-type-taxonomy menu-item-
              object-category menu-item-60"><a href="http://fordummies.local/
              category/uncategorized/">Uncategorized</a>
        </li>

      </ul>

    </li>

    <li id="menu-item-59" class="menu-item menu-item-type-post_type menu-item-
          object-page menu-item-59"><a href="http://fordummies.local/
          contact/">Contact</a>
    </li>

  </ul>
</div>
```

Notice that the first line defines CSS class: `<div class="menu-main-container">`. The class in that line reflects the name that you gave your menu. Because I gave my menu the name Main when I created it in the Dashboard, the CSS ID is `menu-main-container`. If I'd named the menu Foo, the ID would instead be `menu-foo-container`. The ability to assign menu names in the CSS and HTML markup is why WordPress allows you to use CSS to create different styles and formats for different menus.

When you're developing themes for yourself or others to use, you want to make sure that the CSS you define for the menus can do things like account for subpages by creating drop-down-menu effects. Listing 10-1 gives you just one example of a block of simple CSS that you can use to create a nice style for your menu. (This example assumes that you have a menu named Main; therefore, the HTML and CSS markup indicates 'menu-main-container'.)

**TIP**

In Chapter 9, you built an entire simple WordPress theme from scratch, and I gave you a link for the location where you can download the full theme. The menu CSS from this section in this chapter is included in the style.css file in that theme. You can find the full theme file here: https://lisasabin-wilson.com/wpfd/my-theme.zip.

---

**LISTING 10-1:**    **Sample CSS for Drop-Down-Menu Navigation**

---

```
div.menu-main-container {
    display: block;
    width: 100%;
    clear: both;
    height: 55px;
}

div.menu-main-container ul {
    list-style: none;
    margin: 0;
    padding: 0;
}

div.menu-main-container li {
    position: relative;
    line-height: 1.7;
}

div.menu-main-container a {
    color: blue;
    text-decoration: none;
    display: block;
}

div.menu-main-container > ul > li {
    float: left;
}

div.menu-main-container > ul > li > a{
```

```css
    padding: 10px 10px;
}

div.menu-main-container > ul > li:hover > a{
    background: #333;
    color: white;
}

/* Sub/Children Menu */
div.menu-main-container .sub-menu,
div.menu-main-container .children {
    position: absolute;
    left: 0;
    top: 100%;
    z-index: 1;
    background: #333;
    color: white;
    min-width: 200px;
    display: none;
}

div.menu-main-container li:hover > .sub-menu,
div.menu-main-container li:hover > .children {
    display: block;
}

div.menu-main-container .sub-menu a,
div.menu-main-container .children a {
    padding: 5px 10px;
    color:white;
}

div.menu-main-container .sub-menu li:hover a,
div.menu-main-container .children li:hover a {
    background: #444;
    color: white;
}

/* Grandchildren Menu */
div.menu-main-container .sub-menu .sub-menu,
div.menu-main-container .children .children {
    position: absolute;
    left: 100%;
    top: 0;
    background: #444;
}
```

**REMEMBER**

The CSS that you use to customize the display of your menus will differ from Listing 10-1, which is just an example. After you get the hang of using CSS, you can try different methods, colors, and styling to create a custom look.

## Displaying custom menus with widgets

You don't have to use the `wp_nav_menu();` template tag to display the menus on your site, because WordPress also provides Custom Menu widgets that you can add to your theme. Therefore, you can use widgets instead of template tags to display the navigation menus on your site. This feature is especially helpful if you've created multiple menus on your site.

Your first step is registering a special widget area for your theme to handle the Custom Menu widget display. To register this widget, open your theme's `functions.php` file, and add the following code on a new line:

```
// ADD MENU WIDGET
function my_widgets_init() {
  register_sidebar( array (
  'name'    => __( 'Blog Sidebar' ),
  'id'      => 'blog-widget-area',
  ) );
}
add_action('widgets_init', 'my_widgets_init');
```

These few lines of code create a new widget area called Menu Widget Area on the Widgets page of your Dashboard. At this point, you can drag the Custom Menu widget into the Blog Sidebar area to indicate that you want to display a custom menu in that area. Figure 10-4 shows the Available Widgets area with the Blog Sidebar displayed.

To add the widget area to your theme, go to the Theme Editor by choosing Appearance ➪ Editor, and open the `header.php` file (or whatever template file you want to include in the widget area). Then add these lines of code in the area where you want the Menu widget to be displayed:

```
<ul>
  <?php
    if ( !function_exists( 'dynamic_sidebar' ) || !dynamic_sidebar( 'Blog
    Sidebar' ) ) :
    endif;
  ?>
</ul>
```

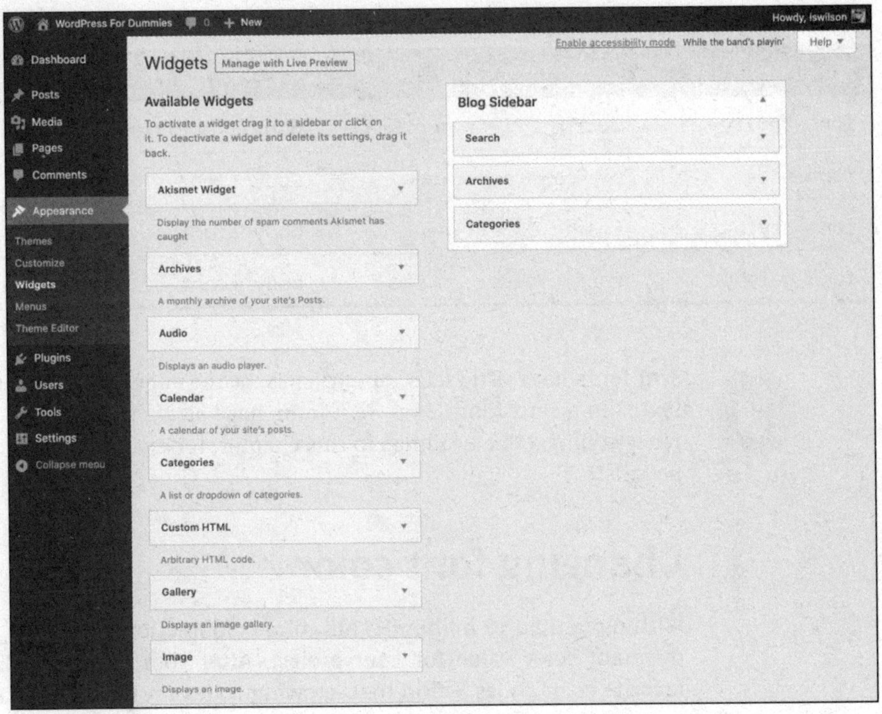

**FIGURE 10-4:**
Widgets page displaying the Blog Sidebar widget in Available Widgets.

These lines of code tell WordPress that you want information contained in the Menu widget area to be displayed on your site.

# Changing Font Family, Color, and Size

You can change the fonts in your theme for style or readability. I've seen typographic (or font) design experts use simple font variations to achieve amazing design results. You can use fonts to separate headlines from body text (or widget headlines and text from the main content) to be less distracting. Table 10-4 lists some examples of often-used font properties.

The web is kind of picky about how it displays fonts, as well as what sort of fonts you can use in the `font-family` property. Not all fonts appear correctly on the web. To be safe, consider sticking to some common font families that appear correctly in most browsers:

>> **Serif fonts:** Times New Roman, Georgia, Garamond, Bookman Old Style

>> **Sans-serif fonts:** Verdana, Arial, Tahoma, Trebuchet MS

**TABLE 10-4**  **Font Properties**

| Font Properties | Common Values | CSS Examples |
|---|---|---|
| font-family | Georgia, Times, serif | body {font-family: Georgia; serif;} |
| font-size | px, %, em, rem | body {font-size: 14px;} |
| font-style | italic, underline | body {font-style: italic;} |
| font-weight | bold, bolder, normal | body {font-weight: normal;} |

**REMEMBER**

*Serif* fonts have little tails, or curlicues, at the edges of letters. (This book's body text is in a serif font.) *Sans-serif* fonts have straight edges and are devoid of any fancy styling. (The headings in this chapter, for example, use a sans-serif font — no tails!)

## Changing font color

With more than 16 million HTML color combinations available, you can find just the right color value for your project. After some time, you'll memorize your favorite color codes. I find that knowing codes for different shades of gray helps me quickly add an extra design touch. I often use the shades of gray listed in Table 10-5 for backgrounds, borders on design elements, and widget headers.

**TABLE 10-5**  **My Favorite CSS Colors**

| Color | Value |
|---|---|
| White | #FFFFFF |
| Black | #000000 |
| Gray | #CCCCCC |
|  | #DDDDDD |
|  | #333333 |
|  | #E0E0E0 |

You can easily change the color of your font by changing the color property of the CSS selector you want to tweak. You can use hex codes to define the colors.

You can define the overall font color in your site by defining it in the body CSS selector like this:

```
body {
  color: #333333;
}
```

# Changing font size

To tweak the size of your font, change the `font-size` property of the CSS selector you want to change. Font sizes are generally determined by units of measurement, as in these examples:

» px: Pixel measurement. Increasing or decreasing the number of pixels increases or decreases the font size. (12px is larger than 10px.)

» rem: The values for these fonts are relative to the root HTML element. For example, if the font size of the root element is 14px, then 1 rem = 14px. If the font size for the root element is not defined, then 1 rem will be equal to the default font size of the browser.

» em: Width measurement. An *em* is a scalable unit of measurement that's equal to the current font size. (Originally, an em space was equal to the width of the capital letter *M*.) If the font size of the body of the site is defined as 12px, 1em is equal to 12px; likewise, 2em is equal to 24px.

» %: Percentage measurement. Increasing or decreasing the percentage number affects the font size. (If the body of the website uses 14px as the default, 50% is the equivalent of 7 pixels, and 100% is the equivalent of 14 pixels.)

In the default template CSS, the font size is defined in the `<body>` tag in pixels, like this:

```
font-size: 12px;
```

When you put the font family, color, and font size elements together in the `<body>` tag, they style the font for the overall body of your site. Here's how they work together in the `<body>` tag of the default template CSS:

```
body {
    font-size: 12px;
    font-family: Georgia, "Bitstream Charter", serif;
    color: #666666;
}
```

When you want to change a font family in your CSS, open the stylesheet (style.css), search for property: font-family, change the values for that property, and save your changes.

In the default template CSS, the font is defined in the `<body>` tag, like this:

```
font-family: Georgia, "Bitstream Charter", serif;
```

**TIP**

Font families, or fonts with multiple names, should appear in quotes in your stylesheet.

## Adding borders

CSS borders can add an interesting, unique flair to elements of your theme design. Table 10-6 lists common properties and CSS examples for borders in your theme design.

**TABLE 10-6** **Common Border Properties**

| Border Properties | Common Values | CSS Examples |
|---|---|---|
| border-size | px, em | body {border-size: 1px;} |
| border-style | solid, dotted, dashed | body {border-style: solid;} |
| border-color | hexadecimal values | body {border-color: #CCCCCC;} |

# Understanding Basic HTML Techniques

HTML can help you customize and organize your theme. To understand how HTML and CSS work together, consider this example: If a website were a building, HTML would be the structure (the studs and foundation), and CSS would be the paint.

HTML contains the elements that CSS provides the styles for. All you have to do to apply a CSS style is use the right HTML element. Here's a basic block of HTML:

```
<body>
  <div id="content">
    <h1>Headline Goes Here</h1>
    <p>
      This is a sample sentence of body text.
```

```
        <blockquote>
            The journey of a thousand miles starts with the first step.
        </blockquote>
        I'm going to continue on this sentence and end it here.
    </p>
    <p>
        Click <a href="http://corymiller.com">here</a> to visit my website.
    </p>
  </div>
</body>
```

All HTML elements must have opening and closing tags. Opening tags are contained in less-than (<) and greater-than (>) symbols. Closing tags are the same, except that they're preceded by a forward slash (/).

Here's an example:

```
<h1>Headline Goes Here</h1>
```

The HTML elements must be properly nested. In the fourth line of the preceding example, a paragraph tag is opened (<p>). Later in that line, a block quote is opened (<blockquote>) and is nesting inside the paragraph tag. When editing this line, you couldn't end the paragraph (</p>) before you end the block quote (</blockquote>). Nested elements must close before the elements within which they're nested close.

Finally, proper *tabbing*, or indenting, is important when writing HTML, mainly for readability so that you can scan code to find what you're looking for quickly. A good rule to follow is that if you didn't close a tag in the line above, indent one tab over. This practice allows you to see where each element begins and ends. It can also be very helpful for diagnosing problems.

You'll use several basic HTML markup practices over and over in designing and building websites. Earlier in this chapter, in the "Styling with CSS: The Basics" section, I discuss how to combine CSS styling with HTML markup to create different display styles (borders, fonts, and so on).

The following sections discuss commonly used HTML markup samples that you can use as references on using HTML in your website code. For more in-depth tutorials on HTML, see the HTML section of https://www.w3schools.com/html/default.asp.

# Inserting images

You'll probably want to insert an image into your website, perhaps within the body of a post or page, on the sidebar by using a widget, or within the template code itself. Modern tools like the Block Editor in WordPress take all the guesswork out of inserting images by allowing you to insert images without having to know anything about HTML. But you should know the basic HTML markup for inserting an image so that you'll recognize it when you see it. The HTML markup looks like this:

```
<img src="/path/to/image-file.jpg" alt="Image File Name"/>
```

The following list breaks down this code to help you understand what's at work here:

» `<img src=`: This HTML markup tells the browser that the website is looking for an image file.

» `"/path/to/image-file.jpg"`: This code is the directory path through which the web browser will find the image file. If you uploaded an image to your web server in the `/wp-content/uploads` directory, for example, the physical path for that image file would be `/wp-content/uploads/image-file.jpg`.

» `alt="Image File Name"`: The `alt` tag, which is part of the HTML markup, provides a description of the image that search engines will pick up and recognize as keywords. The `alt` tag description also displays as text in a browser that can't load the image file. (Perhaps the server's load time is slow, or the user is using a screen reader with images turned off. In that case, the text description loads to tell visitors what the image is.)

» `/>`: This HTML markup tag closes the initial `<img src="` tag, telling the web browser that the call to the image file is complete.

# Inserting hyperlinks

You'll probably want to insert a hyperlink within the body of a website. A *hyperlink* is a line of text that's anchored to a web address (URL) so that clicking that link takes a visitor to another website or page that appears in the browser window.

The HTML markup that inserts a hyperlink looks like this:

```
<a href="https://wiley.com">John Wiley & Sons, Inc.</a>
```

Here's a simple explanation of that code:

>> `<a href=`: This code tells the browser that the text within this tag should be hyperlinked to the web address provided next.

>> `"https://wiley.com"`: This code is the URL to which you intend the text to be anchored. Surround the URL with quotation marks to define it as the intended anchor.

>> `>`: This markup closes the previously opened `<a href=` tag.

>> `John Wiley & Sons, Inc.`: In this example, this text is linked, or anchored, by the URL. This text displays on your website and is clickable by your visitors.

>> `</a>`: This HTML markup tag tells the web browser that the hyperlink is closed. Anything that exists between `<a href="..">` and `</a>` is hyperlinked to the intended anchor.

Commonly, designers use URLs to link words to other websites or pages, but you can also provide hyperlinks to files, such as `.pdf` (Adobe Acrobat) and `.docx` (Microsoft Word) files.

# Inserting lists

Suppose that you need to create a clean-looking format for lists of information that you publish on your website. With HTML markup, you can easily create lists that are formatted as either ordered (numbered) or unordered (bulleted), depending on your needs.

Ordered lists are numbered sequentially. An example is a step-by-step list of things to do, like this:

1. Write my book chapters.

2. Submit my book chapters to my publisher.

3. Panic a little when the book is released to the public.

4. Breathe a sigh of relief when public reviews are overwhelmingly positive!

Ordered lists are easy to create in a program such as Microsoft Word or even in the WordPress post editor (because you can use the block editor to format the list for you). If you want to use HTML to code an ordered list, the experience is a little different. The preceding list looks like this in HTML markup:

```
<ol>
    <li>Write my book chapters.</li>
    <li>Submit my book chapters to my publisher.</li>
    <li>Panic a little when the book is released to the public.</li>
    <li>Breathe a sigh of relief when public reviews are overwhelmingly
    positive!</li>
</ol>
```

The beginning <ol> tells your web browser to display this list as an ordered list, meaning that it's ordered with numbers starting with 1. The list ends with the </ol> HTML tag, which tells your web browser that the ordered list is complete.

Between the <ol> and </ol> are list items, designated as such by the HTML markup <li>. Each list item starts with <li> and ends with </li>, which tells the web browser to display the line of text as one list item.

**WARNING**

If you don't close an open HTML markup tag — perhaps you start an ordered list with <ol> but don't include the closing </ol> at the end — you mess up the display on your website. A web browser considers anything after the initial <ol> to be part of the ordered list until it recognizes the closing tag: </ol>.

Unordered lists are similar to ordered lists, but instead of using numbers, they use bullets, like this:

>> Write my book chapters.

>> Submit my book chapters to my publisher.

>> Panic a little when the book is released to the public.

>> Breathe a sigh of relief when public reviews are overwhelmingly positive!

The HTML markup for an unordered list is just like the markup for an ordered list, except that instead of using the <ol> tag, it uses the <ul> tag (UL = unordered list):

```
<ul>
    <li>Write my book chapters.</li>
    <li>Submit my book chapters to my publisher.</li>
    <li>Panic a little when the book is released to the public.</li>
    <li>Breathe a sigh of relief when public reviews are overwhelmingly
    positive!</li>
</ul>
```

## FINDING ADDITIONAL RESOURCES

There may come a time when you want to explore customizing your theme further. Here are some recommended resources:

- **WordPress Codex** (`https://developer.wordpress.org/`): Official WordPress documentation

- **W3Schools** (`https://www.w3schools.com`): A free, comprehensive HTML and CSS reference

Note that the ordered and unordered lists use the list item tags, `<li>` and `</li>`. The only difference is in the first opening and last closing tags:

>> **Ordered lists:** Use <ol> and </ol>.

>> **Unordered lists:** Use <ul> and </ul>.

>> **List items:** Use <li> and </li>.

# Chapter **11**

# Understanding Parent and Child Themes

U sing a theme exactly how the author released it is great. If a new version is released that fixes a browser compatibility issue or adds features offered by a new version of WordPress, a quick theme upgrade is very easy to do.

But there's a good chance that you'll want to tinker with the design, add features, or modify the theme structure. If you modify the theme, you won't be able to upgrade to a newly released version without modifying the theme again.

If only you could upgrade customized versions of themes with new features when they're released! Fortunately, child themes give you this best-of-both-worlds theme solution.

This chapter explores what child themes are, how to create a child theme–ready parent theme, and how to get the most out of using child themes.

# Customizing Theme Style with Child Themes

A WordPress *theme* consists of a collection of template files, stylesheets, images, and JavaScript files. The theme controls the layout and design that your visitors see on the site. When such a theme is properly set up as a parent theme, it allows a *child theme,* or a subset of instructions, to override its files, ensuring that a child theme can selectively modify the layout, styling, and functionality of the parent theme.

The quickest way to understand child themes is by example. In this section, you create a simple child theme that modifies the style of the parent theme.

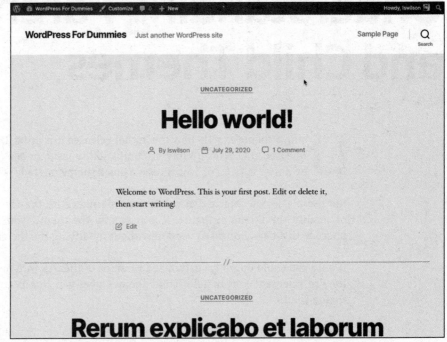

**FIGURE 11-1:**
The Twenty Twenty theme.

Currently, the default WordPress theme is Twenty Twenty. Figure 11-1 shows how the Twenty Twenty theme appears on a sample site.

You likely have Twenty Twenty on your WordPress site, and Twenty Twenty is child theme–ready; therefore, it's a great candidate for a child theme.

Like regular themes, a child theme needs to reside in a directory inside the `/wp-content/themes` directory. The first step in creating a child theme is adding the directory that will hold it. For this example, connect to your hosting account via SFTP, and create a new directory called `twentytwenty-child` inside the `/wp-content/themes` directory.

To register the `twentytwenty-child` directory as a theme and make it a child of the Twenty Twenty theme, create a `style.css` file, and add the appropriate theme headers. To do this, type the following code in your favorite code or plain-text editor (such as Notepad for the PC or TextEdit for the Mac), and save the file as `style.css`:

```
/*
Theme Name: Twenty Twenty Child
Description: My 2020 child theme
Author: Lisa Sabin-Wilson
Version: 1.0
Template: twentytwenty
*/
```

Typically, you find the following headers in a WordPress theme:

>> `Theme Name`: The theme user sees this name in the back end of WordPress.

>> `Description`: This header provides the user any additional information about the theme. Currently, it appears only on the Manage Themes screen (which you access by clicking the Themes link on the Appearance menu).

>> `Author`: This header lists one or more theme authors. Currently, it appears only on the Manage Themes screen (which you access by clicking the Themes link on the Appearance menu).

>> `Version`: The version number is useful for keeping track of outdated versions of the theme. It's always a good idea to update the version number when modifying a theme.

>> `Template`: This header changes a theme into a child theme. The value of this header tells WordPress the directory name of the parent theme. Because your child theme uses Twenty Twenty as the parent, your `style.css` needs to have a `Template` header with a value of `twentytwenty` (the directory/folder name of the Twenty Twenty theme).

Now activate the new Twenty Twenty Child theme as your active theme. (For details on activating a theme on your site, check out Chapter 8.) You should see a site layout similar to the one shown in Figure 11-2.

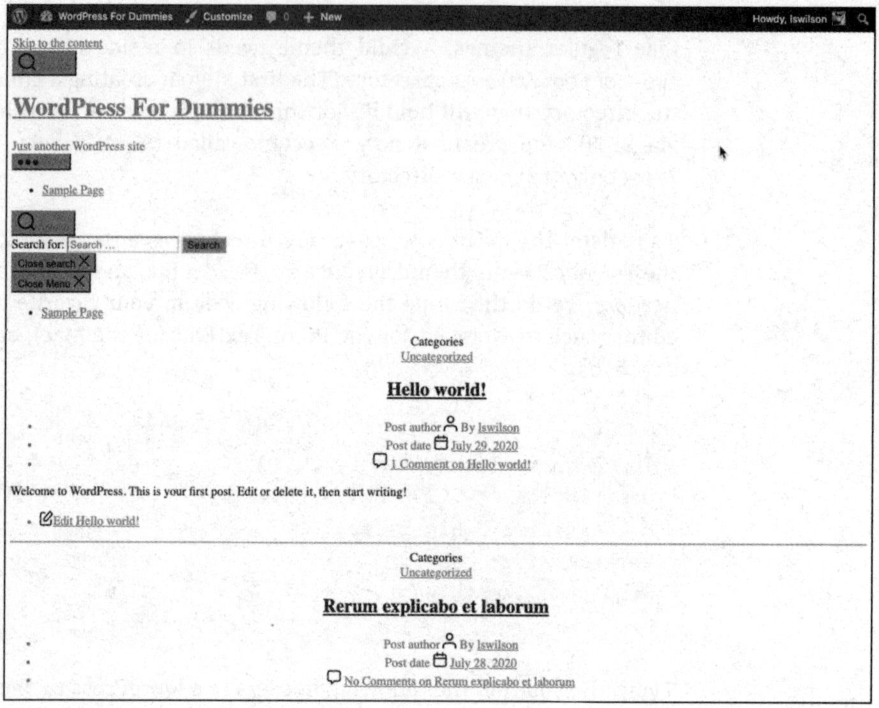

FIGURE 11-2:
The Twenty
Twenty Child
theme.

Figure 11-2 shows that the new theme doesn't look quite right, does it?. The problem is that the new child theme replaced the style.css file of the parent theme, yet the new child theme's style.css file is empty.

You could just copy and paste the contents of the parent theme's style.css file, but that method would waste some of the potential of child themes.

## Loading a parent theme's style

**REMEMBER**

One great thing about CSS is that rules can override one another. If you list the same rule twice in your CSS, the rule that comes last takes precedence.

Consider this example:

```
a {
color: blue;
}
a {
color: red;
}
```

This example is overly simple, but it nicely shows what I'm talking about. The first rule says that all links (tags) should be blue, whereas the second rule says that links should be red. With CSS, the last instruction takes precedence, so the links will be red.

Using this feature of CSS, you can inherit all the styling of the parent theme and selectively modify it by overriding the rules of the parent theme. But how can you load the parent theme's `style.css` file so that it inherits the parent theme's styling? It happens in the Functions file for the theme, `functions.php`, like this:

```
function mytheme_enqueue_styles() {
    wp_enqueue_style('parent-theme', get_template_directory_uri() .'/style.css');
}
add_action('wp_enqueue_scripts', 'mytheme_enqueue_styles');
```

The `parent-theme` parameter in the `wp_enqueue_style` function recognizes the template you have identified in the theme header in the stylesheet, the function tells WordPress to include the stylesheet from the template identified in the stylesheet header.

Figure 11-3 shows how the site appears after the child theme's `style.css` file is updated to match the listing.

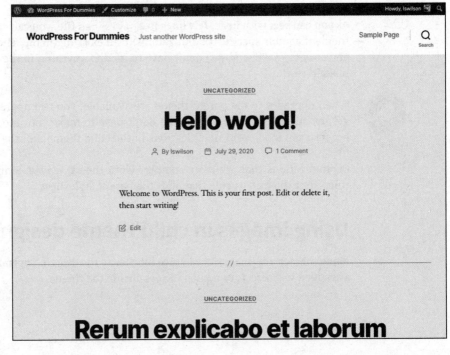

**FIGURE 11-3:**
The updated child theme.

# Customizing the parent theme's styling

Your Twenty Twenty Child theme is set up to match the parent Twenty Twenty theme. Now you can add new styling to the Twenty Twenty Child theme's `style.css` file. To see a simple example of how customizing works, add a style that converts all H1, H2, and H3 headings to uppercase, like so:

```
/*
Theme Name: Twenty Twenty Child
Description: My 2020 child theme
Author: Lisa Sabin-Wilson
Version: 1.0
Template: twentytwenty
*/

h1, h2, h3 {
  text-transform: uppercase;
}
```

Figure 11-4 shows how the child theme looks with the code additions applied. In the preceding code snippet, you targeted the H1, H2, and H3 heading tags for the site and inserted the value `text-transform: uppercase`, which made all of those heading tags appear in uppercase font (or all capital letters) Now the site title and post titles are all uppercase, which differs from Figure 11-3.)

As you can see, with just a few lines in a `style.css` file, you can create a new child theme that adds specific customizations to an existing theme. The change is quick and easy to make, and you don't have to modify anything in the parent theme to make it work.

**TIP**

When upgrades to the parent theme are available, you can upgrade the parent to get the additional features, but you don't have to make your modifications again because you made your modifications in the child theme, not the parent theme.

Customizations that are more complex work the same way. Simply add the new rules after the import rule that adds the parent stylesheet.

## Using images in child theme designs

Many themes use images to add nice touches to the design. Typically, these images are added to a directory named `images` inside the theme.

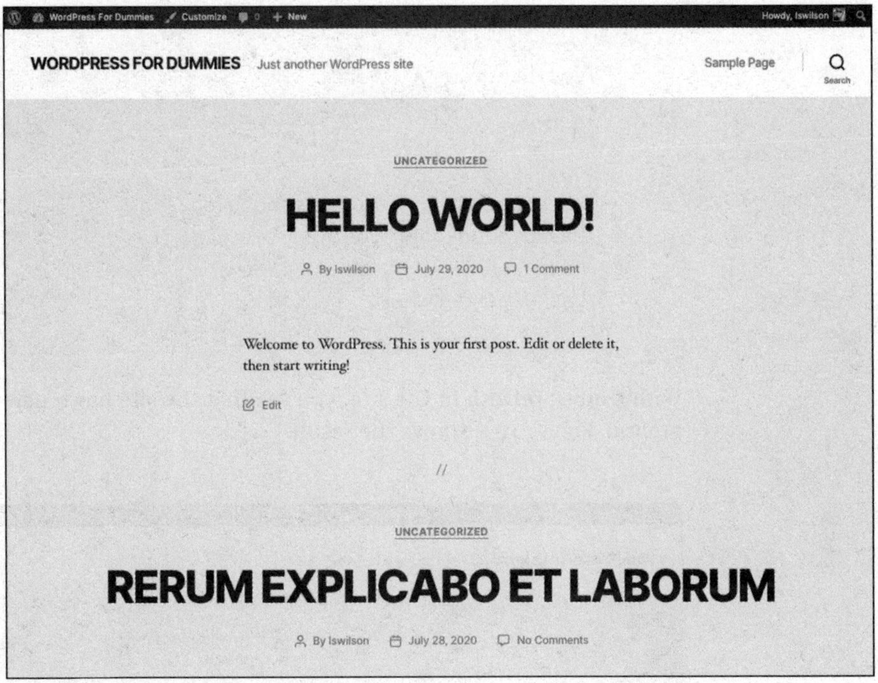

**FIGURE 11-4:**
The updated
child theme
with upper
case headings.

Just as a parent theme may refer to images in its `style.css` file, your child themes can have their own `images` directory. The following sections show examples of how you can use these images.

## Using a child theme image in a child theme stylesheet

Including a child theme image in a child theme stylesheet is common. To do so, simply add the new image to the child theme's `images` directory and refer to it in the child theme's `style.css` file. To get a feel for the mechanics of this process, follow these steps:

1. **Create an `images` directory inside the child theme's directory.**

2. **Add an image to the `images` directory.**

   For this example, add an image called `body-bg.png`.

3. **Add the necessary styling to the child theme's `style.css` file, as follows:**

   ```
   /*
   Theme Name: Twenty Twenty Child
   Description: My 2020 child theme
   Author: Lisa Sabin-Wilson
   ```

```
Version: 1.0
Template: twentytwenty
*/

body {
  background: url('images/body-bg.jpg');
  background-position: center top;
  background-attachment: fixed;
  background-repeat: repeat-x;
}
```

With a quick refresh of the site, you see that the site has a new parchment background. Figure 11-5 shows the result.

FIGURE 11-5:
The Twenty
Twenty Child
theme after the
background
image is edited.

## Using images in a child theme

Child theme images are acceptable for most purposes. You can add your own images to the child theme even if the image doesn't exist in the parent theme folder — and you can accomplish that task without changing the parent theme at all.

In the footer of the Twenty Twenty theme, you can add a WordPress logo to the left of the phrase *Powered by WordPress*, as shown in Figure 11-6. The logo doesn't appear in the footer of the Twenty Twenty theme by default.

**FIGURE 11-6:** The WordPress logo in the Twenty Twenty footer.

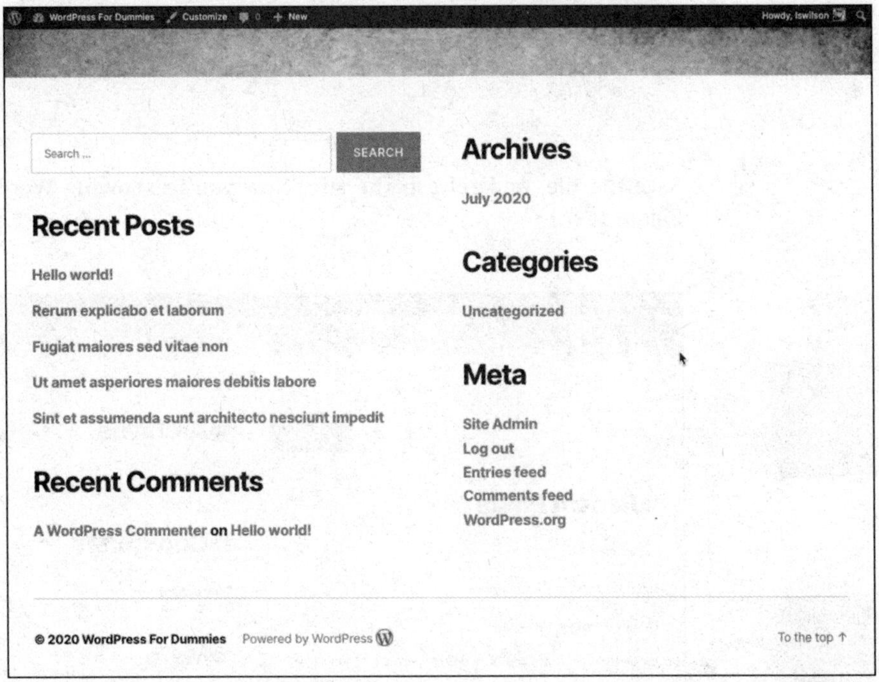

Create a folder in your child theme called `/images`, and add your selected images to that folder. Then you can call those images into your child theme by using the stylesheet (`style.css`) file in your child theme folder.

In this next example, add the same WordPress logo in front of each widget title in the sidebar. Because the logo image already exists inside the child theme's `/images` folder (from the preceding example), you can simply add a customization to the child theme's `style.css` file to make this change, as follows:

```
/*
Theme Name: Twenty Twenty Child
Description: My 2020 child theme
Author: Lisa Sabin-Wilson
Version: 1.0
```

```
Template: twentytwenty
*/

.widget-title {
    background: url(images/wordpress.png);
    background-repeat: no-repeat;
    background-position: left center;
    padding: 0px 30px;
}}
```

Save the file, and refresh the site. Now you're showing WordPress pride! (See Figure 11-7.)

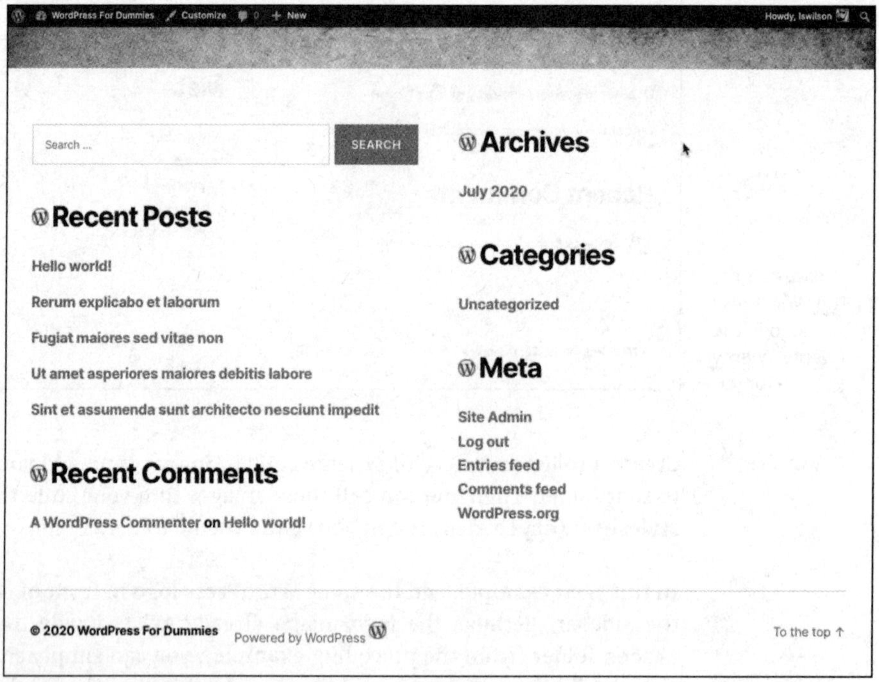

**FIGURE 11-7:**
Showing the WordPress logo before each widget title.

# Modifying Theme Structure with Child Themes

The preceding section shows how to use a child theme to modify the stylesheet of an existing theme. This feature is tremendously powerful. A talented CSS developer can use this technique to create an amazing variety of layouts and designs by using the features and functionality that exist in a parent theme. Creating a child theme removes, or at least reduces, the need to reinvent the wheel by re-creating all the templates in a theme.

This feature is just the beginning of the power of child themes, however. Although every child theme overrides the parent theme's `style.css` file, the child theme can override the parent theme's template (PHP) files, too.

Child themes aren't limited to overriding template files; when needed, child themes also can supply their own template files.

*Template files* are `.php` files that WordPress runs to render different views of your site. A *site view* is the type of content being looked at. Examples of views are home, category archive, individual post, and page content.

Examples of common template files are `index.php`, `archive.php`, `single.php`, `page.php`, `attachment.php`, and `search.php`. (You can read more about available template files, including how to use them, in Chapter 9.)

You may wonder what purpose modifying template files of a parent theme serves. Although modifying the stylesheet of a parent theme can give you powerful control of the design, it can't add new content, modify the underlying site structure, or change how the theme functions. To get that level of control, you need to modify the template files.

## Overriding parent template files

When the child theme and parent theme supply the same template file, WordPress uses the child theme file and ignores the parent theme file. This process of replacing the original parent template file is referred to as *overriding*.

**REMEMBER**

Although overriding each of the theme's template files can defeat the purpose of using a child theme (updates of those template files won't enhance the child theme), sometimes, you must do so to produce a needed result.

The easiest way to customize a specific template file in a child theme is to copy the template file from the parent theme folder to the child theme folder. After copying the file, you can customize it, and the child theme reflects the changes.

A good example of a template file that can be overridden is the `footer.php` file. Customizing the footer allows for adding site-specific branding.

## Adding new template files

A child theme can override existing parent template files, but it also can supply template files that don't exist in the parent. Although you may never need your child themes for this purpose, this option can open possibilities for your designs.

This technique proves to be most valuable with page templates. The Twenty Twenty theme has a template that handles pages named `singular.php`. This template creates a full-width layout for the content and removes the sidebar, as shown in Figure 11-8.

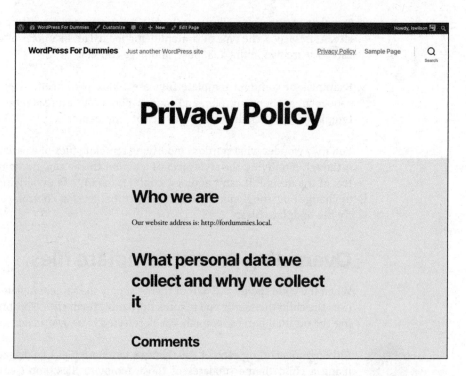

**FIGURE 11-8:**
Page template in Twenty Twenty.

The layout was intentionally set up this way to improve readability on pages where you may not want the distraction of other content in a sidebar. Sometimes, I like to have a full-width layout option so that I can embed a video, add a forum, or add other content that works well at full width; at other times, I want a static page that displays the sidebar. If you want to customize that template and override what the Twenty Twenty theme currently has available, simply create a new page template with the same filename as the one you're replacing (in this case, `page.php`), and add the code to display the sidebar (`get_sidebar();`). Thereafter, WordPress uses the `page.php` template file in your child theme by default, ignoring the one in the Twenty Twenty parent theme folder.

# Removing template files

You may be asking why you'd want to remove a parent's template file. That's a good question. Unfortunately, the Twenty Twenty theme doesn't provide a good example of why you'd want to do this. Therefore, you must use your imagination a bit.

Suppose that you're creating a child theme built from a parent theme called Example Parent. Example Parent is well designed, and a great child theme was built from it quickly. The child theme looks and works exactly the way you want it to, but there's a problem.

The Example Parent theme has a home.php template file that provides a highly customized non-blog home page. What you want, however, is a standard blog home page. If the home.php file didn't exist in Example Parent, everything would work perfectly.

You can't remove the home.php file from Example Parent without modifying the theme, so you have to use a trick. Instead of removing the file, override the home.php file, and have it emulate index.php.

You may think that simply copying and pasting the Example Parent index.php code into the child theme's home.php file is a good approach. Although this technique works, there's a better way: You can tell WordPress to run the index.php file so that the intended index.php file is respected. This single line of code inside the child theme's home.php is all you need to replace home.php with index.php:

```
<?php locate_template( array( 'index.php' ), true ); ?>
```

The locate_template function does a bit of magic. If the child theme supplies an index.php file, that file is used; if not, the parent index.php file is used.

This technique produces the same result as removing the parent theme's home.php file. WordPress ignores the home.php code and respects the changes in index.php.

# Modifying the functions.php file

Like template files, child themes can provide a Theme Functions template, or functions.php file. Unlike in template files, the functions.php file of a child theme doesn't override the file of the parent theme.

When a parent theme and a child theme each have a functions.php file, both the parent and child functions.php files run. The child theme's functions.php file runs first; then the parent theme's functions.php file runs. This setup is intentional, because it allows the child theme to replace functions defined in the parent theme. However, having two function.php files works only if the functions are set up to allow having both.

The Twenty Twenty functions.php file defines a function called twentytwenty_theme_support. This function handles the configuration of many theme options and activates some additional features. Child themes can replace this function to change the default configuration and features of the theme, too.

The following lines of code summarize how the functions.php file allows the child theme to change the default configuration and features of the theme:

```
if ( ! function_exists( 'twentytwenty_theme_support' ) ):
function twentytwenty_theme_support() {
// removed code
}
endif;
```

REMEMBER

Wrapping the function declaration in the if statement protects the site from breaking in the event of a code conflict and allows a child theme to define its own version of the function.

In the Twenty Twenty Child theme, you can see how modifying this function affects the theme. Add a new twentytwenty_setup function that adds post thumbnails support to the Twenty Twenty Child theme's functions.php file, as follows:

```
<?php
function twentytwenty_theme_support() {
    add_theme_support( 'post-thumbnails' );
}
```

As a result of this change, the child theme no longer supports other special WordPress features, such as custom editor styling, automatic feed link generation, and internationalization and location.

The takeaway from this example is that a child theme can provide its own custom version of the function, because the parent theme wraps the function declaration in an if block that checks for the function first.

# Preparing a Parent Theme

WordPress makes it easy for you to make parent themes. WordPress does most of the hard work, but you must follow some rules for a parent theme to function properly.

I've used the *stylesheet* and *template* numerous times in this book in many contexts. Typically, *stylesheet* refers to a CSS file in a theme, and *template* refers to a template file in the theme. But these words also have specific meaning in the context of parent and child themes. You must understand the difference between a stylesheet and a template when working with parent and child themes.

**REMEMBER**

In WordPress, the active theme is the stylesheet, and the active theme's parent is the template. If the theme doesn't have a parent, the active theme is both the stylesheet and the template.

**TECHNICAL STUFF**

Originally, child themes could replace only the `style.css` file of a theme. The parent provided all the template files and `functions.php` code. Thus, the child theme provided style, and the parent theme provided the template files. The capabilities of child themes expanded in subsequent versions of WordPress, making the use of these terms for parent and child themes somewhat confusing.

Imagine two themes: parent and child. The following code is in the parent theme's `header.php` file and loads an additional stylesheet provided by the theme:

```
<link type="text/css" rel="stylesheet" media="all" href="<?php
bloginfo( 'stylesheet_directory' ) ?>/reset.css"/>
```

The `bloginfo` function prints information about the blog configuration or settings. This example uses the function to print the URL location of the stylesheet directory. The site is hosted at `http://example.com`, and the parent is the active theme. It produces the following output:

```
<link type="text/css" rel="stylesheet" media="all"
href="http://example.com/wp-content/themes/Parent/reset.css"/>
```

If the child theme is activated, the output would be

```
<link type="text/css" rel="stylesheet" media="all"
href="http://example.com/wp-content/themes/Child/reset.css"/>
```

Now the location refers to the `reset.css` file in the child theme. This code could work if every child theme copies the `reset.css` file of the parent theme, but requiring child themes to add files in order to function isn't good design. The solution is simple, however. Instead of using the `stylesheet_directory` in the `bloginfo` call, use `template_directory`. The code looks like this:

```
<link type="text/css" rel="stylesheet" media="all" href="<?php
bloginfo( 'template_directory' ) ?>/reset.css"/>
```

Now all child themes properly load the parent `reset.css` file.

When developing, use `template_directory` in stand-alone parent themes and `stylesheet_directory` in child themes.

TIP

You can download a copy of my Twenty Twenty Child Theme for practice at this link: `https://lisasabin-wilson.com/wpfd/twentytwenty-child.zip`.

IN THIS CHAPTER

» Creating templates for static pages, post categories, and sidebars

» Working with custom styles for sticky posts, categories, and tags

» Using Custom Post Types

» Creating New Taxonomies

» Creating post thumbnails

» Optimizing for search engine success

Chapter **12**

# WordPress As a Content Management System

I f you've avoided using WordPress as a solution for building your own website because you think it's only a blogging platform and you don't want to have a blog (not every website owner does, after all), it's time to rethink your position. WordPress is a powerful content management system that's flexible and extensible enough to run an entire website — with no blog at all, if you prefer.

A *content management system* (CMS) is a system used to create and maintain your entire site. It includes tools for publishing and editing, as well as for searching and retrieving information and content. A CMS lets you maintain your website with little or no knowledge of HTML. You can create, modify, retrieve, and update your content without ever having to touch the code required to perform those tasks.

This chapter shows you a few ways that you can use the WordPress platform to power your entire website, with or without a blog. It covers different template configurations that you can use to create separate sections of your site. This chapter also dips into a feature in WordPress called custom post types, which lets you control how content is displayed on your website.

TIP

This chapter touches on working with WordPress templates and themes — a concept that's covered in depth in Chapter 9. If you find templates and themes to be intimidating, check out chapters 8 and 9 first.

You can do multiple things with WordPress to extend it beyond the blog. I use the WordPress theme called Twenty Twenty to show you how to use WordPress to create a fully functional website that has a CMS platform — anything from the smallest personal site to a large business site.

# Creating Different Page Views Using WordPress Templates

A *static page* contains content that doesn't appear on the blog page, but as a separate page within your site. You can have numerous static pages on your site, and each page can have a different design, based on the template you create. (Flip to Chapter 8 to find out all about choosing and using templates on your site.) You can create several static-page templates and assign them to specific pages within your site by adding code to the top of the static-page templates.

Here's the code that appears at the top of the static-page template my company used for an About page template:

```php
<?php
/*
Template Name: About
*/
?>
```

Using a template on a static page is a two-step process: Upload the template, and tell WordPress to use the template for a specific page by tweaking the page's code.

TIP

In Chapter 10, you can discover information about the WordPress Custom Menus feature, including how to create navigation menus for your website. You can create a menu of links that includes all the pages you created on your WordPress Dashboard, for example, and display that menu on your website by using the WordPress Menus feature.

# Uploading the template

To use a page template, you have to create one. You can create this file in a text-editor program, such as Notepad or TextEdit. (To see how to create a template, turn to Chapter 9, which gives you extensive information on WordPress templates and themes.) To create an About page, for example, you can save the template with the name about.php.

**TIP**

For beginners, the best way to get through this step is to make a copy of your theme's page.php file, rename the file about.php, and then make your edits (outlined in this section) in the new about.php file. As you gain more confidence and experience with WordPress themes and template files, you'll be able to create these files from scratch without having to copy other files to make them — or you can keep copying from other files. Why re-create the wheel?

For the purposes of this chapter, I'm using the Twenty Twenty theme, which is the default theme included in every new installation of WordPress. You should have installed WordPress by now, so chances are good that you already have the Twenty Twenty theme.

**WARNING**

Normally, when I practice coding on a website, I use a code repository with version control so that I can keep automatic backups of my template files; this practice allows me to reverse any changes I've made in the files easily and quickly if I've made a mistake or want to undo something I've done. The use of a code repository with version control is a topic for a whole other book. (You could try *Professional Git,* by Brent Laster, from Wrox Publishing.) For the purposes of the coding exercises in the book, you'll be downloading, editing and re-uploading the files via Secure FTP (SFTP), so I strongly recommend that you keep a copy of the files on your computer in a different folder, safe from any alterations. That way, you can restore the original files to your web server if needed.

When you have your template created, follow these steps to make it part of WordPress:

1. **Upload the template file to your WordPress theme folder.**

   You can find that folder on your web server in /wp-content/themes. (See Chapter 3 for more information about SFTP.)

2. **Log in to your WordPress Dashboard, and click the Editor link on the Appearance menu.**

   The Edit Themes screen opens.

3. **Click the about.php template link located on the right side of the page.**

**4.** **Type the Template Name tag directly above the** `get_header()` **template tag.**

The header tag looks like this:

```
get_header(); ?>
```

If you're creating an About Page, the code to create the Template Name tag contains code that looks like this:

```
<?php
/*
Template Name: About
*/
get_header();
?>
```

**5.** **Click the Update File button.**

The file is saved, and the page refreshes. If you created an About page template, the about.php template is now called About Page in the template list on the right side of the page.

Figure 12-1 shows the Page template and displays the code needed to define a specific name for the template.

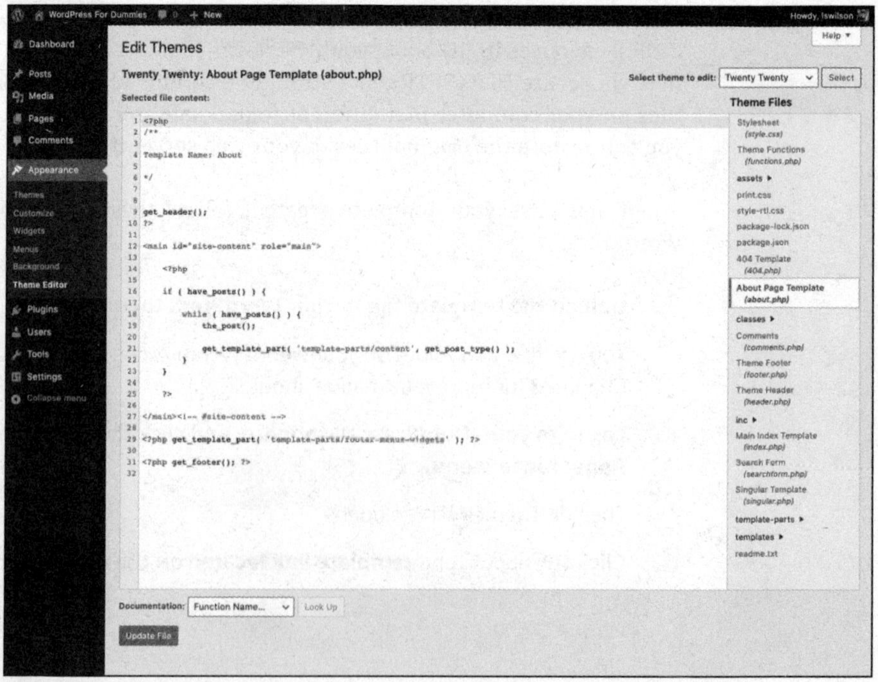

**FIGURE 12-1:** Naming a static-page template.

# Assigning the template to a static page

After you create the template and name it the way you want, assign that template to a page by following these steps:

1. **Click the Add New link on the Pages menu of the Dashboard.**

   The Add New Page screen opens, allowing you to write a new page for your WordPress site.

2. **Type the title in the Title text box and the page content in the large text box.**

3. **Choose the page template from the Template drop-down menu.**

   By default, the Template drop-down menu in the Page Attributes module appears on the right side of the page.

4. **Click the Publish button to save and publish the page to your site.**

Figure 12-2 shows the layout of my home page on my business site at https://webdevstudios.com and the information it contains. Figure 12-3 shows the layout and information provided on the Team page at https://webdevstudios.com/about/team. Both pages are on the same site, in the same WordPress installation, with different page templates to provide different looks, layouts, and sets of information.

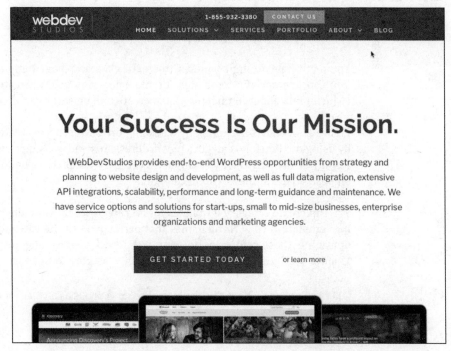

**FIGURE 12-2:** Lisa's business home page at WebDev Studios.com.

FIGURE 12-3:
The Team page
at WebDev
Studios.com.

# Creating a Template for Each Post Category

You don't have to limit yourself to creating a static-page template for your site. You can use specific templates for the categories you've created on your blog (which I talk about in Chapter 5) and create unique sections for your site.

Figure 12-4 shows my company's design portfolio. You can create a page like this by using the Portfolio category in WordPress. Instead of using a static page for the display of the portfolio, you can use a category template to handle the display of all posts made in the Portfolio category.

You can create category templates for all categories on your site simply by creating template files with filenames that correspond to the category slug and then uploading those templates to your WordPress `themes` directory via SFTP (see Chapter 3). Here's the logic behind creating category templates:

>> A template that has the filename `category.php` is a catch-all for the display of categories.

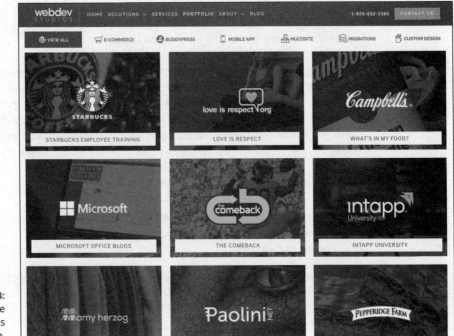

**FIGURE 12-4:**
A Portfolio page on a WordPress website.

>> Add a dash and the category slug to the end of the filename (as shown in Table 12-1) to specify a template for an individual category.

>> If you don't have a `category.php` or `category-slug.php` file, the category display gets defined from the Main Index template (`index.php`).

Table 12-1 shows three examples of the category template naming requirements.

**TABLE 12-1**

## WordPress Category Template Naming Conventions

| If the Category Slug Is . . . | The Category Template Filename Is . . . |
| --- | --- |
| portfolio | category-portfolio.php |
| books | category-books.php |
| music-i-like | category-music-i-like.php |

WordPress makes it possible to pull in very specific types of content on your website through the use of the `WP_Query` class. If you include `WP_Query` before The Loop (see Chapter 9), WordPress lets you specify which category you want to pull information from. If you have a category called WordPress and want to display the last three posts from that category — on your front page, on your sidebar, or somewhere else on your site — you can use this class.

The WP_Query class accepts several parameters that let you display different types of content, such as posts in specific categories and content from specific pages/posts or dates in your blog archives. The WP_Query class lets you pass so many variables and parameters that I just can't list all the possibilities. Instead, you can visit the WordPress Code Reference and read about the options available with this class: https://developer.wordpress.org/reference/classes/wp_query.

Here are two parameters that you can use with WP_Query:

>> posts_per_page=X: This parameter tells WordPress how many posts you want to display. If you want to display only three posts, enter **posts_per_page=3**.

>> category_name=slug: This parameter tells WordPress that you want to pull posts from the category with a specific slug. If you want to display posts from the WordPress category, enter **category_name=wordpress**.

Following is a code snippet that you can use in any theme template to display posts from a single category anywhere on your site. This snippet of code uses the WP_Query class to construct a query that pulls content from posts in a category named wordpress. With the snippet in place, your site displays the title of the post, hyperlinked to the URL of the post, surrounded by HTML <p> .. </p> tags for some formatting. Using techniques covered in Chapter 10 about using CSS to customize the style of HTML tags on your site, you can create a unique look for these posts. You may need to change the wordpress portion of this snippet to pull content from a category that you use on your own site.

```php
<?php
  $args = array( 'category_name' => 'wordpress' );
  $query = new WP_Query( $args );
    if( $query->have_posts() ) :
    while( $query->have_posts() ) :
      $query->the_post(); {
        echo '<p><a href="';
        echo the_permalink();
        echo '">';
        echo the_title();
        echo '</a></p>';
      }
    endwhile; endif;
?>
```

**TECHNICAL STUFF**

In past versions of WordPress, people used the `query_posts();` function to pull content from a specific category, but the `WP_Query` class is more efficient. Although the `query_posts();` function provides the same result, it increases the number of calls to the database and also increases page load and server resources, so please don't use `query_posts();` (no matter what you see written on the Internet!).

# Using Sidebar Templates

You can create separate sidebar templates for different pages of your site by using an `include` statement. When you write an `include` statement, you're simply telling WordPress that you want it to include a specific file on a specific page.

The code that pulls the usual Sidebar template (`sidebar.php`) into all the other templates, such as the Main Index template (`index.php`), looks like this:

```php
<?php get_sidebar(); ?>
```

What if you create a page and want to use a sidebar that has different content from what you already have in the Sidebar template (`sidebar.php`)? You can accomplish that by following these steps to create a new Sidebar template and include it in any template file:

1. **Create a new sidebar template in a text editor such as Notepad or TextEdit.**

   See Chapter 9 for information on template tags and themes. My recommendation is to make a copy of the existing `sidebar.php` file in your existing theme and rename it.

2. **Save the file as `sidebar2.php`.**

   In Notepad (or Text Edit for Mac), choose File ➪ Save. When you're asked to name the file, type **sidebar2.php** and then click Save.

3. **Upload `sidebar2.php` to your Themes folder on your web server.**

   See Chapter 3 for SFTP information, and see Chapter 9 for information on how to locate the Themes folder.

   The template is now in your list of theme files on the Edit Themes screen. (Log in to your WordPress Dashboard and click the Editor link on the Appearance menu.)

**4.** **To include the** `sidebar2.php` **template in one of your page templates, replace** `<?php get_sidebar(); />` **with this code:**

```
<?php get_template_part( 'sidebar2' ); ?>
```

This code calls in a template you've created within your theme.

**TIP**

By using that `get_template_part` function, you can include virtually any file in any of your WordPress templates. You can use this method to create footer templates for pages on your site, for example. First, create a new template that has the filename `footer2.php`. Then locate the following code in your template

```
<?php get_footer(); ?>
```

and replace it with this code:

```
<?php get_template_part( 'footer2' ); ?>
```

**TECHNICAL STUFF**

Every single WordPress theme is different, so you'll find that one theme may have a Sidebar template (`sidebar.php`) in the main directory of its theme folder. Likewise, you may find that another theme puts its Sidebar template in a folder called `/includes` or a folder called `/content`. The default Twenty Twenty theme, for example, doesn't use a Sidebar template at all, whereas one of the default themes used before the latest version of WordPress, called Twenty Seventeen, does have a Sidebar template in the main directory of the theme folder. Each theme you encounter is different, with a different method of structuring the file directory of the theme, so you may have to dig around a little to discover the mysteries of the theme you are working with. Or you can visit Chapter 9 to discover how to create your own theme from scratch.

# Creating Custom Styles for Sticky, Category, and Tag Posts

In Chapter 9, you can find the method for putting together a basic WordPress theme, which includes a Main Index template that uses the WordPress Loop. You can use a custom tag to display custom styles for sticky posts, categories, and tags on your blog. That special tag looks like this:

```
<article id="post-<?php the_ID(); ?>" <?php post_class(); ?> >
```

The post_class() section is the coolest part of the template. This template tag tells WordPress to insert specific HTML markup in your template that allows you to use Cascading Style Sheets (CSS) to make custom styles for sticky posts, categories, and tags.

**REMEMBER**

In Chapter 5, I tell you all about how to publish new posts to your site, including the options you can set for your blog posts, such as categories, tags, and publishing settings. One of those settings is Stick This Post to the Front Page. In this chapter, I show you how to custom-style those sticky posts. The process isn't as messy as it sounds!

Suppose that you publish a post with the following options set:

>> Stick this post to the front page.

>> File it in a category called WordPress.

>> Tag it News.

When the post_class() tag is in the template, WordPress inserts HTML markup that allows you to use CSS to style *sticky posts* (posts assigned top priority for display) with different styling from the rest of your posts. WordPress inserts the following HTML markup for your post:

```
<article class="post sticky category-wordpress tag-news">
```

In Chapter 10, you can discover CSS selectors and HTML markup, and see how they work together to create style and format for your WordPress theme. With the post_class() tag in place, you can go to the CSS file (style.css) in your theme and define styles for the following CSS selectors:

>> .post: Use this tag as the generic style for all posts on your site. The CSS for this tag is

```
.post {
   background: #ffffff;
   border: 1px solid silver;
   padding: 10px;
}
```

A style is created for all posts to have a white background with a thin silver border and 10 pixels of padding space between the post text and the border of the post.

» `.sticky`: You stick a post to your front page to call attention to that post, so you may want to use different CSS styling to make it stand out from the rest of the posts on your site:

```
.sticky {
    background: #ffffff;
    border: 4px solid red;
    padding: 10px;
}
```

This CSS style creates a style for all posts that have been designated as sticky in the settings panel on the Edit Post page to appear on your site with a white background, a thick red border, and 10 pixels of padding space between the post text and border of the post.

» `.category-wordpress`: Because I blog a lot about WordPress, my readers may appreciate it if I give them a visual cue as to which posts on my blog are about that topic. I can do that through CSS by telling WordPress to display a small WordPress icon in the top-right corner of all my posts in the WordPress category:

```
.category-wordpress {
    background: url(wordpress-icon.jpg) top right no-repeat;
    height: 100px;
    width: 100px;
}
```

This CSS style inserts a graphic — `wordpress-icon.jpg` — that's 100 pixels in height and 100 pixels in width in the top-right corner of every post I assign to the WordPress category on my site.

» `.tag-news`: I can style all posts tagged with News the same way I style the WordPress category:

```
.tag-news {
    background: #f2f2f2;
    border: 1px solid black;
    padding: 10px;
}
```

This CSS styles all posts tagged with News with a light gray background and a thin black border with 10 pixels of padding between the post text and border of the post.

You can easily use the `post-class()` tag, combined with CSS, to create dynamic styles for the posts on your site!

# Working with Custom Post Types

Custom post types and custom taxonomies have expanded the CMS capabilities of WordPress and are likely to become a big part of plugin and theme features as more developers become familiar with their use. *Custom post types* allow you to create new content types separate from posts and pages, such as movie reviews or recipes. *Custom taxonomies* allow you to create new types of content groups separate from categories and tags, such as genres for movie reviews or seasons for recipes.

Posts and pages are nice generic containers of content. A *page* is timeless content that has a hierarchal structure; a page can have a parent (forming a nested, or hierarchal, structure of pages). A *post* is content that is listed in linear (not hierarchal) order based on when it was published and organized into categories and tags. What happens when you want a hybrid of these features? What if you want content that doesn't show up in the post listings, displays the posting date, and doesn't have either categories or tags? Custom post types are created to satisfy this desire to customize content types.

By default, WordPress has different post types built into the software, ready for you to use. The default post types include

>> Blog posts

>> Pages

>> Menus

>> Attachments

>> Revisions

Custom post types give you the ability to create new, useful types of content on your website, including a smart, easy way to publish those content types to your site.

The possibilities for the use of custom post types are endless. To kick-start your imagination, here are some of the most popular and useful ideas that developers have implemented on sites:

>> Photo gallery

>> Podcast or video

>> Book reviews

>> Coupons and special offers

>> Events calendar

To create and use Custom Post Types on your site, you need to be sure that your WordPress theme contains the correct code and functions. In the following steps, I create a basic Custom Post Type called Generic Content. Follow these steps to create the Generic Content basic Custom Post Type:

1. **Log in to your web server via SFTP.**

   To see how to connect to your website via SFTP, refer to Chapter 3.

2. **Navigate to the Twenty Twenty theme folder.**

   On your web server, that folder is /wp–content/themes/twentytwenty/.

3. **Locate the `functions.php` template file.**

4. **Download the `functions.php` template file to your computer.**

5. **Open the `functions.php` file in a text editor.**

   Use Notepad if you're on a PC or TextEdit if you're using a Mac.

6. **Add the Custom Post Types code below this line of code**

   ```
   function twentytwenty_theme_support() {.
   ```

   to add a Generic Content Custom Post Type to your site:

   ```
   // Add a Custom Post Type called: Generic Content
   add_action( 'init', 'create_my_post_types' );
      function create_my_post_types() {
         register_post_type( 'generic_content', array(
            'label' => ( 'Generic Content' ),
            'singular_label' => ( 'Generic Content' ),
            'description' => ( 'Description of Generic Content
   type' ),
            'public' => true,
         )
      );
   }
   ```

7. **Save the Functions file, and upload it to your web server again.**

Figure 12-5 shows you the WordPress Dashboard screen with a new menu item on the left menu; below the Comments link, you see a link for Generic Content. This link is the sample Custom Post Type that you added to WordPress in the preceding steps. You can see that the Generic Content post type has been successfully registered in my installation of WordPress and is ready to use.

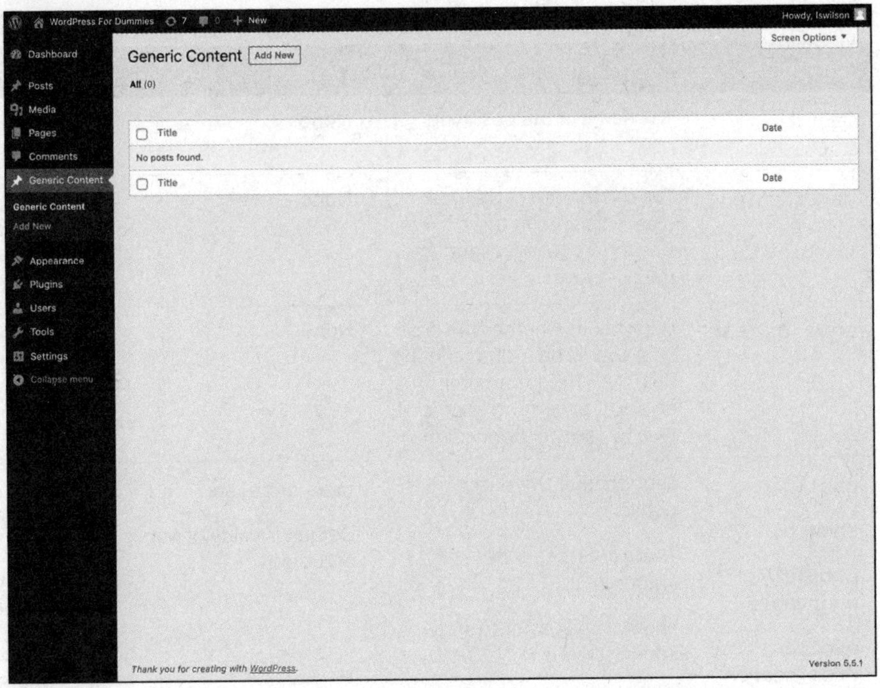

FIGURE 12-5:
The new Generic
Content post type.

**TECHNICAL STUFF**

The function `register_post_type` can accept several arguments and parameters, which are detailed in Table 12-2. You can use a variety of combinations of arguments and parameters to create a specific post type. You can find more information on Custom Post Types and on using the `register_post_type` function in the official WordPress Code Reference at `https://developer.wordpress.org/reference/functions/register_post_type`.

After you add a Generic Content Custom Post Type to your site, a new post type labeled Generic appears on the left navigation menu of the Dashboard.

You can add and publish new content by using the new Custom Post Type, just as you do when you write and publish posts or pages. (See Chapter 4.) The published content isn't added to the chronological listing of posts; rather, it's treated like separate content from your blog (just like static pages).

View the permalink for the published content, and you see that it adopts the custom post type name Generic Content and uses it as part of the permalink structure, creating a permalink that looks like `https://yourdomain.com/generic-content/new-article`.

**TABLE 12-2**     **Arguments and Parameters for register_post_type();**

| Parameter | Information | Default | Example |
|---|---|---|---|
| label | The name of the post type. | None | `'label' => __( 'Generic Content' ),` |
| description | The description of the post type, displayed in the Dashboard to represent the post type. | None | `'description' => __( 'This is a description of the Generic Content type' ),` |
| show_in_rest | Determines whether the post type should be included in the REST API. This parameter must be set to true for this post type to use the Block editor. | None | `'show_in_rest' => true,` |
| public<br><br>show_ui<br><br>publicly_queryable<br><br>exclude_from_search | Sets whether the post type is public.<br><br>There are three other arguments:<br><br>show_ui: whether to show admin screens<br><br>publicly_queryable: whether to query for this post type from the front end<br><br>exclude_from_search: whether to show post type in search results | true or false<br><br>Default is value of public argument | `'public' => true,`<br><br>`'show_ui' => true,`<br><br>`'publicly_queryable' => true,`<br><br>`'exclude_from_search' => false,` |
| menu_position | Sets the position of the post type menu item on the Dashboard navigation menu. | Default: null<br><br>By default, appears after the Comments menu on the Dashboard<br><br>Set integer in intervals of 5 (5, 10, 15, 20, and so on) | `'menu_position' => 25,` |
| menu_icon | Defines a custom icon (graphic) to the post type menu item on the Dashboard navigation menu.<br><br>Creates and uploads the image to the images directory of your theme folder. | Posts icon | `'menu_icon' => get_stylesheet_directory_uri() . '/images/generic-content.png',` |
| hierarchical | Tells WordPress whether to display the post type content list in a hierarchical manner. | true or false<br><br>Default is false | `'hierarchical' => true,` |

| Parameter | Information | Default | Example |
|---|---|---|---|
| query_var | Controls whether this post type can be used with a query variable such as query_posts or WP_Query. | true or false<br><br>Default is true | `'query_var' => true,` |
| capability_type | Defines permissions for users to edit, create, or read the Custom Post Type. | post (default)<br><br>Gives the same capabilities for those users who can edit, create, and read blog posts | `'query_var' => post,` |
| supports | Defines what meta boxes, or modules, are available for this post type in the Dashboard. | title: Text box for the post title<br><br>editor: Text box for the post content<br><br>comments: Check boxes to toggle comments on/off<br><br>trackbacks: Check boxes to toggle trackbacks and pingbacks on/off<br><br>revisions: Allows post revisions to be made<br><br>author: Drop-down menu to define post author<br><br>excerpt: Text box for the post excerpt<br><br>thumbnail: The featured image selection<br><br>custom-fields: Custom Fields input area<br><br>page-attributes: The page parent and page template drop-down menus<br><br>post-formats: adds Post Formats | `'supports' => array( 'title', 'editor', 'excerpt', 'custom-fields', 'thumbnail' ),` |

*(continued)*

**TABLE 12-2** *(continued)*

| Parameter | Information | Default | Example |
|-----------|-------------|---------|---------|
| rewrite | Rewrites the permalink structure for the post type. | true or false<br><br>The default is true with the post type as the slug<br><br>Two other arguments are available:<br><br>slug: Permalink slug to use for your Custom Post Types<br><br>with_front: If you've set your permalink structure with a specific prefix, such as /blog | `'rewrite' => array( 'slug' => 'my-content', 'with_front' => false ),` |
| taxonomies | Uses existing WordPress taxonomies (category and tag). | Category<br><br>post_tag | `'taxonomies' => array( 'post_tag', 'category'),` |

TIP

A very helpful plugin for building Custom Post Types quickly in WordPress is one called Custom Post Type UI. Written by my team at WebDevStudios, this plugin (https://wordpress.org/plugins/custom-post-type-ui) gives you a clean interface within your WordPress Dashboard that can help you build Custom Post Types on your website easily and quickly. It eliminates the need to add the code to your functions.php file by giving you options and settings to configure the Custom Post Type that you want. Figure 12-6 shows the Custom Post Type UI options page on the Dashboard.

TIP

To add custom post types to the Menus options on the Menus page that are accessible from the Dashboard (choose Appearance ⇨ Menus), click the Screen Options tab in the top-right corner of that screen. Select the check box next to Post Types to enable your Custom Post Types in the menus you create. (Post Types appears in the screen options only if you have Custom Post Types enabled in your theme.)

By default, a Custom Post Type uses the single.php template in your theme unless you create a specific template for that post type. You might find the regular WordPress single.php template to be limiting for your post type, depending on the type of content you want to include and whether you want to apply different formats and styles with HTML and CSS markup.

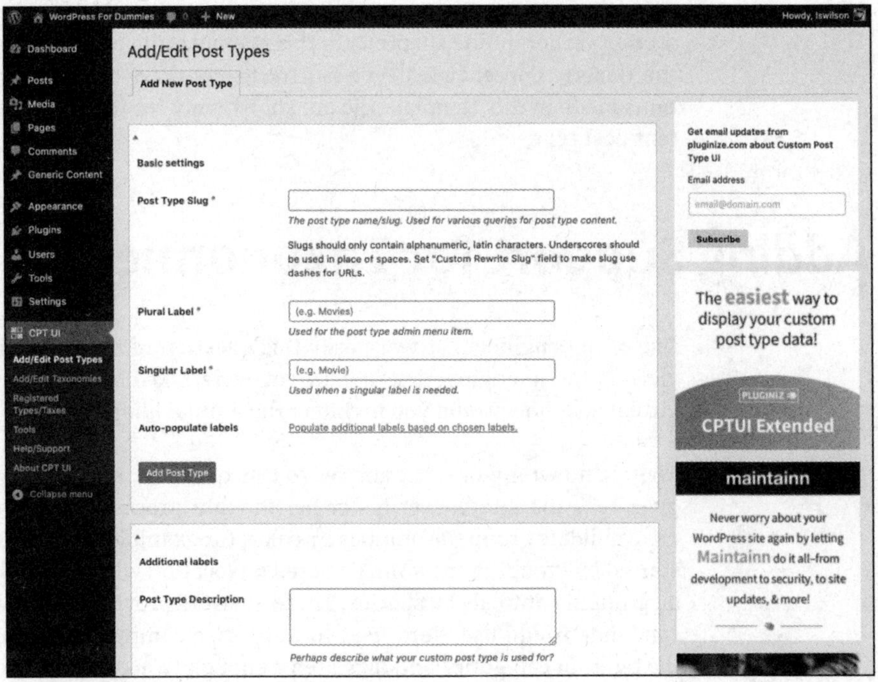

FIGURE 12-6:
The Custom Post
Type UI plugin
options page.

Earlier in this chapter in the "Working with Custom Post Types" section, I share the code to build a simple Generic Content custom post. After you add it, the Generic Content menu appears on the WordPress Dashboard. Choose Generic Content ⇨ Add New and then publish a new post with some content for testing. Add a new Generic Content type with a Test title and a test slug, for example. Because the Generic Content type has no specific template yet, it uses the single. php template, and resulting posts look no different from standard posts.

**TIP**

If you get a Not Found page when you try to go to a new custom post type entry, reset your permalink settings. Click the Permalinks link on the Settings menu on the WordPress Dashboard and then click the Save Changes button. This action causes WordPress to reset the permalinks, which adds the new custom post type link formats in the process.

To build a template specifically for the Generic Content post type, add a new single-posttype.php template. (posttype is the first argument passed to the register_post_type function from the "Working with Custom Post Types" section earlier in this chapter.) In this example, the template file that's specific to the Generic Content post type is `single-generic-content.php`. Any modifications made in this template file are shown only for instances of the Generic Content post type.

# Adding Support for Taxonomies

One of my engineers at work asks this question of every candidate who's interviewing for an engineering job: "If you were asked to design a class system for a virtual zoo, how would you architect the animal kingdom?"

There's no wrong or right answer to this question, but the answer the candidate gives tells the interviewer his or her thought process on architecting data. Does the candidate group the animals by color, for example, or by breed? One candidate started by grouping the animals based on location within the zoo; another started by grouping animals by species. These answers provide helpful insight into how a candidate would use WordPress to categorize complex data. Sometimes, though, the built-in categories and tags aren't enough to accomplish a complex task such as architecting a virtual zoo.

On a website where content about different types of zoo animals is published, the zoo content might need a variety of new taxonomies or grouping options. Organizing zoo animals by color, breed, species, location in the zoo, and size allows visitors to the site to view different groups of animals that might interest them.

To register this new taxonomy, use the `register_taxonomy` function. Adding the following code to the bottom of your theme's `functions.php` file registers the Color taxonomy that you could use to categorize the zoo animals by color:

```
register_taxonomy( 'color', 'post', array(
  'label' => 'Color' )
);
```

This function call gives the new custom taxonomy an internal name of `color`, assigns the new taxonomy to Posts, and gives the taxonomy a human-readable name of Color. When you've successfully registered the new taxonomy, you see a link for it on the Posts menu of the Dashboard. In Figure 12-7, you see the Color taxonomy screen of the Dashboard, where you can add new terms in the same manner that you add Categories on the Category screen.

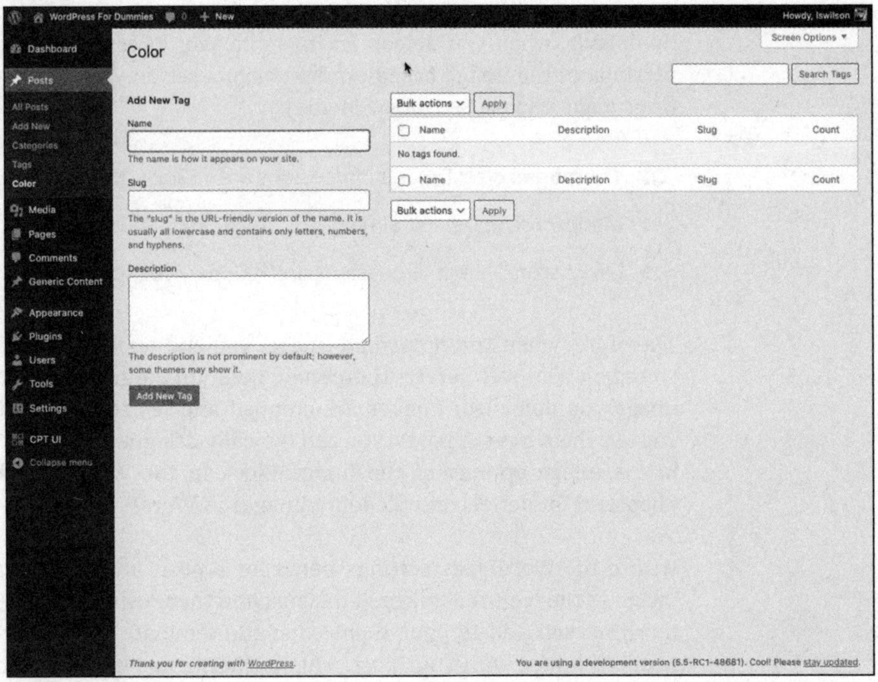

**FIGURE 12-7:**
A new taxonomy, called Color, registered in WordPress.

# Adding Support for Post Thumbnails

The WordPress feature called post thumbnails (also known as featured images) takes a lot of the work out of associating an image with a post and using the correct size each time. A popular way to display content in WordPress themes includes a thumbnail image with a snippet (short excerpt) of text; the thumbnail images are consistent in size and placement within your theme.

Before the inclusion of post thumbnails in WordPress, users had to open their images in an image-editing program (such as Adobe Photoshop) and then crop and resize their images to the desired size or use fancy scripts (which tend to be resource-intensive on web servers) to resize images on the fly. Neither solution is optimal. How about using a CMS that crops and resizes your images to the exact dimensions that you specify? Yep, WordPress does all that with just a few adjustments.

By default, when you upload an image in WordPress, the software creates three versions of the image based on dimensions set on your Dashboard's Media Settings page (choose Settings ⇨ Media):

>> **Thumbnail size:** Default dimensions are 150px × 150px.

>> **Medium size:** Default dimensions are 300px × 300px.

>> **Large size:** Default dimensions are 1024px × 1024px.

Therefore, when you upload an image, you end up with four sizes of that image stored on your web server: thumbnail, medium, large, and the original (full-size) image you uploaded. Images are cropped and resized proportionally, and when you use them in your posts, you can typically designate which size you want to use in the image options of the Image Block in the WordPress Block Editor. (See Chapter 6 for details on uploading images in WordPress.)

Within the WordPress settings panel for a post, you can designate a particular image as the featured image of the post and then, using the featured images function that you add to your theme, include template tags to display your chosen featured image with your post. This technique is helpful for creating the magazine- or news-style themes that are popular on WordPress sites. Figure 12-8 shows Post Thumbnails and featured images on my business website's blog at `https://webdevstudios.com/blog`.

In Chapter 6, I cover the default image sizes that are set on the Media Settings page of your Dashboard.

## Adding the post thumbnails to a theme

Adding support for post thumbnails includes adding a single line of code to your theme functions (`functions.php`) file:

```
add_theme_support( 'post-thumbnails' );
```

**TIP**

Many themes, including the default themes that come packaged in WordPress, already have support for post thumbnails. If a theme you're using (or creating) doesn't offer support, you can add it with the preceding line of code.

After you add this line of code to your Theme Functions template file, you can use the post thumbnails feature for your posts. You can designate featured images in the Featured Image section of the settings panel on the Edit Post screen.

**FIGURE 12-8:**
Post thumbnails
in use at
`https://`
`webdevstudios.`
`com/blog.`

After you add featured images to your post, make sure that you add the correct tag in your template(s) so that the featured images display on your site in the area where you want them to display. Open your `index.php` template, and add the following line of code to include the default thumbnail-size version of your chosen featured image in your posts:

```
<?php if ( has_post_thumbnail() ) { the_post_thumbnail( 'thumbnail' ); } ?>
```

The first part of that line of code (`if ( has_post_thumbnail()`) checks whether a featured image is associated with the post. If so, the image is attached to the post. The second part of that code snippet (`the_post_thumbnail('thumbnail')`) displays the thumbnail-size version of the image. If a featured image doesn't exist for the post, the second part of the code snippet is ignored, and the code returns nothing. You also can include the other default image sizes set in the Media Settings screen of the Dashboard for medium, large, and full-size images by using these tags:

```
<?php if ( has_post_thumbnail() ) { the_post_thumbnail( 'medium' ); } ?>
```

```
<?php if ( has_post_thumbnail() ) { the_post_thumbnail( 'large' ); } ?>
```

```
<?php if ( has_post_thumbnail() ) { the_post_thumbnail( 'full' ); } ?>
```

# Adding custom image sizes for post thumbnails

If the predefined, default image sizes in WordPress (thumbnail, medium, large, and full) don't satisfy you, and you want to display images with nondefault dimensions, WordPress makes it relatively easy to extend the functionality of the post thumbnail feature by defining custom image sizes for your images in your Theme Functions template file. Then, you use the `the_post_thumbnail` function to display it in your theme.

There's no limit to what sizes you can use for your images. The following example shows how to add a new image size of 600px × 300px. Add this line to your Theme Functions template file (`functions.php`) below the `add_theme_support( 'post-thumbnails' )` function:

```
add_image_size( 'custom', 600, 300, true );
```

This code tells WordPress to create an additional version of the images you upload and to crop and resize them to 600px × 300px. Notice the four parameters in the `add_image_size` function:

» **Name** ($name): Gives the image size a unique name that you can use later in your template tag. The image size in this example uses the name `'custom'`.

» **Width** ($width): Gives the image size a width dimension in numbers. In the example, the width is 600.

» **Height** ($height): Gives the image size a height dimension in numbers. In the example, the height is 300.

» **Crop** ($crop): Tells WordPress whether it should crop the image to exact dimensions or do a soft proportional resizing of the image. Rather than cropping to exact dimensions, WordPress automatically corrects dimensions based on the width of the thumbnail settings. The accepted arguments are true and false. In the example, the crop setting is true (accepted arguments: true or false).

Adding a custom image size to your template to display the image you've designated as featured is the same as adding default image sizes, except that the name of the image is set in the parentheses of the template tag. To add an example custom image size, use this tag:

```
<?php if ( has_post_thumbnail() ) { the_post_thumbnail( 'custom' ); ?>.
```

# Optimizing Your WordPress Site

*Search engine optimization* (SEO) is the practice of preparing your site to make it as easy as possible for the major search engines to crawl and cache your data in their systems so that your site appears as high as possible in the search returns. This section gives you a brief introduction to SEO practices with WordPress.

If you search for the keywords *WordPress website design and development* in Google, my business site at WebDevStudios is in the top-ten search results for those keywords (at least, it is while I'm writing this chapter). Those results can change from day to day, so by the time you read this book, someone else may very well have taken over that coveted position. The reality of chasing those high-ranking search engine positions is that they're here today, gone tomorrow. The goal of SEO is to make sure that your site ranks as high as possible for the keywords that you think people will use to find your site. After you attain those high-ranking positions, the next goal is to keep them.

WordPress is equipped to create an environment that's friendly to search engines, giving them easy navigation through your archives, categories, and pages. WordPress provides this environment with a clean code base, content that's easily updated through the WordPress interface, and a solid navigation structure.

To extend SEO even further, you can tweak five elements of your WordPress posts, pages, and templates:

>> **Custom permalinks:** Use custom permalinks, rather than the default WordPress permalinks, to fill your post and page URLs with valuable keywords. Check out Chapter 5 for information on WordPress permalinks.

>> **Posts and page titles:** Create descriptive titles for your posts and pages to provide rich keywords in your site.

>> **Text:** Fill your posts and pages with keywords for search engines to find and index. Keeping your site updated with descriptive text and phrases helps the search engines find keywords to associate with your site.

>> **Category names:** Use descriptive names for the categories you create in WordPress to place great keywords right in the URL for those category pages if you use custom permalinks.

>> **Images and <ALT> tags:** Place <ALT> tags in your images to further define and describe the images on your site. You can accomplish this task easily by using the description field in the WordPress image uploader.

# Planting keywords on your website

If you're interested in a higher ranking for your site, use custom permalinks. By using custom permalinks, you're automatically inserting keywords into the URLs of your posts and pages, letting search engines include those posts and pages in their databases of information on those topics. If a provider that has the Apache `mod_rewrite` module enabled hosts your site, you can use the custom permalink structure for your WordPress-powered site.

Keywords are the first step on your journey to great search engine results. Search engines depend on keywords, and people use keywords to look for content.

The default permalink structure in WordPress is pretty ugly. When you're looking at the default permalink for any post, you see a URL something like this:

```
http://yourdomain.com/p?=105
```

This URL contains no keywords of worth. If you change to a custom permalink structure, your post URLs automatically include the titles of your posts to provide keywords, which search engines absolutely love. A custom permalink may appear in this format:

```
http://yourdomain.com/2020/12/01/your-post-title
```

I explain setting up and using custom permalinks in full detail in Chapter 4.

# Optimizing your post titles for search engine success

Search engine optimization doesn't completely depend on how you set up your site. It also depends on you, the site owner, and how you present your content.

You can present your content in a way that lets search engines catalog your site easily by giving your blog posts and pages titles that make sense and coordinate with the actual content being presented. If you're doing a post on a certain topic, make sure that the title of the post contains at least one or two keywords about that particular topic. This practice gives the search engines even more ammunition to list your site in searches relevant to the topic of your post.

**REMEMBER**

As your site's presence in the search engines grows, more people will find your site, and your readership will increase as a result.

A post with the title A Book I'm Reading doesn't tell anyone *what* book you're reading, making it difficult for people who are searching for information on that particular book to find the post. If you give the post the title *WordPress For Dummies: My Review,* however, you provide keywords in the title, and (if you're using custom permalinks) WordPress automatically inserts those keywords into the URL, giving the search engines a triple keyword play:

» Keywords exist in your blog post title.

» Keywords exist in your blog post URL.

» Keywords exist in the content of your post.

## Writing content with readers in mind

When you write your posts and pages and want to make sure that your content appears in the first page of search results so that people will find your site, you need to keep those people in mind when you're composing the content.

When search engines visit your site to crawl through your content, they don't see how nicely you've designed your site. They're looking for words to include in their databases. You, the site owner, want to make sure that your posts and pages use the words and phrases you want to include in search engines.

If your post is about a recipe for fried green tomatoes, for example, you need to add a keyword or phrase that you think people will use when they search for the topic. If you think people would use the phrase *recipe for fried green tomatoes* as a search term, you may want to include that phrase in the content and title of your post.

A title such as A Recipe I Like isn't as effective as a title such as A Recipe for Fried Green Tomatoes. Including a clear, specific title in your post or page content gives the search engines a double-keyword whammy.

## Creating categories that attract search engines

One little-known SEO tip for WordPress users: The names you give the categories you create for your site provide rich keywords that attract search engines like honey attracts bees. Search engines also see your categories as keywords that are relevant to the content on your site. Make sure that you're giving your categories names that are relevant to the content you're providing.

If you sometimes write about your favorite recipes, you can make it easier for search engines to find your recipes if you create categories specific to the recipes you're blogging about. Instead of having one Favorite Recipes category, you can create multiple category names that correspond to the types of recipes you blog about: Casserole Recipes, Dessert Recipes, Beef Recipes, and Chicken Recipes, for example.

**REMEMBER**

Creating specific category titles not only helps search engines, but also helps your readers discover content that's related to topics they're interested in.

You also can consider having one category called Favorite Recipes and creating subcategories (also known as *child categories*) that give a few more details on the types of recipes you've written about. (See Chapter 5 for information on creating categories and child categories.)

Categories use the custom permalink structure, just as posts do, so links to your WordPress categories also become keyword tools within your site to help the search engines — and, ultimately, search engine users — find the content. Using custom permalinks creates category page URLs that look something like this:

```
http://yourdomain.com/category/category_name
```

The `category_name` portion of that URL puts the keywords right in the hands of search engines.

## Using the <ALT> tag for images

When you use the WordPress Media Library to upload and edit images, an ALT Text text field appears; in it, you can enter a description of the image. (I cover uploading and inserting images into your posts and pages in great detail in Chapter 6.) This text automatically becomes what's referred to as the ⟨ALT⟩ tag, otherwise known as alternative text.

The purpose of the ⟨ALT⟩ attribute is to provide a description of the image to screen readers for people with visual impairments. In a text-based browser that doesn't display images, for example, visitors see the description, or ⟨ALT⟩ text, telling them what image would be there if they could see it. Also, the tag helps people who have impaired vision and rely on screen-reading technology because the screen reader reads the ⟨ALT⟩ text from the image. You can read more about website accessibility for people with disabilities at `https://www.w3.org/WAI/people-use-web`.

An extra benefit of ‹ALT› tags is that search engines gather data from them to further classify the content of your site. The following code inserts an image with the ‹ALT› tag of the code in bold to demonstrate what I'm talking about:

```
<img src="http://yourdomain.com/image.jpg" alt="This is an ALT tag"/>
```

Search engines harvest those ‹ALT› tags as keywords. The WordPress image uploader gives you an easy way to include those ‹ALT› tags without having to worry about inserting them into the image code yourself. Just fill out the Description text box before you upload and add the image to your post. Chapter 6 provides in-depth information on adding images to your site content, including how to add descriptive text for the ‹ALT› tag and keywords.

A great resource for further reading on ALT text for images can be found from Google at https://support.google.com/webmasters/answer/114016?hl= en#descriptive-alt-text.

Chapter **13**

# Hosting Multiple Sites with WordPress

I n this chapter, I introduce you to the network feature that's built into the WordPress software. The network feature allows you, the site owner, to add and maintain multiple blogs within one installation of WordPress. In this chapter, you discover how to set up the WordPress network feature, explore settings and configurations, gain an understanding of the Network Administrator role, determine which configuration is right for you (do you want subdirectories or subdomains?), and find some great resources to help you on your way.

With the network features enabled, users of your network can run their own sites within your installation of WordPress. They also have access to their own Dashboards with the same options and features covered in the first 12 chapters of this book. Heck, it would probably be a great idea to buy a copy of this book for every member of your network so that everyone can become familiar with the WordPress Dashboard and features, too. At least have a copy on hand so people can borrow yours!

# Deciding When to Use the Multisite Feature

Usually, for multiple users to post to one site, the default WordPress setup is sufficient. The *Multi* part of the WordPress Multisite feature's name doesn't refer to how many users were added to your WordPress website; it refers to the ability to run multiple sites in one installation of the WordPress software. *Multisite* is a bit of a misnomer and an inaccurate depiction of what the software actually does. *Network of sites* is a much closer description.

Determining whether to use the Multisite feature depends on user access and publishing activity. Each site on the network shares a codebase and users but is a self-contained unit. Users still have to access the back end of each site to manage options or post to that site. A limited number of general options are available networkwide, and posting isn't one of those options.

You can use multiple sites on a network to give the appearance that only one site exists. Put the same theme on each site, and the visitor doesn't realize that the sites are separate. This technique is a good way to separate sections of a magazine site, using editors for complete sections (sites) but not letting them access other parts of the network or the back ends of other sites.

Another factor to consider is how comfortable you are with editing files directly on the server. Setting up the network involves accessing the server directly, and ongoing maintenance and support for your users often leads to the network owner's doing the necessary maintenance, which is not for the faint of heart.

Generally, you should use a network of sites in the following cases:

>> **You want multiple sites and one installation.** You're a blogger or site owner who wants to maintain another site, possibly with a subdomain or a separate domain, with both sites on one web host. You're comfortable with editing files; you want to work with one codebase to make site maintenance easier; and most of your plugins and themes are accessible to all the sites. You can have one login across the sites and manage each site individually.

>> **You want to host blogs or sites for others.** This process is a little more involved. You want to set up a network in which users sign up for their own sites or blogs below (or on) your main site and you maintain the technical aspects.

Because all files are shared, some aspects are locked down for security purposes. One of the most puzzling security measures for new users is suppression of errors. Most PHP errors (such as those that occur when you install a faulty plugin or incorrectly edit a file) don't output messages to the screen. Instead, WordPress displays what I like to call the White Screen of Death.

Finding and using error logs and doing general debugging are necessary skills for managing your own network. Even if your web host sets up the ongoing daily or weekly tasks for you, managing a network can involve a steep learning curve.

**REMEMBER**

When you enable the Multisite feature, the existing WordPress site becomes the main site in the installation.

Although WordPress can be quite powerful, in the following situations, managing multiple sites has limitations:

>> **One web account is used for the installation.** You can't use multiple hosting accounts.

>> **You want to post to multiple blogs at the same time.** WordPress doesn't allow this practice by default.

>> **If you choose subdirectory sites, the main site regenerates permalinks with /blog/ in them to prevent collisions with subsites.** Plugins are available that prevent this regeneration.

The best example of a large blog network with millions of blogs and users is the hosted service at WordPress.com (https://wordpress.com). At WordPress.com, people are invited to sign up for an account and start a blog by using the Multisite feature within the WordPress platform on the WordPress server. When you enable this feature on your own domain and enable the user registration feature, you're inviting users to do the following:

>> Create an account

>> Create a blog in their WordPress installations (on your domain)

>> Create content by publishing blog posts

>> Upload media files such as photos, audio, and video

>> Invite friends to view their blogs or sign up for their own accounts

# Understanding the Difference between Sites and Blogs

Each additional blog in a WordPress Multisite network is a *site* instead of a *blog*. What's the difference?

Largely, the difference is one of perception. Everything functions the same way, but people see greater possibilities when they no longer think of each site as being "just" a blog. WordPress can be much more:

>> With the built-in domain mapping feature, you can manage multiple sites that have different, unique domain names. None of these sites even has to be a blog. The sites can have blog elements, or they can be static sites that use only pages.

>> The built-in options let you choose between subdomains and subfolder sites when you install the network. If you install WordPress in the root of your web space, you get subdomain.*yourdomain.com* (if you choose subdomains) or yourdomain.com/subfolder (if you choose subfolders).

REMEMBER

After you choose the kind of sites you want to host and create those sites, you can't change them. These sites are served virtually, meaning that they don't exist as files or folders anywhere on the server; they exist only in the database. The correct location is served to the browser by means of rewrite rules in the .htaccess file.

>> The main, or parent, site on the network can also be a landing page for the entire network of sites, showcasing content from other sites on the network and drawing in visitors further.

# Considering Web-Hosting Services

This chapter assumes that you already have the WordPress software installed and running correctly on your web server, and that your web server meets the minimum requirements to run WordPress (see Chapter 2).

Before you enable the WordPress network feature, you need to determine how you're going to use the feature. You have a couple of options:

>> Manage just a few of your own WordPress blogs or websites

>> Run a full-blown network with several hundred sites and multiple users

If you're planning to run just a few of your own sites with the WordPress network feature, your current hosting situation is probably well suited to the task. (See Chapter 3 for information on web-hosting services.) If, however, your plan is to host a large network with hundreds of blogs and multiple users, you should consider contacting your host and increasing your bandwidth, as well as the disk-space limitations on your account.

**WARNING**

Beyond the necessary security measures, time, and administrative tasks that go into running a community of sites, you have a few more things to worry about. Creating a community increases resource use, bandwidth use, and disk space on your web server. In many cases, if you go over the limits assigned to you by your web host, you'll incur great cost. Make sure that you anticipate your bandwidth and disk-space needs before running a large network on your website. (Don't say I didn't warn you!)

Many WordPress Multisite communities start with grand dreams of being large, active communities. Be realistic about how your community will operate so that you can make the right hosting choice for yourself and your community.

Small multisite communities can be easily handled via a shared-server solution, whereas larger, more active communities really need a dedicated-server solution for operation. The difference between the two lies in their names:

>> **Shared-server solution:** You have one account on one server that has several other accounts on it. Think of this solution as apartment living. One apartment building has several apartments for multiple people to live in, all under one roof.

>> **Dedicated-server solution:** You have one account. You have one server. That server is dedicated to your account, and your account is dedicated to the server. Think of this solution as owning a home and not sharing your living space with anyone else.

A dedicated-server solution is a more expensive investment for your blog community, whereas a shared-server solution is more economical. Your decision about which solution to go with for your WordPress Network blogging community should be based on your realistic estimates of how big and how active your community will be. You can move from a shared-server solution to a dedicated-server solution if your community gets larger than you expected, but it's easier to start with the right solution for your community from day one.

# Enabling the WordPress Network Feature

WordPress makes it pretty easy to enable the network feature, but doing so does require opening a file on your web server called `wp-config.php` and making a small alteration. Follow these steps to get the process started:

1. **Download the file called `wp-config.php` from the WordPress installation on your web server.**

   It's easiest to use an SFTP program to download a copy of this file from your web server to your computer. (Chapter 3 provides the information you need about using SFTP.)

2. **Using your preferred text editor, open the `wp-config.php` file.**

   Windows users can use Notepad to edit the file; Mac users can use TextEdit.

3. **Click at the end of the line that reads `define( 'DB_COLLATE', '' );` and then press Enter to create a new blank line.**

4. **Type the following on the new blank line:**

   ```
   define( 'WP_ALLOW_MULTISITE', true );
   ```

5. **Save the file to your computer as `wp-config.php`.**

6. **Upload the new file to your web server in your WordPress installation directory.**

7. **Go to your WordPress Dashboard in your browser.**

   You see a new item, labeled Network Setup, on the Tools menu.

8. **Click the Network Setup link on the Tools menu.**

   A page called Create a Network of WordPress Sites loads in your browser window, as shown in Figure 13-1.

**WARNING**

If you have any plugins installed and activated on your WordPress installation, deactivate them before you proceed with the network setup. WordPress won't allow you to continue until you deactivate all your plugins.

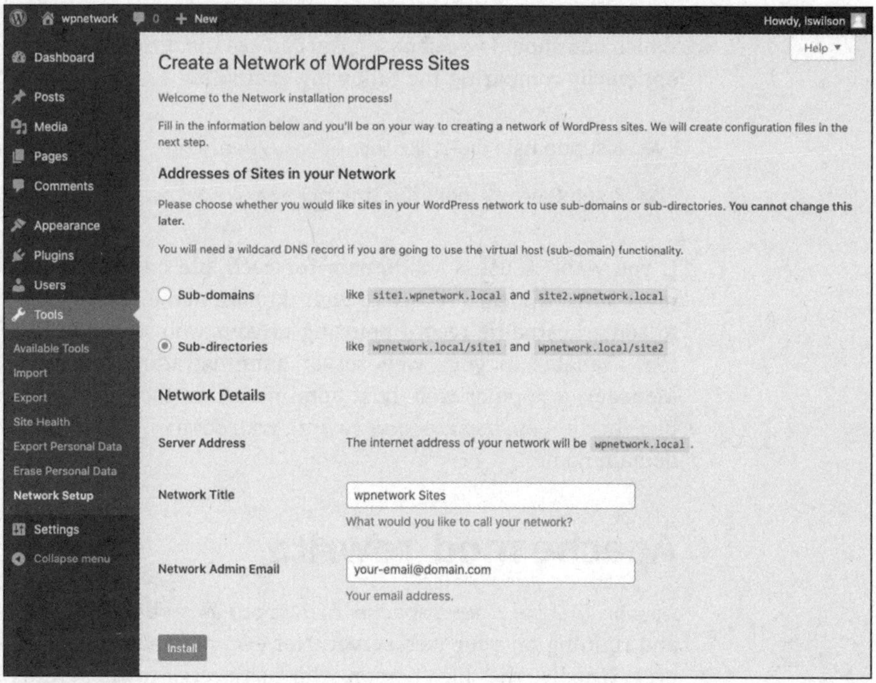

**FIGURE 13-1:**
The Create a
Network of
WordPress Sites
page in your
WordPress
Dashboard.

Before proceeding with the final steps in enabling the WordPress network feature, you need to get the items in the next section in order on your web server. You also need to decide how to handle the multiple sites within your network. These configurations need to be in place so that you can run the WordPress network successfully. If you can perform the configurations in this section yourself (and if you have access to the Apache configuration files), this section is for you. If you don't know how, are uncomfortable with adjusting these settings, or don't have access to change the configurations in your web server software, ask your hosting provider for help, or hire a consultant to perform the configurations for you.

# DNS

The WordPress Multisite feature gives you two ways to run a network of sites on your domain: the subdomain option and the subdirectory option. The most popular option (and recommended structure) sets up subdomains for the sites created within a WordPress network. With the subdomain option, the username of the site appears first, followed by your domain name. With the subdirectory option, your domain name appears first, followed by the username of the site.

Which one should you choose? You can see the difference in the URLs of these two options by comparing the following examples:

>> A subdomain looks like this: http://username.yourdomain.com

>> A subdirectory looks like this: http://yourdomain.com/username

If you want to use a subdomain for each site on your WordPress network, you must add a wildcard record to your domain name server (DNS) records. You need to add a hostname record pointing to your web server in the DNS configuration tool available in your web-server administration software (such as WebHost Manager, a popular web-host administration tool). The hostname record looks like this: *.yourdomain.com (where yourdomain.com is replaced by your actual domain name).

## Apache mod_rewrite

Apache (https://www.apache.org/free) is web-server software that's loaded and running on your web server. Not everyone has access to Apache files, however. Usually, the only person who has access to those files is the web-server administrator (which usually is your web host). Depending on your own web server account and configuration, you may not have access to the Apache software files.

The Apache module you need to create nice permalink URLs (see Chapter 5 for information on pretty permalinks) for your WordPress network is called mod_rewrite. This module must be configured so that it's active and installed on your server.

You (or your web host) can make sure that the Apache mod_rewrite module is activated on your server. To do so, open the httpd.conf file, and verify that it includes the following line:

```
LoadModule rewrite_module /libexec/mod_rewrite.so
```

If not, type that line on its own line, and save the file. You'll probably need to restart Apache to make the change take effect.

**WARNING**

Remember that the Apache mod_rewrite module is required for WordPress multisites. If you don't know whether your current hosting environment has this module in place, drop an email to your hosting provider and ask. The provider can answer that question for you (in addition to installing the module for you in the event that your server doesn't have it).

Networks also work well on Nginx and Lightspeed servers, but many users have reported having much difficulty on IIS (Windows) servers. Therefore, I don't recommend setting up WordPress with Multisite features in a Windows server environment.

Subdomain sites work by way of a virtual host entry in Apache, also known as a wildcard subdomain. On shared hosts, your web hosting provider's support team has to enable this entry for you (or may already have done so for all accounts). It's best to ask your hosting provider before you begin. In these situations, the domain you use for your install must be the default domain in your account. Otherwise, the URLs of your subsites will fail to work properly or won't have a folder name in the URL.

Some hosts may require you to have a dedicated IP address, but this requirement isn't a specific software requirement for a WordPress network to function.

Before proceeding with the final steps in enabling the WordPress Multisite feature, you need to get a few items in order on your web server. You also need to decide how the multiple sites within your network will be handled. You need to have these configurations in place to run the WordPress network successfully. The next section takes you through some of these configurations and items, including virtual host and PHP considerations.

## Virtual host

In the same httpd.conf file that I discuss in the preceding section ("Apache mod_rewrite"), you need to make some adjustments in the <VirtualHost> section of that file.

In this section, you edit and configure Apache server files. If you can perform the configurations in this section yourself (and if you have access to the Apache configuration files), this section is for you. If you don't know how, are uncomfortable with adjusting these settings, or don't have access to configurations in your web-server software, ask your hosting provider for help or hire a consultant to perform the configurations for you. I can't stress enough that you shouldn't edit the Apache server files yourself if you aren't comfortable with it or don't fully understand what you're doing. Web-hosting providers have support staff to help you with these things if you need it; take advantage!

Follow these steps to edit your httpd.conf file:

1.  **Find the <VirtualHost> section in the httpd.conf file.**

    This line of the file provides directives, or configurations, that apply to your website.

**2.** Find a line in the `<VirtualHost>` section of the `httpd.conf` that looks like this:

```
AllowOverride None
```

**3.** Replace that line with this line:

```
AllowOverride FileInfo Options
```

**4.** On a new line, type ServerAlias *.*yourdomain*.com.

Replace *yourdomain*.com with whatever your domain is. This line defines the host name for your network site and is essential for the virtual host to work correctly.

**5.** Save the `httpd.conf` file, and close it.

You also need to add a wildcard subdomain DNS record. Depending on how your domain is set up, you can do this at your registrar or your web host. If you simply pointed to your web host's nameservers, you can add more DNS records at your web host in the web server administration interface, such as WHM (Web Host Manager).

You also should add a CNAME record with a value of *. *CNAME*, which stands for *Canonical Name*, is a record stored in the DNS settings of your Apache web server that tells Apache you want to associate a new subdomain with the main-account domain. Applying the value of * tells Apache to send any subdomain requests to your main domain. From there, WordPress looks up that subdomain in the database to see whether it exists.

Networks require a great deal more server memory (RAM) than typical WordPress sites (those that don't use the Multisite feature) because multisites generally are bigger, have a lot more traffic, and use more database space and resources because multiple sites are running (as opposed to one with regular WordPress). You aren't simply adding instances of WordPress; you're also multiplying the processing and resource use of the server when you run the WordPress Multisite feature. Although smaller instances of a network run fine on most web hosts, you may find that when your network grows, you need more memory. I generally recommend that you start with a hosting account with access to at least 256MB of RAM (memory).

For each site created, nine tables are added to the single database. Each table has a prefix similar to `wp_BLOG-ID_tablename` (where `BLOG-ID` is a unique ID assigned to the site).

The only exception is the main site: Its tables remain untouched and remain the same. With WordPress multisites, all new installations leave the main blog tables untouched and number additional site tables sequentially when every new site is added to the network.

## PHP

In this section, you edit the PHP configuration on your web server. PHP needs to have the following configurations in place in the php.ini file on your web server to run WordPress Multisite on your server:

>> Set your PHP to *not* display any error messages in the visitor's browser window. (Usually, this setting is turned off by default; just double-check to be sure.)

>> Find out whether your PHP is compiled with memory-limit checks. You can find out by looking for the text memory_limit in the php.ini file. Usually, the default limit is 8MB. Increase the memory limit to at least 32MB, or even 64MB, to prevent PHP memory errors while running WordPress Multisite.

**TIP**

The default memory limit for WordPress is 40MB or 64MB for a Multisite setup. As an alternative to editing the php.ini file on your web server to increase the PHP memory limit, you can add this line to the wp-config.php file of your WordPress installation:

```
define( 'WP_MEMORY_LIMIT', '64M' );
```

The 64M portion of that line of code defines the memory limit in megabytes, and you can set it to any value that doesn't exceed 512MB.

# Installing the Network on Your Site

The Network Details section of the Create a Network of WordPress Sites page (refer to Figure 13-1 earlier in this chapter) has options that are filled in automatically. The server address, for example, is pulled from your installation and can't be edited. The network title and administrator email address are pulled from your installation database, too, because your initial WordPress site is the main site on the network.

Follow these steps to complete the installation (and be sure to have your preferred text-editor program handy):

1. **Click the Install button at the bottom of the Create a Network of WordPress Sites page of your WordPress Dashboard.**

   The Enabling the Network page opens (not shown).

2. **Add the required network-related configuration lines to the `wp-config.php` file following the `define( 'WP_ALLOW_MULTISITE', true );` you added earlier.**

   On the Enabling the Network page, WordPress gives you up to six lines of configuration rules that need to be added to the `wp-config.php` file. The lines of code you add may look like this:

   ```
   define('MULTISITE', true);
   define('SUBDOMAIN_INSTALL', false);
   define('DOMAIN_CURRENT_SITE', domain.com);
   define('PATH_CURRENT_SITE', '/');
   define('SITE_ID_CURRENT_SITE', 1);
   define('BLOG_ID_CURRENT_SITE', 1);
   ```

   These lines of code provide configuration settings for WordPress by telling WordPress whether it's using subdomains, what the base URL of your website is, and what your site's current path is. This code also assigns a unique ID of 1 to your website for the main installation site of your WordPress Multisite network.

**WARNING**

   The lines of code that appear on the Enabling the Network screen are unique to *your* installation of WordPress. Make sure that you copy the lines of code that are given to you on the Create a Network of WordPress Sites page on *your* installation because they are specific to your site's setup.

**TIP**

   My WordPress installation sets up my network to use subdomains instead of subdirectories. If you'd like to use subdirectories, change `define( 'SUBDOMAIN_INSTALL', true );` to `define( 'SUBDOMAIN_INSTALL', false );`. Make sure that you have the `<VirtualHost>` and Apache `mod_rewrite` configurations on your server in place first; I cover both configurations earlier in this chapter.

3. **Add the necessary rewrite rules to the `.htaccess` file on your web server.**

   WordPress gives you several lines of code that you need to add to a file called `.htaccess` on your web server. (You find that file in the WordPress installation directory.) These lines look something like this:

```
RewriteEngine On
RewriteBase /
RewriteRule ^index\.php$ - [L]

# add a trailing slash to /wp-admin
RewriteRule ^wp-admin$ wp-admin/ [R=301,L]

RewriteCond %{REQUEST_FILENAME} -f [OR]
RewriteCond %{REQUEST_FILENAME} -d
RewriteRule ^ - [L]
RewriteRule ^(wp-(content|admin|includes).*) $1 [L]
RewriteRule ^(.*\.php)$ $1 [L]
RewriteRule . index.php [L]
```

**REMEMBER**

Earlier in this chapter, I discuss the required Apache module mod_rewrite, which you must have installed on your web server to run WordPress Multisite. The rules that you add to the .htaccess file on your web server are mod_rewrite rules, and they need to be in place so that your web server tells WordPress how to handle things like permalinks for blog posts, pages, media, and other uploaded files. If these rules aren't in place, the WordPress multisite feature won't work correctly.

4. **Copy the lines of code that you entered in Step 3, open the** .htaccess **file, and paste the code there.**

   Replace the rules that already exist in that file.

5. **Save the** .htaccess **file, and upload it to your web server again.**

6. **Return to your WordPress Dashboard, and click the Log In link at the bottom of the Enabling the Network screen.**

   You're logged out of WordPress because by following these steps, you changed some of the browser cookie-handling rules in the wp-config.php and .htaccess files.

7. **Log in to WordPress by entering your username and password in the login form.**

# Exploring the Network Admin Dashboard Menu

When the Multisite feature enabled, you see the link to My Sites. If you hover your mouse pointer over that link, the Network Admin link appears in the drop-down menu in the top-left section of the Dashboard, as shown in Figure 13-2.

My Sites

**FIGURE 13-2:**
Link to the
Network Admin
Dashboard.

WordPress has separated the Network Admin menu features from the regular (Site Admin) Dashboard menu features to make it easier for you to know which part of your site you're managing. If you're performing actions that maintain your main website — publishing posts or pages, creating or editing categories, and so on — you work in the regular Dashboard (Site Admin). If you're managing any of the network sites, plugins, and themes for the network sites or registered users, you work in the Network Admin section of the Dashboard.

**REMEMBER**

Keep in mind the distinct differences between the Site Admin and Network Admin Dashboards, as well as their menu features. WordPress does its best to know which features you're attempting to work with, but if you find yourself getting lost on the Dashboard, or if you're not finding a menu or feature that you're used to seeing, make sure that you're working in the correct section of the Dashboard.

The Network Admin Dashboard (see Figure 13-3) is similar to the regular WordPress Dashboard, but as you may notice, the modules pertain to the network of sites. Options include creating a site, creating a user, and searching existing sites and users. Obviously, you won't perform this search if you don't have any users or sites yet. This function is extremely useful when you have a community of users and sites within your network, however.

FIGURE 13-3:
The Network
Admin
Dashboard.

**TIP**

The Network Admin Dashboard is configurable, just as the regular Dashboard is, and you can move the modules around and edit the settings of the modules. See Chapter 4 for more information about arranging the Dashboard modules to suit your taste.

The Search Users feature allows you to search usernames and users' email addresses. If you search for the user *Lisa*, for example, your results include any user whose username or email address contains *Lisa*, so you can receive multiple returns from just one search word or phrase. The Search Sites feature returns any sites within your network that match the terms you've searched for.

The Network Admin Dashboard has two useful links near the top of the screen:

>> **Create a New Site:** Click this link to create a new site within your network. When you click the link, the Add New Site screen appears. Find out how to add a new site in the upcoming "Sites" section.

>> **Create a New User:** Click this link to create a new user account within your community. When you click this link, the Add New User screen appears. Find out how to add a new user to your community in the "Users" section, later in the chapter.

Additionally, the Network Admin Dashboard gives you a real-time count of how many sites and users you have in your network, which is nice-to-know information for any network admin.

# Managing Your Network

As I mention earlier in this chapter, the Network Admin Dashboard has its own set of menus, separate from those of the Site Admin Dashboard. Those menus are located on the left side of the Network Admin Dashboard. This section goes through the menu items, providing explanations; instructions on working with the settings; and configurations to help you manage your network, sites, and users.

These menus are available on the Network Admin Dashboard:

>> **Sites:** View a list of the sites in your network, along with details about them.

>> **Users:** See detailed info about current users in your network.

>> **Themes:** View all the currently available themes to enable or disable them for use on your network.

>> **Plugins:** Manage (activate or deactivate) themes for use on all sites within your network.

>> **Settings:** Configure global settings for your network.

All the items on the Network Admin Dashboard are important, and you'll use them frequently throughout the life of your network. Normally, I'd take you through the menu items in order so that you can follow along on your Dashboard, but it's important to perform some preliminary configurations on your network. Therefore, I start the following sections with the Settings menu; then I take you through the other menu items in the order of their appearance on the Network Admin Dashboard.

## Settings

Click the Settings menu link on the Network Admin Dashboard. The Network Settings screen appears (see Figure 13-4), displaying several sections of options for you to configure to set up your network the way you want.

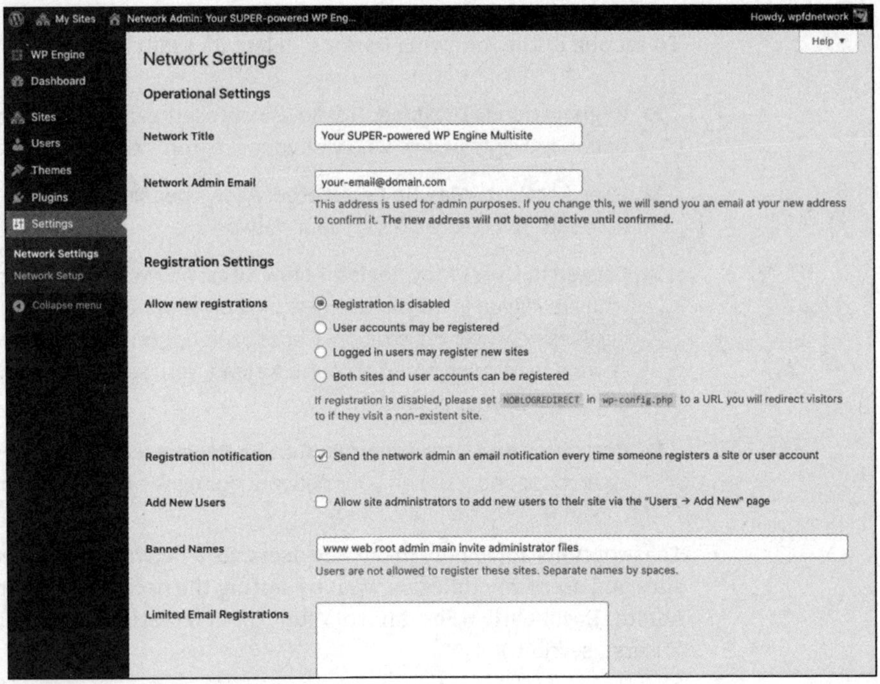

**FIGURE 13-4:**
The Network
Settings screen.

## Operational Settings

The Operational Settings section has these settings:

>> **Network Title:** This setting is the title of your overall network of sites. This
name is included in all communications regarding your network, including
emails that new users receive when they register a new site within your
network. Type your desired network title in the text box.

>> **Network Admin Email:** This setting is the email address that all correspon-
dence from your website is addressed from, including all registration and
signup emails that new users receive when they register a new site and/or
user account within your network. In the text box, type the email that you
want to use for these purposes.

## Registration Settings

The Registration Settings section allows you to control several aspects of allowing
users to sign up on your network. The most important option is whether to allow
open registration.

To set one of the following options, select its radio button:

>> **Registration Is Disabled:** Disallows new user registration. When selected, this option prevents people who visit your site from registering for a user account.

>> **User Accounts May Be Registered:** Allows people to create only user accounts, not create sites on your network.

>> **Logged In Users May Register New Sites:** Allows only existing users (those who are logged in) to create a new blog on your network. This setting also disables new user registration. Choose this option if you don't want just anyone registering for an account. Instead, you, as the site administrator, can add users at your discretion.

>> **Both Sites and User Accounts Can Be Registered:** Allows users to register an account and a site on your network during the registration process.

These options apply only to outside users. As a network admin, you can create new sites and users any time you want by setting the necessary options on the Network Admin Dashboard. (For information about creating users, see the upcoming "Users" section.)

The remaining options in the Registration Settings section are as follows:

>> **Registration Notification:** When this option is selected, an email is sent to the network administrator every time a user or a site is created on the system, even if the network administrator is the person who created the site.

>> **Add New Users:** When this option is selected, your community blog owners (individual site administrators) can add new users to their own community sites via the Users page of their individual Dashboards.

>> **Banned Names:** By default, WordPress bans several usernames from being registered within your community, including *www, web, root, admin, main, invite,* and *administrator*. This ban exists for good reason: You don't want a random user to register a username such as *admin*, because you don't want that person misrepresenting himself as an administrator of your site. You can enter an unlimited amount of usernames in the Banned Names text box.

>> **Limited Email Registrations:** You can limit signups based on email domains by filling in this text box, entering one email domain per line. If you have open registrations but limited email addresses, only the people who have email domains that are in the list can register. This option is an excellent one to use in a school or corporate environment in which you're providing email addresses and sites to students or employees.

>> **Banned Email Domains:** This feature, which is the reverse of Limited Email Registration, blocks all signups from a particular domain and can be useful in stopping spammers. You can enter **gmail.com** in the text box, for example, to ban anyone who tries to sign up with a Gmail address.

## New Site Settings

The New Site Settings section (see Figure 13-5) is a configurable list of items that WordPress populates with default values when a new site is created. These values include the ones that appear in welcome emails, on a user's first post page, and on a new site's first page.

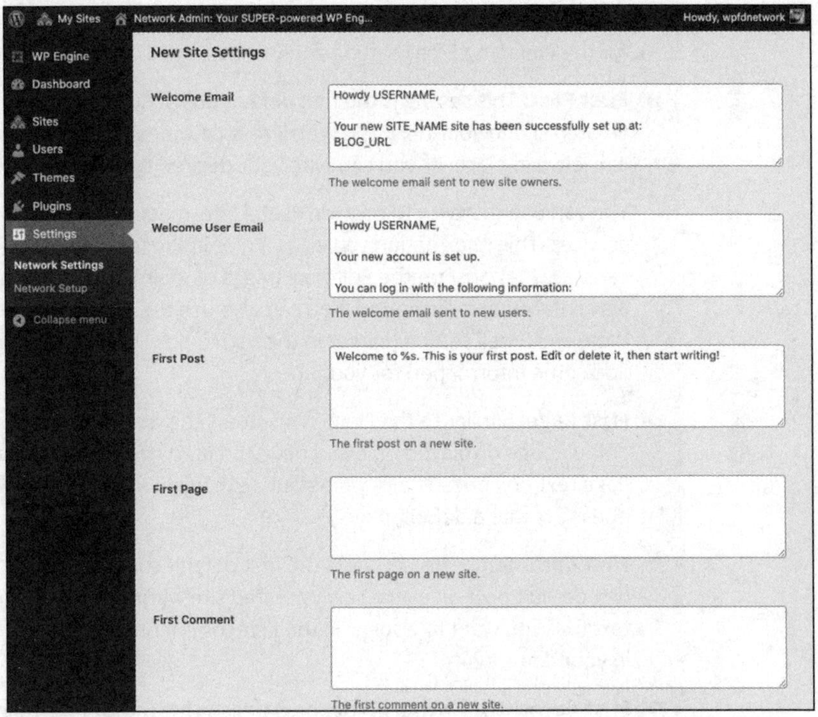

**FIGURE 13-5:**
New Site Settings section of the Network Settings page.

The configurable list of items includes

>> **Welcome Email:** This setting is the text of the email that owners of newly registered sites on your network receive when their registration is complete. You can leave the default message in place, if you like, or you can type the text of the email you want new site owners to receive when they register a new site within your network.

A few variables you can use in this email aren't explained entirely on the Network Settings screen, including

- `SITE_NAME`: Inserts the name of your WordPress site

- `BLOG_URL`: Inserts the URL of the new member's blog

- `USERNAME`: Inserts the new member's username

- `PASSWORD`: Inserts the new member's password

- `BLOG_URLwp-login.php`: Inserts the hyperlinked login URL of the new member's blog

- `SITE_URL`: Inserts the hyperlinked URL of your WordPress site

>> **Welcome User Email:** This setting is the text of the email that newly registered users receive when they complete the registration process. The variables used for the Welcome Email setting apply to this email configuration as well.

>> **First Post:** This setting is the first default post displayed on every newly created site in your network. WordPress provides some default text that you can leave in place, or you can type your desired text in the text box.

You can use this area to provide useful information about your site and services. This information serves as a nice guide for new users, because they can view that post on the Edit Post pages of their Dashboards and see how it was entered and formatted. You can also use the variables described for the Welcome Email setting (earlier in this list) to have WordPress automatically add some information for you.

>> **First Page:** Similar to the First Post setting, this setting is the default text for a default page displayed on every newly created site on your network. (The First Page text box doesn't include default text; if you leave it blank, WordPress doesn't create a default page.)

>> **First Comment:** This setting is the first default comment displayed for the first default post on every newly created site within your network. Type the text that you want to appear in the first comment on every site that's created in your community.

>> **First Comment Author:** Type the name of the author of the First Comment on new sites in your network.

>> **First Comment Email:** Type the email of the author of the first comment on new sites in your network.

>> **First Comment URL:** Type the web address (URL) of the author of the First Comment. WordPress hyperlinks the First Comment Author's name to the URL you type.

# Upload Settings

The Upload Settings section (see Figure 13-6) defines global values pertaining to the type of files you allow site owners within your network to upload by using the file upload feature of the WordPress Add Media window (see Chapter 6).

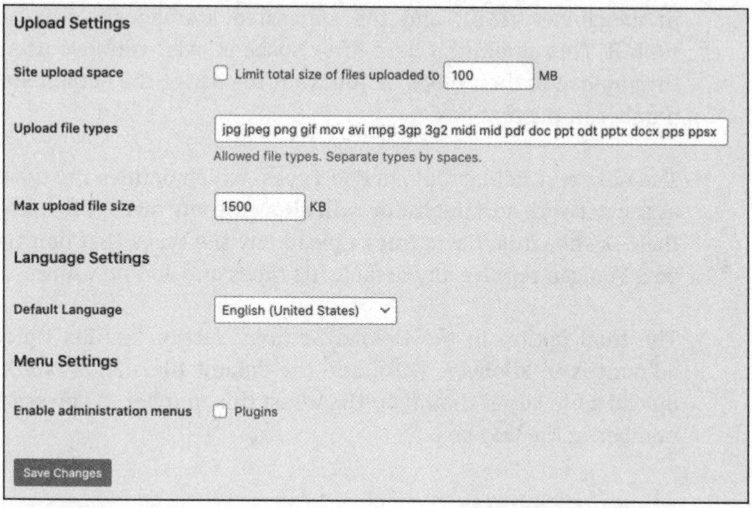

**FIGURE 13-6:**
Upload Settings
section of
the Network
Settings page.

The types of files that site owners can upload include images, videos, documents, and music. The fields in the Upload Settings section have default settings already filled in:

» **Site Upload Space:** If you leave this check box deselected, users are allowed to use all the space they want for uploads; they have no limits. Select the check box to limit the available space per site and then fill in the amount in megabytes (MB); the suggested, default storage space is 100MB. This amount of hard drive space is what you give users within your network for the storage of files they upload to their sites. If you want to change the default storage space, type a number in the text box.

» **Upload File Types:** This text field defines the types of files that you, as the network admin, allow site owners to upload to their sites on their Dashboards. Users can't upload any file types that don't appear in this field. By default, WordPress includes the following file types: `.jpg`, `.jpeg`, `.png`, `.gif`, `.mov`, `.avi`, `.mpg`, `.3gp`, `.3g2`, `.midi`, `.mid`, `.pdf`, `.doc`, `.ppt`, `.odt`, `.pptx`, `.docx`, `.pps`, `.ppsx`, `.xls`, `.xlsx`, `.key`, `.mp3`, `.ogg`, `.flac`, `.m4a`, `.wav`, `.mp4`, `.m4v`, `.webm`, `.ogv`, and `.flv`. You can remove any default file types and add new ones.

>> **Max Upload File Size:** This amount is in kilobytes (KB), and the default file size is 1500KB. This setting means that a user can't upload a file larger than 1500KB. Adjust this number as you see fit by typing a new number in the text box.

The first option in the Upload Settings section is Site Upload Space. The amount is in megabytes (MB), and the suggested storage space is set, by default, to 100MB. This amount of hard drive space is what you give users to store the files they upload to their blogs. If you want to change the default storage space, type a number in the text box.

The next text field is Upload File Types, which defines the types of files that you, as the network administrator, will allow the site owners to upload to their sites on their Dashboards. Users can't upload any file types that don't appear in this text box. You can remove any default file types and add new ones.

The final option in the Upload Settings section is Max Upload File Size. This amount is in kilobytes (KB), and the default file size is 1500KB, so a user can't upload a file larger than 1500KB. Adjust this number as you see fit by typing a new number in the text box.

## Menu Settings

The Plugins administration menu is disabled on the Dashboards of all network site owners, but the network administrator always has access to the Plugins menu. If you leave this check box unselected (refer to Figure 13-6), the Plugins page is visible to users on their own site's Dashboard. Select the box to enable the Plugins administration menu for your network users. For more information about using plugins with WordPress, see Chapter 7.

REMEMBER

When you finish configuring the settings on the Network Settings screen, don't forget to click the Save Changes button at the bottom of the page, below the Menu Settings section. If you navigate away from the Network Settings page without clicking the Save Changes button, none of your configurations will be saved, and you'll need to go through the entire process again.

## Sites

Clicking the Sites menu item on the Network Admin Dashboard takes you to the Sites screen, where you can manage your individual sites. Although each site on the network has its own Dashboard for basic tasks, such as posting and changing themes, the Sites page is where you create and delete sites, as well as edit the properties of the sites on your network. Editing information from this page is handy when you have problems accessing a site's Dashboard.

The Sites screen also lists all the sites within your network and shows the following statistics for each site:

>> **URL:** The site's path in your network. In Figure 13-7, you see a site listed with the path newsite. This path means that the site's domain is newsite.*yourdomain*.com (if you're using a subdomain setup) or yoursite.com/newsite (if you're using a subdirectory setup). I discuss subdomains and subdirectories in "DNS" earlier in this chapter.

>> **Last Updated:** The date when the site was last updated (or published to).

>> **Registered:** The date when the site was registered in your network.

>> **Users:** The username and email address associated with the user(s) of that site.

>> **ID:** The unique ID number assigned to the site. This ID number corresponds to the database tables where the data for this site are stored.

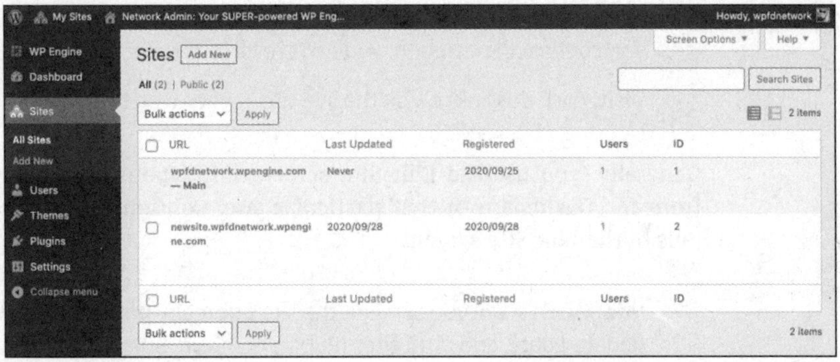

**FIGURE 13-7:**
Site-management options on the Sites screen.

When you hover your mouse pointer over the path name of a site on your network, you see a handy list of links that helps you manage the site. (These links also appear as tabs on the Edit Site screen, shown in Figure 13-8.) The options that appear below a site listing when you hover on the site name include

>> **Edit:** Click this link to go to the Edit Site screen (see Figure 13-8), where you can change aspects of each site.

>> **Dashboard:** Click this link to go to the Dashboard of the site.

>> **Deactivate:** Click this link to mark the site for deletion from your network. A message appears in a pop-up window, asking you to confirm your intention to deactivate the site. Click the Yes button to confirm. The user's site displays a message stating that the site has been deleted.

You can reverse this action by revisiting the Sites screen and clicking the Activate link that appears below the site pathname. (The Activate link appears only for sites that are marked as Deactivated.)

>> **Archive:** Click this link to archive the site on your network and prevent visitors from viewing it. The user's site displays a message stating This site has been archived or suspended.

You can reverse this action by revisiting the Sites screen and clicking the Unarchive link that appears below the site's pathname. (The Unarchive link appears only for sites that are marked as Archived.)

>> **Spam:** Click this link to mark the site as spam and block users from accessing the Dashboard. WordPress displays the message This site has been archived or suspended.

You can reverse this action by revisiting the Sites screen and clicking the Not Spam link that appears below the site's pathname. (The Not Spam link appears only for sites that are marked as Spam.)

>> **Delete:** Click this link to delete the site from your network of sites. Although a confirmation screen asks you to confirm your intention to delete the site, after you confirm the deletion, you can't reverse this decision.

>> **Visit:** Click this link to visit the live site in your web browser.

Generally, you use the Edit Site screen only when the settings are unavailable from the Dashboard of that particular site. Configure these options on the four tabs of the Edit Site screen:

>> **Info:** On this tab, you can edit the site's domain, path, registered date, updated date, and attributes (Public, Archived, Spam, Deleted, or Mature).

>> **Users:** On this tab, you can manage the users who are assigned to the site, as well as add users to the site.

>> **Themes:** On this tab, you can enable themes for the site. This capability is particularly useful if you have themes that aren't network-enabled (see the "Themes" section later in this chapter). All the themes that aren't enabled within your network are listed on the Themes tab, which allows you to enable themes on a per-site basis.

>> **Settings:** The settings on this tab cover all the database settings for the site that you're editing. You rarely, if ever, need to edit these settings because as the network administrator, you have access to each user's Dashboard and should be able to make any changes in the site's configuration settings there.

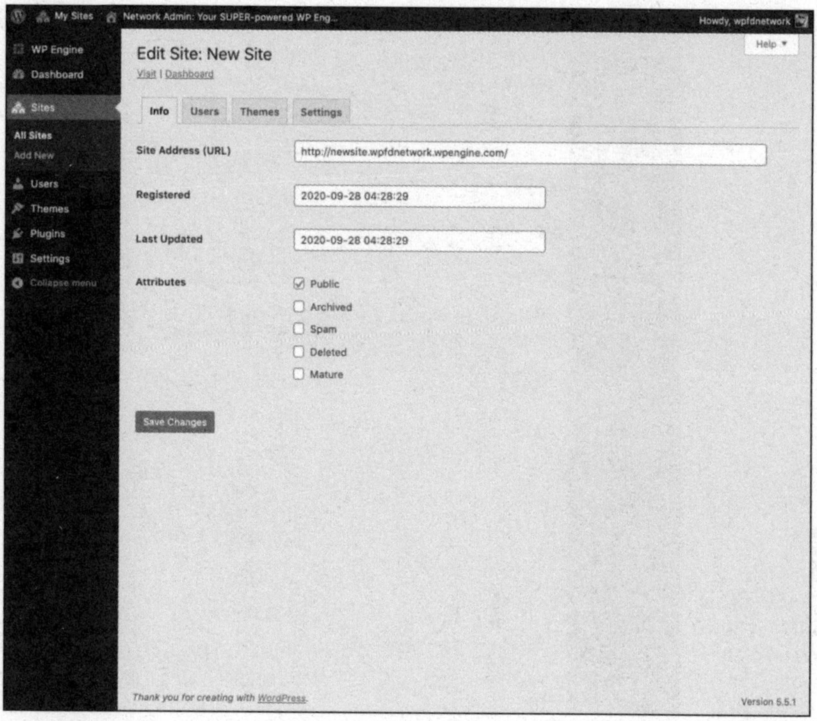

**FIGURE 13-8:**
The Edit Site
screen.

The Sites menu also includes a link called Add New. Click that link to load the Add New Site screen (see Figure 13-9) in your Network Admin Dashboard. Fill in the Site Address (URL), Site Title, Site Language, and Admin Email fields and then click the Add Site button to add the site to your network. If the Admin Email you entered is associated with an existing user, the new site is assigned to that user on your network. If the user doesn't exist, WordPress creates a new user and sends a notification email. The site is immediately accessible. The user receives an email containing a link to his site, a login link, and his username and password.

## Users

Clicking the Users link on the Network Admin Dashboard takes you to the Users screen (see Figure 13-10), where you see a full list of members, or users, within your network.

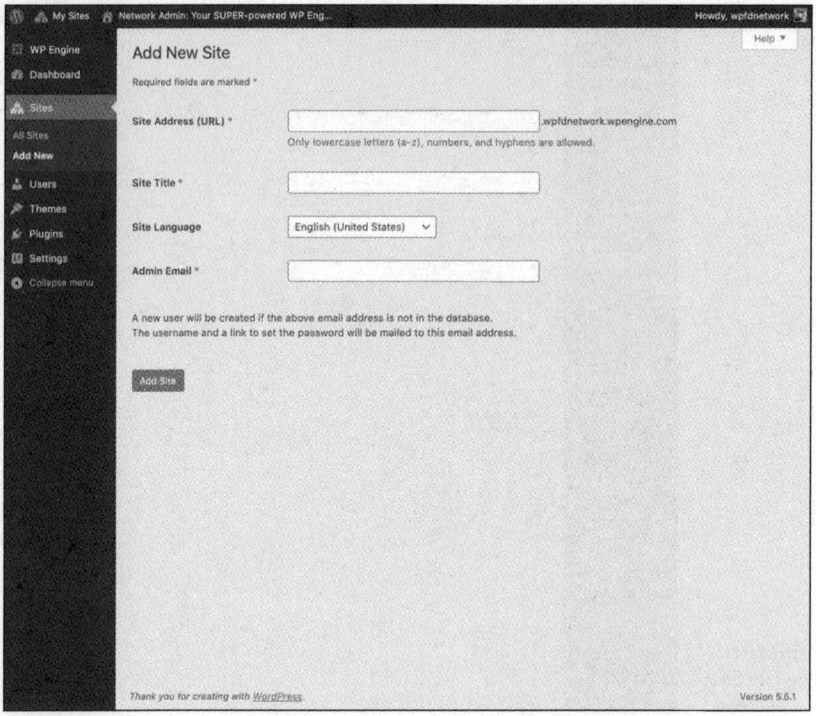

**FIGURE 13-9:**
The Add New Site screen of the Network Admin Dashboard.

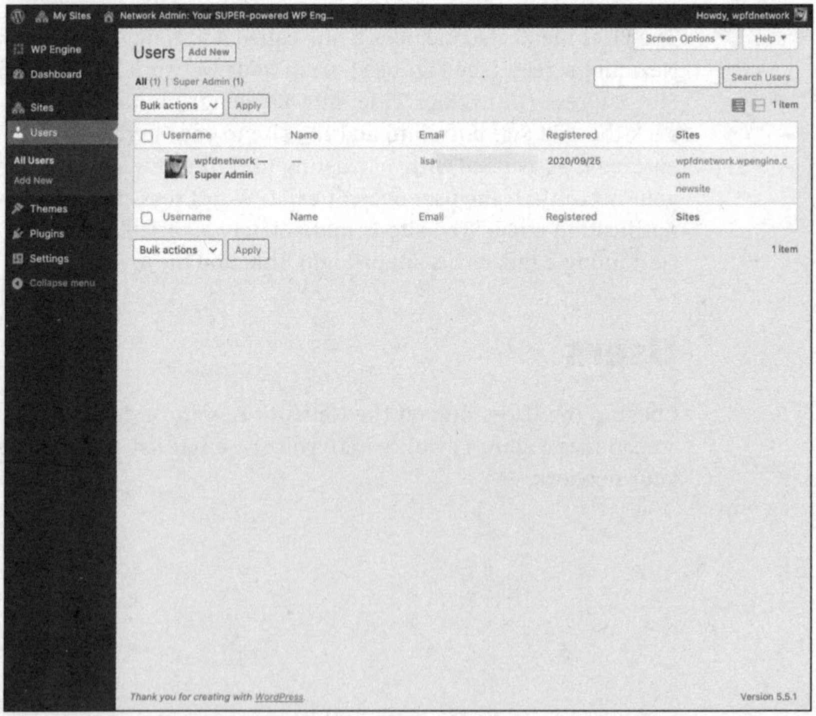

**FIGURE 13-10:**
The Users screen.

The Users screen lists the following information about each user:

>> **Username:** This setting is the login name the member uses when they log in to their account in your community.

>> **Name:** This setting is the user's real name, taken from their profile. If the user hasn't provided their name in their profile, this column is blank.

>> **Email:** This setting is the email address the user entered when they registered on your site.

>> **Registered:** This setting is the date when the user registered.

>> **Sites:** If you enable sites within your WordPress Network, this setting lists any sites of which the user is a member.

You can add and delete users to the network, as well as manage users, by clicking the Edit or Delete link that appears below their names when you hover over them with your mouse pointer.

To delete a user, simply hover over the username in the list that appears on the Users screen. Click the Delete link. A new screen appears, telling you to transfer this user's posts and links to another user account (yours, most likely). Then click the Confirm Deletion button. WordPress removes the user from the network.

WARNING

This action is irreversible, so be certain about your decision before you click that button!

You can also edit a user's profile information by clicking the Edit link that appears below their name when you hover your mouse pointer over it on the Users screen. Clicking that link takes you to the Profile screen (see Figure 13-11).

Also, on the Users menu of the Network Admin Dashboard is a link called Add New. Click that link to load the Add New User screen (see Figure 13-12).

You can add a new user by filling in the Username and Email fields and then clicking the Add User button. WordPress sends the new user an email notification of the new account, along with the site URL, their username, and their password (randomly generated by WordPress at the time the user account is created).

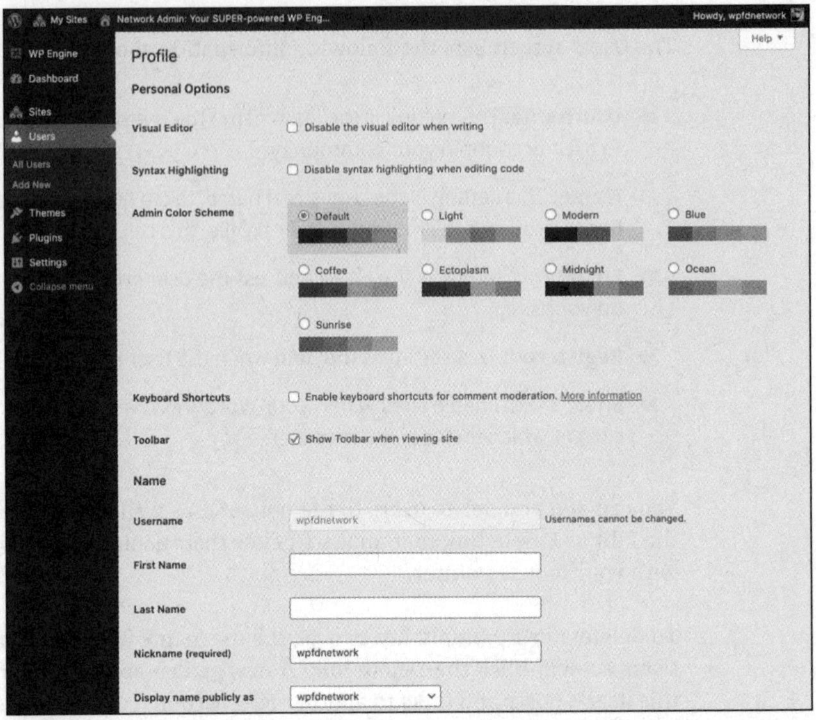

**FIGURE 13-11:**
The Profile
screen.

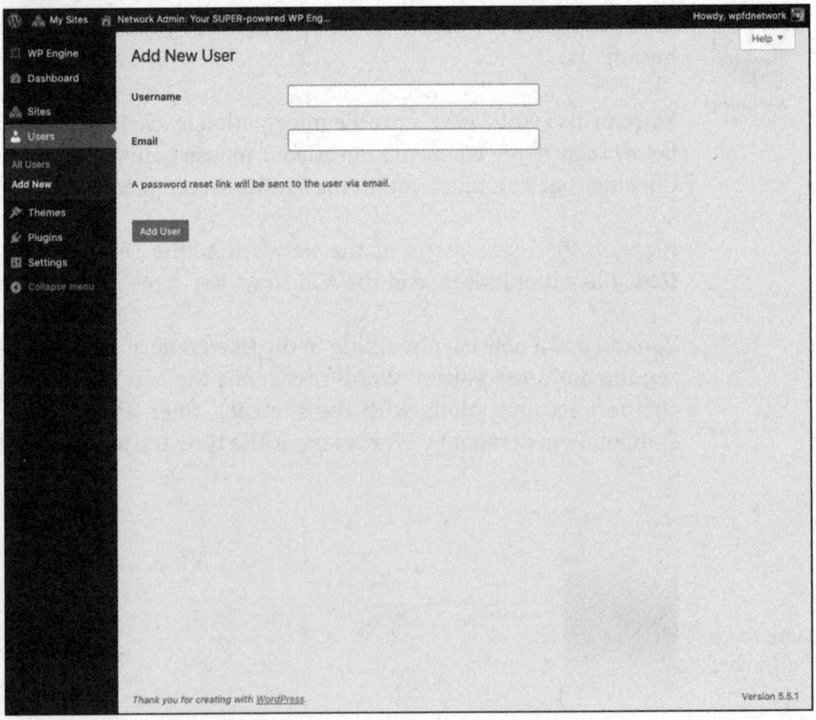

**FIGURE 13-12:**
The Add New
User screen of
the Network
Admin
Dashboard.

## SUPER ADMIN VERSUS NETWORK ADMIN

At this writing, the terms *super admin* and *network admin* are interchangeable. When WordPress first merged the WordPress MU codebase with the regular WordPress software, the term used was *super admin*. Right now, *network admin* is the standard term, but *super admin* is still used in some areas of the Network Admin and regular Dashboards. That situation may change in the very near future, when the folks at WordPress realize the discrepancy and make the updates in later versions of the software.

## Themes

When a network is enabled, only users with Network Admin access have permission to install themes, which are shared across the network. You can see how to find, install, and activate new themes in your WordPress installation in Chapter 8. After you install a theme, you must enable it on your network to have the theme appear on the Appearance menu of each site. To access the Themes screen (shown in Figure 13-13), click the Themes link on the Themes menu of the Network Admin Dashboard.

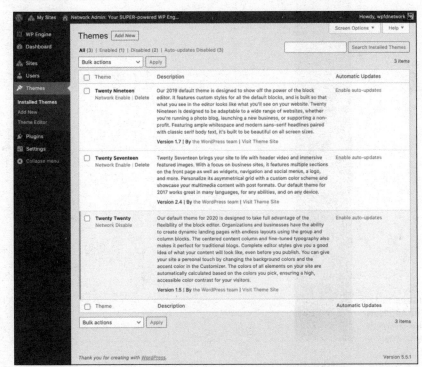

**FIGURE 13-13:** The Themes screen.

# Plugins

By and large, all WordPress plugins work on your network. Some special plugins exist, however, and using plugins on a network involves some special considerations.

TIP

For details on finding, installing, and activating plugins in WordPress, see Chapter 7.

Browse to the Plugins screen of your WordPress Network Admin Dashboard by clicking the Plugins link. The Plugins screen is almost the same as the one shown in Chapter 7, but if you don't know where to look, you can easily miss one very small, subtle difference. Check out Figure 13-14, and look below the name of the plugin. Do you see the Network Activate link? That link is the big difference between plugins listed on the regular Dashboard and those listed on the Network Admin Dashboard. As the network administrator, you can enable certain plugins to be activated across your network. All sites on your network will have the network-activated plugin features available, in contrast to plugins that you activate on the regular Dashboard (Site Admin), which are activated and available only for *your* main website.

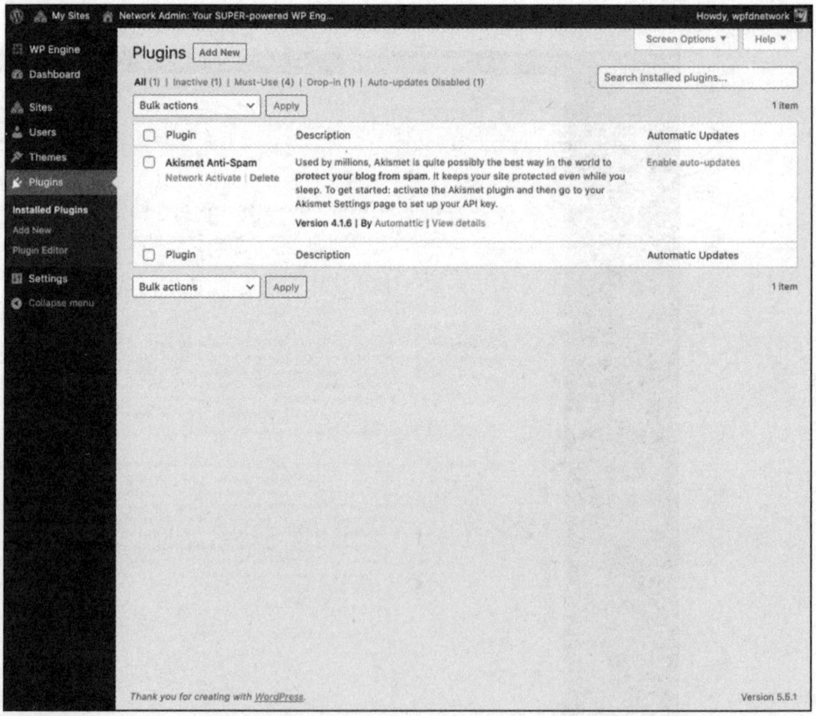

**FIGURE 13-14:**
The network Plugins screen.

**TIP**

If you select the Plugins administration menu (see "Menu Settings" earlier in this chapter) on the Network Settings page, users see the plugins listed on their Plugins pages of their Dashboard. In their list of plugins, they see only the plugins that you haven't network-activated — that is, all the plugins you installed in your WordPress installation but not activated on that user's site. Users can activate and deactivate those plugins as they desire.

Only network administrators have access to install new plugins on the site; regular users within the network don't have that kind of access (unless you've made them network administrators in their User settings).

Also, located on the Plugins menu of the Network Admin Dashboard are two other links: Add New and Editor. The Add New link lets you add and install new plugins by searching WordPress plugins within your Dashboard, and the Editor link gives you access to the Plugin Editor. I cover these topics in detail in Chapter 7.

# Stopping Spam Signups and Splogs

If you choose to have open signups, allowing any member of the public to register and create a new site on your network, at some point, automated bots run by malicious users and spammers will visit your network signup page and attempt to create sites on your network. They do so by automated means, hoping to create links to their sites or fill their site on your network with spam posts. This kind of spam blog or site is a *splog*.

Spam bloggers don't hack your system to take advantage of this feature; they call aspects of the signup page directly. You can do a few simple things to slow them or stop them.

In the "Registration Settings" section earlier in this chapter, I go over a few options, including areas in which you can specify email addresses to allow or block. The Add New Users check box (refer to Figure 13-4) stops many spammers when it's deselected. When spammers access the system to set up a spam site, they often use the Add New Users feature to programmatically (through the use of programs built into the bots) create many other sites.

Spammers often find your site via Google, and that's where they find the link to the signup page. You can stop Google and other search engines from crawling your signup page by adding `rel=nofollow,noindex` to the signup page link. To do so, wherever you add a link to your signup page, inviting new users to sign up, the HTML code you use to add the `nofollow,noindex` looks like this:

```
<a href="http://yoursite.com/wp-signup.php" rel="nofollow,noindex">Get your own
  site here</a>
```

You can add this code to any page or widget as a normal link to instruct legitimate visitors to sign up for a site on your network.

IN THIS CHAPTER

» Finding the Dashboard upgrade notification

» Backing up your database before upgrading

» Upgrading WordPress automatically and manually

» Moving to WordPress from a different platform

» Transferring your website from one host to another

# Chapter **14**

# Upgrading, Backing Up, and Migrating

At some point, you may need to move your site to a different home on the web, either to a new web host or to a different account on your current hosting account. Or, you may be reading this book because you're moving your blog from a different platform to WordPress. You'll find that during your time as a WordPress user, upgrading the WordPress software is important, as is maintaining backups of your site so that you don't lose precious months or years of content.

In this chapter, you discover the WordPress upgrade notification system and find out what to do when WordPress notifies you that a new version of the software is available. This chapter covers the best practices for upgrading the WordPress platform on your site to ensure the best possible outcome (that is, not break your website after a WordPress upgrade).

This chapter also covers how to migrate a blog that exists within a different blogging platform (such as Movable Type or TypePad) to WordPress. Finally, it takes you through how to back up your WordPress files, data, and content and then move it to a new hosting provider or different domain.

# Getting Notified of an Available Upgrade

After you install WordPress and log in for the first time, you can see the version number on the WordPress Dashboard. Therefore, if anyone asks what version you're using, you know exactly where to look to find out.

Suppose that you have WordPress installed, and you've been happily publishing content to your website with it for several weeks or maybe even months. Then one day, you log in and see a message at the top of your Dashboard screen that you've never seen before: WordPress 5.5.1 is available! Please update now. You can see such a message displayed in Figure 14-1.

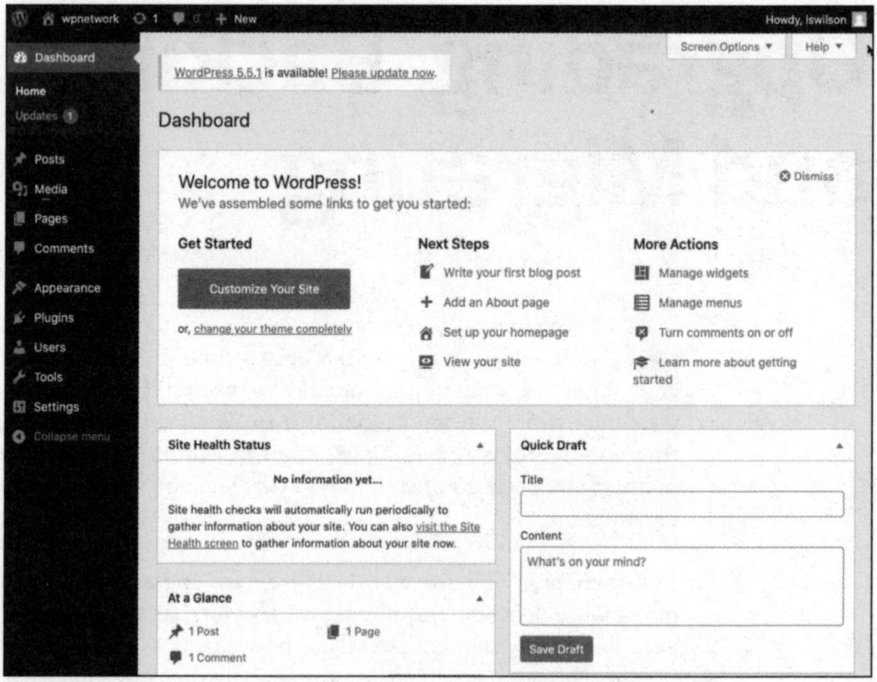

FIGURE 14-1:
Notification
of an available
WordPress
upgrade.

Both the message at the top of the screen and the notification bubble on the Dashboard menu are visual indicators that you're using an outdated version of WordPress and that you can (and need to) upgrade the software.

The message at the top of your Dashboard has two links that you can click for more information. The first, in Figure 14-1, is a link titled `WordPress 5.5.1`. Clicking it takes you to the WordPress Codex page titled Version 5.5.1, which is filled with information about the version upgrade, including

>> Installation/upgrade information

>> Summary of the development cycle for this version

>> List of revised files

The second link, Please Update Now, takes you to the WordPress Updates screen of the WordPress Dashboard, shown in Figure 14-2.

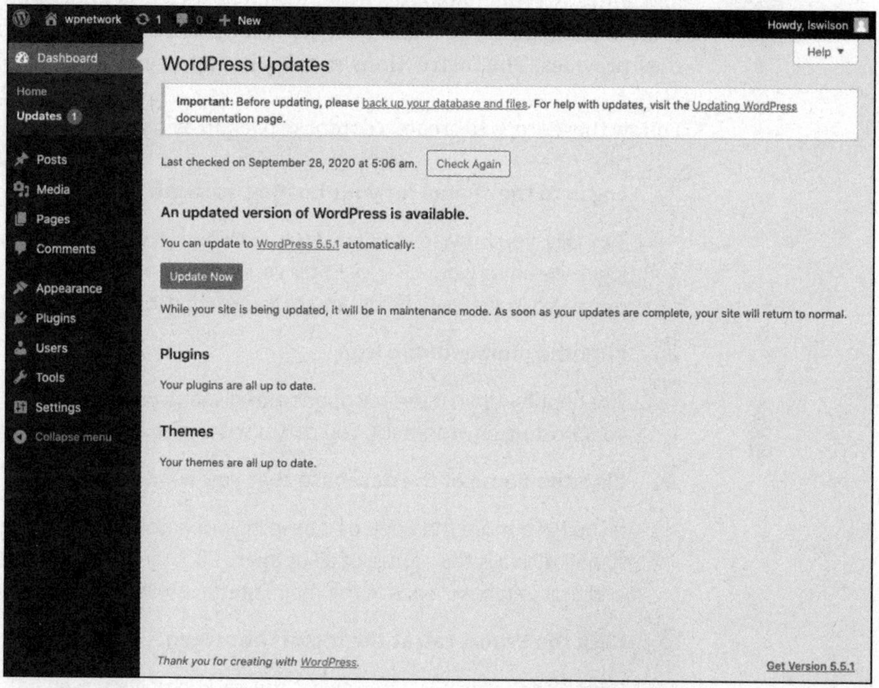

FIGURE 14-2:
Get the latest
version of
WordPress
through the
WordPress
Updates screen.

At the top of the WordPress Updates screen is another important message for you (see Figure 14-2): `Important: Before updating, please back up your database and files` (`https://wordpress.org/support/article/wordpress-backups`). `For help with updates, visit the Updating WordPress documentation page` (`https://wordpress.org/support/article/updating-wordpress`). The WordPress Updates screen tells you that an updated version of WordPress is available.

# Backing Up Your Database

Before upgrading your WordPress software installation, make sure that you back up your database. This step isn't required, of course, but it's a smart step to take to safeguard your website and make absolutely sure that — should the upgrade go wrong for some reason — you have a complete copy of all your website data that can be restored if necessary.

The best way to back up your database is to use the MySQL administration interface provided by your web-hosting provider.

**TIP**

cPanel is a web-hosting interface provided by many web hosts as an account management tool, containing phpMyAdmin as the preferred tool for managing and administering databases. Not all web hosts use cPanel or phpMyAdmin, however; if yours doesn't, consult the user documentation for the tools that your web host provides. The instructions in this chapter use cPanel and phpMyAdmin.

Follow these steps to create a database backup by using the phpMyAdmin interface:

1.  **Log in to the cPanel for your hosting account.**

    Typically, you browse to `http://yourdomain.com/cpanel` to bring up the login screen for your cPanel. Enter your hosting account username and password in the login fields, and click OK to log in.

2.  **Click the phpMyAdmin icon.**

    The phpMyAdmin interface opens and displays your database. (Depending on your hosting environment, you may also have to log in to phpMyAdmin.)

3.  **Click the name of the database that you want to back up.**

    If you have more than one database in your account, the left menu in phpMy-Admin displays the names of all of them. Click the one you want to back up, and that database loads in the main interface window at the side of the screen.

4.  **Click the Export tab at the top of the screen.**

    The screen refreshes and displays the backup utility screen.

5.  **Select the Save As File check box.**

6.  **Select the "zipped" option.**

    This option compiles the database backup file in a `.zip` file and prepares it for download.

7. **Click the Go button.**

   A pop-up window appears, allowing you to select a location on your computer to store the database backup file.

8. **Click the Save button to download the backup file and save it to your computer.**

With your WordPress website data safely backed up, you can proceed to upgrading the WordPress software. If anything happens to go wrong, you have a full copy of the content from your website, which can be restored at a future date if necessary.

# Upgrading WordPress Automatically

WordPress provides you an easy, quick, and reliable method to update the core software from within your Dashboard. I recommend using this option whenever possible to make sure that you're accurately updating the WordPress software.

To update WordPress automatically, follow these steps:

**WARNING**

1. **Back up your WordPress website.**

   Do not skip this step!

   Backing up your website includes taking a backup of your database (covered in the previous section, "Backing Up Your Database") and downloading the crucial elements via SFTP (see Chapter 3). The crucial elements include the entire / wp–content folder, which contains your plugins, themes, and all media (images, videos, documents, and so on) that you've uploaded to your website. Also, download a backup of your wp–config.php file, located in the root install.

2. **Deactivate all plugins.**

   This step ensures that any plugin conflicts caused by the upgraded version of WordPress can't affect the upgrade process. It also ensures that your website won't break when the upgrade is complete. You can find more information on working with and managing plugins in Chapter 7. For the purpose of this step, you can deactivate plugins by following this procedure:

   (a) *Click the Plugins link on the Plugins menu of the Dashboard to load the Plugins screen.*

   (b) *Select all plugins by selecting the check box to the left of the Plugin column.*

   (c) *From the drop-down menu at the top, choose Deactivate.*

   (d) *Click the Apply button.*

**3.** **Click the Updates link in the Dashboard menu and then click the Update Now button on the WordPress Updates screen.**

The Dashboard refreshes to display the welcome page for the latest version. This page displays a list of new features in the version you just installed, as shown in Figure 14-3. You're now using the latest version of WordPress.

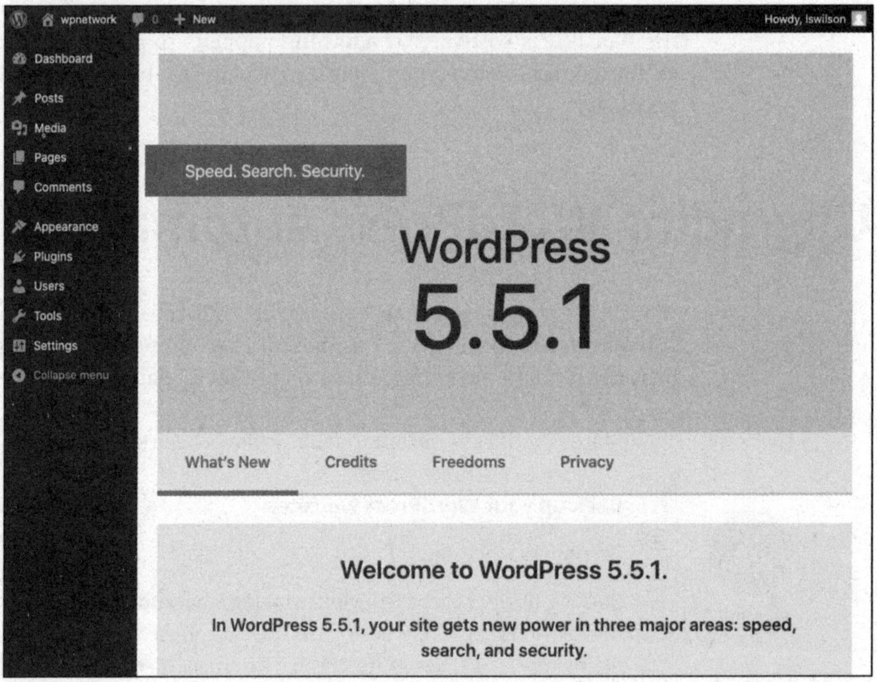

**FIGURE 14-3:**
The welcome page.

When you complete the WordPress software upgrade, you can revisit the Plugins screen and reactivate the plugins you deactivated in step 2 of the preceding list.

# Upgrading WordPress Manually

The second, less-used method of upgrading WordPress is the manual method. The method is less used mainly because the automatic method (discussed in the preceding section, "Upgrading WordPress Automatically") is so quick and easy. In certain circumstances, however — probably related to the inability of your hosting environment to accommodate the automatic method — you have to upgrade WordPress manually.

The steps to upgrade WordPress manually are as follows:

**1.** **Back up your WordPress website, and deactivate all plugins.**

See steps 1 and 2 of "Upgrading WordPress Automatically" earlier in this chapter.

**2.** **Navigate to the WordPress website, and click the Get WordPress button.**

This step takes you to the Get WordPress page.

**3.** **Click the Download button.**

This step opens a dialog box that allows you to save the `.zip` file of the latest WordPress download package to your computer (see Figure 14-4).

**4.** **Select a location to store the download package, and click Save.**

The `.zip` file downloads to the selected location on your computer.

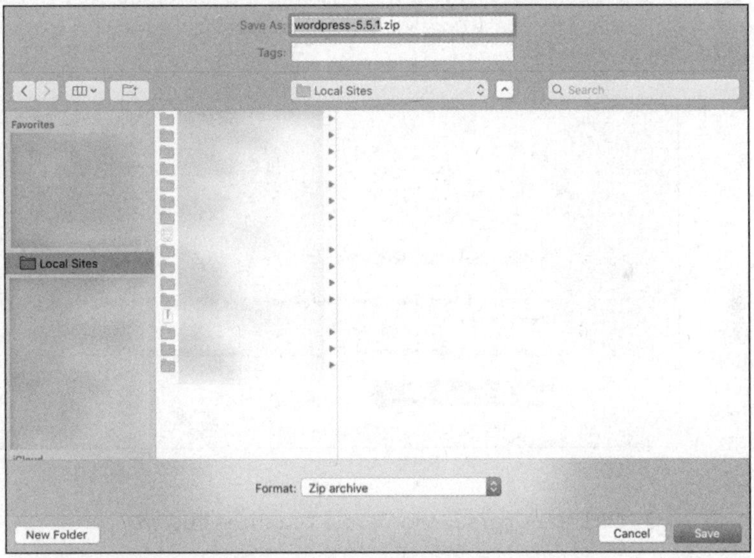

**FIGURE 14-4:** Downloading the WordPress files to your computer.

**5.** **Browse to the `.zip` file on your computer.**

**6.** **Unzip the file.**

Use a program such as WinZip (`https://www.winzip.com`) to unzip a `.zip` file.

**7.** **Connect to your web server via SFTP.**

See Chapter 3 for information on how to use SFTP.

8. **Delete all the files and folders in your existing WordPress installation directory** *except* **the following:**

- /wp-content folder

- .htaccess

- wp-config.php

9. **Upload the contents of the** /wordpress **folder — not the folder itself — to your web server.**

   Most SFTP client software lets you select all the files and drag and drop them to your web server. Other programs have you select the files and click a Transfer button.

10. **Navigate to the following URL on your website:** http://yourdomain.com/wp-admin.

    Don't panic: Your database still needs to be upgraded to the latest version, so instead of seeing your website on your domain, you see a message telling you that a database upgrade is required (see Figure 14-5).

**FIGURE 14-5:** Click the button to upgrade your WordPress database.

11. **Click the Upgrade WordPress Database button.**

    This action causes WordPress to initiate the upgrade of the MySQL database associated with your website. When the database upgrade is complete, the page refreshes and displays a message saying so.

12. **Click the Continue button.**

    Your browser loads the WordPress login page. The upgrade is complete, and you can continue using WordPress with all its newly upgraded features.

During your time as a WordPress user, you'll upgrade on a regular basis, at least three to four times per year. For some users, the need to update is a frustrating reality of using WordPress. Because of the very active development environment of the WordPress project, however, WordPress is the most popular platform available today. WordPress is always adding great new features and functions to the platform, so upgrading always ensures that you're on top of the game and using the latest tools and features available.

If you're uncomfortable performing administrative tasks such as upgrading and creating database backups, you can hire someone to perform these tasks for you — a member of your company, if you run a business, or a WordPress consultant who's skilled in these tasks.

# Migrating Your Existing Site to WordPress

So, you have a site on a different content management system (CMS) and want to move your site to WordPress? This chapter helps you accomplish that task. WordPress makes it relatively easy to pack up your data and archives from one platform and move to a new WordPress site.

By default, WordPress lets you move your site from such platforms as Blogger, TypePad, and Movable Type. It also gives you a nifty way to migrate from any platform via RSS feeds, as long as the platform you're importing from has an RSS feed available. Some platforms, such as Medium (https://medium.com), have some limitations on RSS feed availability, so be sure to check with your platform provider. In this section, you discover how to prepare your site for migration and how to move from the specific platforms for which WordPress provides importer plugins.

**TECHNICAL STUFF**

For each platform, the WordPress.org platform provides a quick, easy way to install plugins so that you can import and use your content right away. The importers are packaged in plugin format because most people use an importer only once, and some people don't use the importer tools at all. The plugins are there for you to use if you need them. WordPress.com, on the other hand, has the importers built into the software. Note the differences for the version you're using.

## Movin' on up

Website owners have a variety of reasons to migrate away from one system to WordPress:

>> **Simple curiosity:** WordPress currently powers more than 35 percent of all websites today. People are naturally curious to check out popular content-generating software.

>> **More control of your website:** This reason applies particularly to those who have a site on Medium, TypePad, or any other hosted service. Hosted programs limit what you can do, create, and mess with. When it comes to plugins, add-ons, and theme creation, hosting a WordPress blog on your own web server wins hands down. In addition, you have complete control of your data, archives, and backup capability when you host your blog on your own server.

>> **Ease of use:** Many people find WordPress to be easier to use, more understandable, and a great deal more user-friendly than many of the other blogging platforms available today.

**REMEMBER**

In the WordPress software, the importers are added to the installation as plugins. The importer plugins included in this chapter are the plugins packaged within the WordPress software; you can also find them by searching the Plugins page at https://wordpress.org/plugins/tags/importer. You can import content from several other platforms by installing other plugins that aren't available from the official WordPress Plugins page, but you may have to do an Internet search to find them.

## Preparing for the big move

Depending on the size of your site (that is, how many posts and comments you have), the migration process can take as little as 5 minutes in some cases to more than 30 minutes in others. As with any major change or update you make, no matter where your site is hosted, the first thing you need to do is create a backup of your site. You should back up the following:

>> **Archives:** Posts, pages, comments, and trackbacks. Back up by making a backup of your database.

>> **Template:** Template files and image files. Back up by transferring the /wp-content/themes folder from your hosting server to your local computer via SFTP.

>> **Plugins:** Plugin files. Back up by transferring the /wp-content/plugins folder from your hosting server to your local computer via SFTP.

>> **Media:** Any images, video, audio, or documents you use in your blog. Back up by transferring from your hosting server to your local computer via SFTP. (Images uploaded to posts or pages are generally stored in /wp-content/uploads.)

Table 14-1 gives you tips on creating the export data for your blog on a few major blogging platforms. *Note:* This table assumes that you're logged in to your blog software.

**TABLE 14-1**     **Backing Up Your Website Data on Major Platforms**

| Content Platform | Backup Information |
| --- | --- |
| Movable Type | Click the Import/Export button on the menu of your Movable Type Dashboard; then click the Export Entries From link. When the page stops loading, save it on your computer as a .txt file. |
| TypePad | Click the name of the site you want to export; then click the Import/Export link on the Overview menu, and click the Export link at the bottom of the Import/Export page. When the page stops loading, save it on your computer as a .txt file. |
| Blogger | Back up your template by copying the text of your template to a text editor such as Notepad; then save it on your computer as a .txt file. |
| Live Journal | Browse to https://livejournal.com/export.bml, and enter your information; choose XML as the format. Save this file on your computer. |
| Tumblr | Browse to https://www.tumblr.com/oauth/apps, and follow the directions to create a Tumblr app. When you're done, copy the OAuth Consumer Key and Secret Key, and paste them into a text file on your computer. Use these keys to connect your WordPress site to your Tumblr account. |
| WordPress | Click the Export link on the Tools menu of the Dashboard; the Export page opens. Choose your options on the Export page, click the Download Export File button, and then save this file on your computer. |
| RSS feed | Point your browser to the URL of the RSS feed you want to import. Wait until the feed loads fully. (You may need to set your feed to display all posts.) View the source code of the page, copy and paste that source code into a .txt file, and save the file on your computer. |

**TIP**

The WordPress import script allows for a maximum file size of 128MB. If you get an "out of memory" error, try dividing the import file into pieces and uploading the pieces separately. The import script is smart enough to ignore duplicate entries, so if you need to run the script a few times to get it to take everything, you can do so without worrying about duplicating your content.

## Converting templates

Every platform has a unique way of delivering content and data to your blog. Template tags vary from program to program; no two tags are the same, and each template file requires conversion if you want to use *your* template with your new WordPress blog. In such a case, two options are available to you:

» **Convert the template yourself.** To accomplish this task, you need to know WordPress template tags and HTML. If you have a template that you're using on another blogging platform and want to convert it for use with WordPress, you need to swap the original platform tags for WordPress tags. The information provided in chapters 8 through 11 gives you the rundown on working with themes as well as basic WordPress template tags; you may find that information useful if you plan to attempt a template conversion yourself.

» **Hire an experienced WordPress consultant to do the conversion for you.** You can find a list of WordPress consultants, assembled by the folks at WP Engine (a managed hosting provider for WordPress), at https://wpengine.com/partners/agencies/.

To use your own template, make sure that you've saved *all* the template files, the images, and the stylesheet from your previous site setup. You need them to convert the template(s) for use in WordPress.

**REMEMBER**

Thousands of free themes are available for use with WordPress, so it may be a lot easier to abandon the theme you're working with and find a free WordPress theme that you like. If you've paid to have a custom design done for your site, contact the designer of your theme, and hire her to perform the theme conversion for you. Also, you can hire several WordPress consultants to perform the conversion for you — including yours truly.

## Moving your website to WordPress

You've packed all your stuff, and you have your new place prepared. Moving day has arrived! This section takes you through the steps for moving your site from one platform to WordPress. The section assumes that you already have the WordPress software installed and configured on your own domain.

Find the import function that you need by following these steps:

1. **On the Dashboard, choose Tools ⇨ Import.**

   The Import screen opens, listing platforms from which you can import content (such as Blogger and Movable Type). Figure 14-6 shows the Import screen of the WordPress Dashboard.

2. **Find the platform you're working with.**

3. **Click the Install Now link to install the importer plugin and begin using it.**

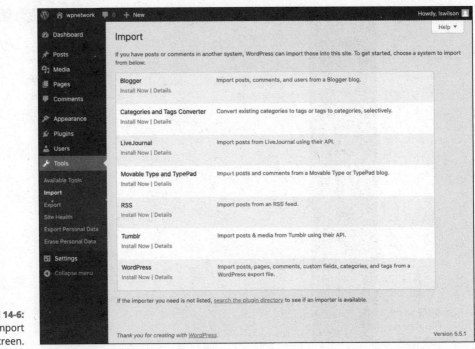

**FIGURE 14-6:**
The Import
screen.

The following sections provide some import directions for a few of the most popular CMSes (other than WordPress, that is). Each platform has its own content export methods, so be sure to check the documentation for the platform that you're using.

## Importing from Blogger

Blogger (formerly called Blogspot) is the blogging application owned by Google.

**1.** **On the Dashboard, choose Tools ⇨ Import.**

The Import screen opens, listing platforms from which you can import content (refer to Figure 14-6).

**2.** **Find the platform you're working with.**

**3.** **Click the Install Now link below the Blogger heading on the Import screen, and install the plugin for importing from Blogger.**

**4.** **Click the Run Importer link.**

The Import Blogger page loads, with instructions for importing your file, as shown in Figure 14-7. If you haven't already prepared a file in Blogger to import to WordPress, you need to do that now.

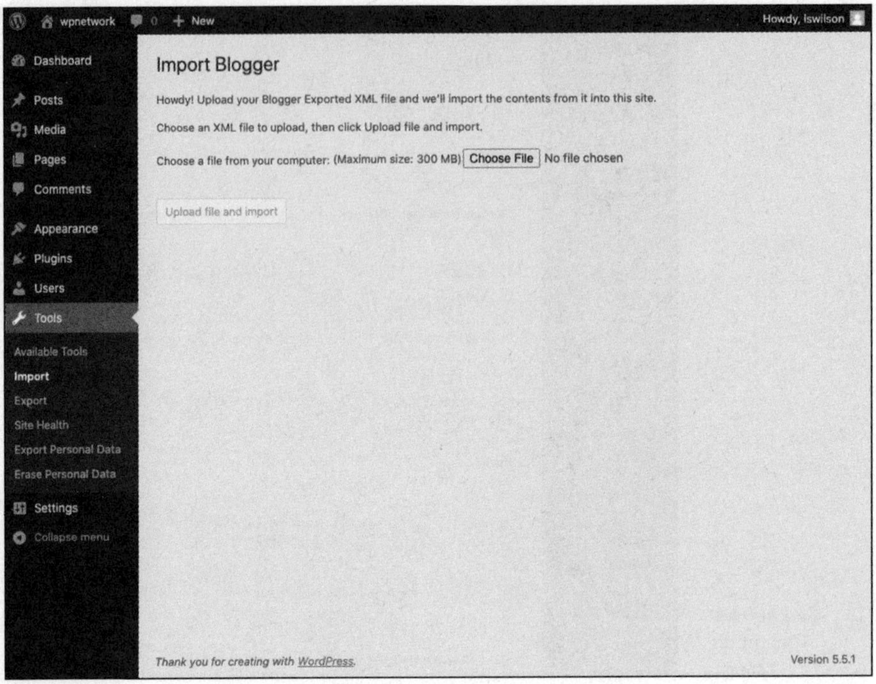

**FIGURE 14-7:**
The Import
Blogger page.

5. **Log in to your Blogger account.**

6. **In your Blogger account, click the blog you want to import.**

7. **Choose Settings ⇨ Other in your Blogger account.**

   This link is on the left menu.

8. **In your Blogger account, choose Back up Content ⇨ Save to Your Computer to save the** `.xml` **file.**

9. **On the Import Blogger screen of your WordPress Dashboard, click the Choose File button to upload the Blogger file.**

10. **Click the Upload File and Import button.**

    This step uploads the file, and the screen refreshes to the Assign Authors section of the Import Blogger screen.

11. **Click the Set Authors button to assign the authors to the posts.**

    The Blogger username appears on the left side of the page; a drop-down menu on the right side of the page displays the WordPress login name.

12. **Assign authors by using the drop-down menu.**

    If you have only one author on each blog, the process is especially easy: Use the drop-down menu on the right to assign the WordPress login to your

Blogger username. If you have multiple authors on both blogs, each Blogger username is listed on the left side with a drop-down menu to the right of each username. Select a WordPress login for each Blogger username to make the author assignments.

**13.** **Click Save Changes.**

You're done!

# Importing from LiveJournal

Both WordPress.com and WordPress.org offer an import script for LiveJournal users, and the process of importing from LiveJournal to WordPress is the same for each platform.

To export your blog content from LiveJournal, log in to your LiveJournal blog and then type this URL in your browser's address bar: `https://www.livejournal.com/export.bml`.

LiveJournal lets you export the `.xml` files one month at a time, so if you have a blog with several months' worth of posts, be prepared to be at this process for a while. First, you have to export the entries one month at a time, and then you have to import them into WordPress — yep, you guessed it — one month at a time.

TIP

To speed the process a little, you can save all the exported LiveJournal files in one text document by copying and pasting each month's `.xml` file into one plain-text file (created in a text editor such as Notepad), thereby creating one long .xml file with all the posts from your LiveJournal blog. Then you can save the file as an `.xml` file to prepare it for import into your WordPress blog.

After you export the `.xml` file from LiveJournal, return to the Import page of your WordPress Dashboard, and follow these steps:

**1.** **Click the Install Now link below the LiveJournal heading, and install the plugin for installing from LiveJournal.**

**2.** **Click the Run Importer link.**

The Import LiveJournal screen loads, with instructions for importing your file, as shown in Figure 14-8.

**3.** **In the LiveJournal Username field, type the username for your LiveJournal account.**

**4.** **In the LiveJournal Password field, type the password for your LiveJournal account.**

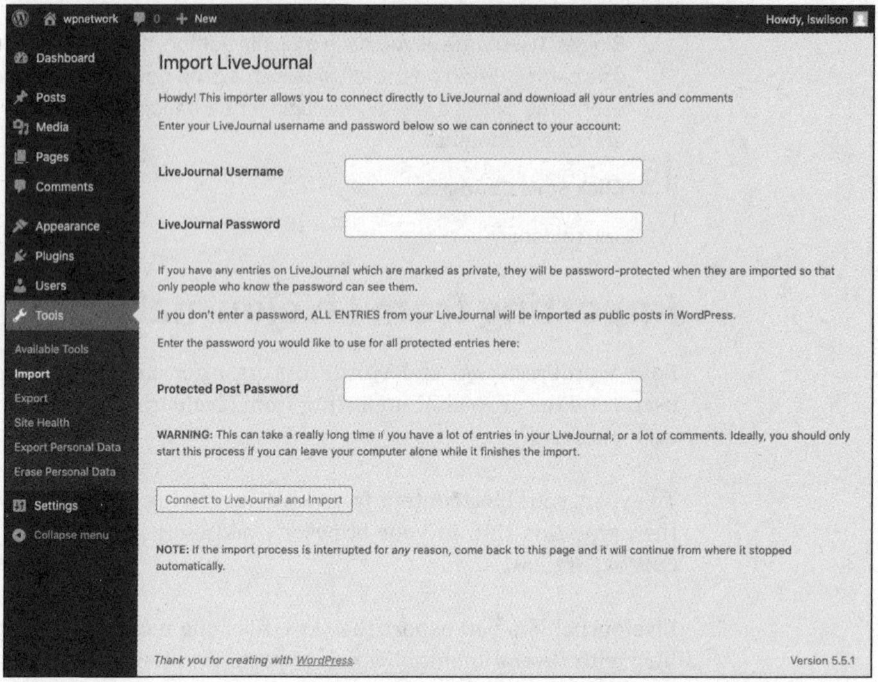

FIGURE 14-8:
The Import
LiveJournal page
of the WordPress
Dashboard.

5. **In the Protected Post Password field, enter the password you want to use for all protected entries in your LiveJournal account.**

    If you don't complete this step, every entry you import into WordPress will be viewable by anyone. Be sure to complete this step if any entry in your LiveJournal account is password-protected (or private).

    **WARNING**

6. **Click the Connect to LiveJournal and Import button.**

    This step connects your WordPress site to your LiveJournal account and automatically imports all entries from your LiveJournal into your WordPress installation. If your LiveJournal site has a lot of entries, this process could take a long time, so be patient.

## Importing from Movable Type and TypePad

Movable Type and TypePad were created by the same company, Six Apart. These two platforms run on essentially the same code base, so the import/export procedure is basically the same for both. Refer to Table 14-1 earlier in this chapter for details on how to run the export process in both Movable Type and TypePad. This import script moves all your posts, comments, and trackbacks to your WordPress site. Follow these steps to import your Movable Type or TypePad content:

1. **On the Dashboard, choose Tools ⇨ Import.**

   The Import screen opens, listing platforms from which you can import content (refer to Figure 14-6).

2. **Find the platform you're working with.**

3. **Click the Install Now link below the Movable Type and TypePad heading, and install the plugin for importing from Movable Type and TypePad.**

4. **Click the Run Importer link.**

   The Import Movable Type or TypePad screen loads, with instructions for importing your file, as shown in Figure 14-9.

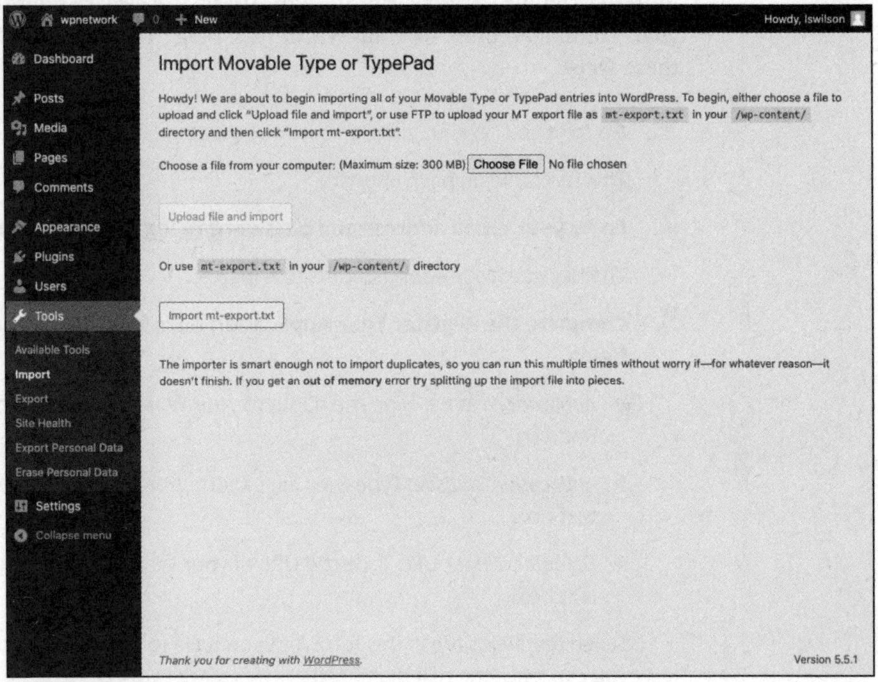

**FIGURE 14-9:**
The Import
Movable Type or
TypePad screen
of the WordPress
Dashboard.

5. **Click the Choose File button.**

6. **Double-click the name of the export file you saved from your Movable Type or TypePad blog.**

7. **Click the Upload File and Import button.**

   Sit back and let the import script do its magic. When the script finishes, it reloads the page with a message confirming that the process is complete.

8. **When the import script finishes, assign users to the posts, matching the Movable Type or TypePad usernames with WordPress usernames.**

   If you have only one author on each blog, this process is easy; you simply assign your WordPress login to the Movable Type or TypePad username by using the drop-down menu. If you have multiple authors on both blogs, match the Movable Type or TypePad usernames with the correct WordPress login names.

9. **Click Save Changes.**

# Importing from Tumblr

With the Tumblr import script for WordPress, it's easy to import the content from your Tumblr account to your WordPress blog. To complete the import, follow these steps:

1. **Go to** `https://www.tumblr.com/oauth/apps.`

   The Tumblr login page appears.

2. **Enter your email address and password to log in to your Tumblr account.**

   The Register Your Application page appears.

3. **Complete the Register Your Application form by filling in the following fields:**

   - *Application Name:* Type the name of your WordPress website in the text box.

   - *Application Website:* Type the URL of your WordPress website in the text box.

   - *Default Callback URL:* Type the URL of your WordPress website in the text box.

   Seven text fields are in this form, but you have to fill in only these three fields; you can leave the rest blank.

4. **Click the check box titled I'm Not a Robot to prove that you're human and not a spammer.**

5. **Click the Register button.**

   The Applications page refreshes and displays your registered app information at the top.

6. **Copy the OAuth Consumer Key, and paste it into a text file on your computer.**

**7.** Copy the Secret Key, and paste it into the same text file where you placed the OAuth Consumer Key in step 6.

**8.** On your Dashboard, choose Tool ⇨ Import and then click the Tumblr link.

The Import Tumblr screen of your Dashboard opens, as shown in Figure 14-10.

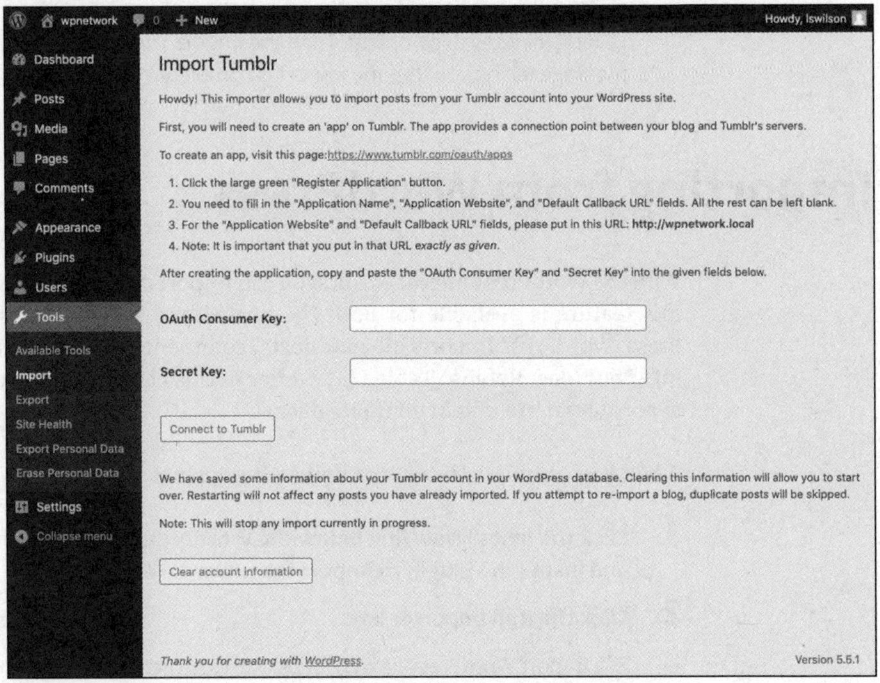

FIGURE 14-10:
The Import
Tumblr screen of
the WordPress
Dashboard.

**9.** Insert the OAuth Consumer Key into the indicated text box.

Use the OAuth Consumer Key you saved to a text file in step 6.

**10.** Insert the Secret Key into the indicated text box.

Use the Secret Key you saved to a text file in step 7.

**11.** Click the Connect to Tumblr button.

The Import Tumblr screen appears, with a message instructing you to authorize Tumblr.

**12.** Click the Authorize the Application link.

The Authorization page on the Tumblr website asks you to authorize your WordPress site access to your Tumblr account.

**13.** Click the Allow button.

The Import Tumblr page opens in your WordPress Dashboard and displays a list of your sites from Tumblr.

**14.** Click the Import This blog button in the Action/Status section.

The content from your Tumblr account is imported into WordPress. Depending on how much content you have on your Tumblr site, this process may take several minutes to complete. Then the Import Tumblr page refreshes with a message telling you that the import is complete.

# Importing from WordPress

With the WordPress import script, you can import one WordPress site into another; this feature is available for both the hosted and self-hosted versions of Word-Press. WordPress imports all your posts, comments, custom fields, and categories into your blog. Refer to Table 14-1 earlier in this chapter to find out how to use the export feature to obtain your site data.

When you complete the export, follow these steps:

**1.** Click the Install Now link below the WordPress title on the Import page, and install the plugin to import from WordPress.

**2.** Click the Run Importer link.

The Import WordPress screen loads, with instructions for importing your file, as shown in Figure 14-11.

**3.** Click the Choose File button.

A window opens, listing the files on your computer.

**4.** Double-click the export file you saved earlier from your WordPress blog.

**5.** Click the Upload File and Import button.

The import script gets to work, and when it finishes, it reloads the page with a message confirming that the process is complete.

## Importing from an RSS feed

If all else fails, or if WordPress doesn't provide an import script that you need for your current site platform, you can import your site data via the RSS feed for the site you want to import. With the RSS import method, you can import only posts; you can't use this method to import comments, trackbacks, categories, or users.

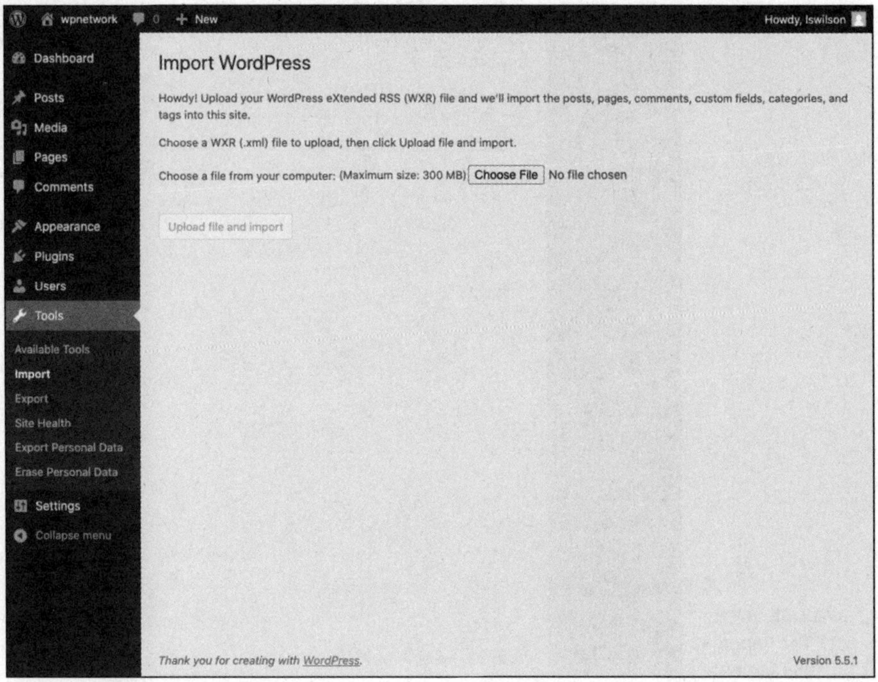

**FIGURE 14-11:**
The Import
WordPress
screen.

Refer to Table 14-1 for information on creating the file you need to import via RSS. Then follow these steps:

1. **On the Import page of the WordPress Dashboard, click the Install Now link below the RSS heading, and install the plugin to import from an RSS feed.**

2. **Click the Run Importer link.**

   The Import RSS screen loads, with instructions for importing your RSS file, as shown in Figure 14-12.

3. **Click the Choose File button on the Import RSS page.**

4. **Double-click the export file you saved earlier from your RSS feed.**

5. **Click the Upload File and Import button.**

   The import script does its magic and then reloads the page with a message confirming that the process is complete.

## Finding other import resources

WordPress Support has a long list of scripts, plugins, workarounds, and outright hacks for importing from other platforms. You can find that information at https://wordpress.org/support/article/importing-content.

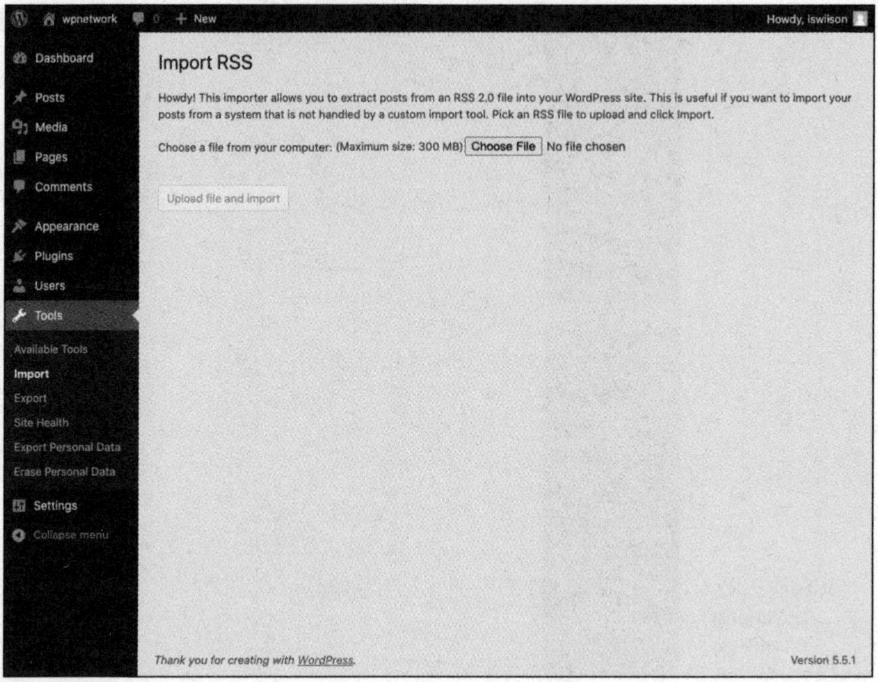
wpnetwork
0
New
Howdy, iswilson
Dashboard
Posts
Media
Pages
Comments
Appearance
Plugins
Users
Tools
Available Tools
Import
Export
Site Health
Export Personal Data
Erase Personal Data
Settings
Collapse menu
Import RSS
Howdy! This importer allows you to extract posts from an RSS 2.0 file into your WordPress site. This is useful if you want to import your posts from a system that is not handled by a custom import tool. Pick an RSS file to upload and click Import.
Choose a file from your computer: (Maximum size: 300 MB)
Choose File
No file chosen
Upload file and import
Thank you for creating with WordPress.
Version 5.5.1

**FIGURE 14-12:** The Import RSS screen.

REMEMBER

Note that volunteers run the WordPress Support site. When you refer to it, be aware that not everything listed in it is necessarily up to date or accurate, including import information (or any other information about running your WordPress site).

# Moving Your Website to a Different Host

You may decide at some point that you need to switch from your current hosting provider to a new one. You may have to switch for several reasons. Maybe you're unhappy with your current provider and want to move to a new one, or perhaps your current provider is going out of business, forcing you to move.

Transferring from one host to another is a reality that some website owners must face. Transferring an existing website, with all its content, files, and data, from one host to another can be a daunting task. This section of the chapter should make the task easier for you to face, however.

You have two ways to go about it:

- Manually, by backing up your database and downloading essential files
- Using a plugin to automate as much of the process as possible

Obviously, using a tool to automate the process is the easier and more desirable way to go, but just in case you need to do the job manually, the next section of this chapter, "Creating a backup and moving manually," provides instructions.

# Creating a backup and moving manually

Earlier in this chapter, in "Backing Up Your Database," I provide step-by-step instructions for making a backup of your database with phpMyAdmin. Follow the steps in that section, and you'll have a backup of your database with all the recent content you've published to your blog. By *content*, I mean the content that you (or others) have written or typed in your blog via the WordPress Dashboard, including

>> Blog posts, pages, and custom post types

>> Links, categories, and tags

>> Post and page options such as excerpts, time and date, custom fields, categories, tags, and passwords

>> WordPress settings you configured on the Settings menu of the Dashboard

>> All widgets that you've created and configured

>> All plugin options that you've configured for the plugins you installed

Other elements of your website aren't stored in the database, which you need to download (via SFTP) from your web server. Following is a list of those elements, including instructions on where to find them and how to download them to your local computer:

>> **Media files:** These files are the ones you uploaded by using the WordPress media upload feature; they include images, videos, audio files, and documents. Media files are located in the /wp-content/uploads/ folder; connect to your web server via SFTP, and download that folder to your computer.

>> **Plugin files:** Although all the plugin settings are stored in the database, the actual plugin files that contain the programming code are not. The plugin files are located in the /wp-content/plugins/ folder; connect to your web server via SFTP, and download that folder to your computer.

>> **Theme files:** Widgets and options you've set for your current theme are stored in the database, but the theme template files, images, and stylesheets are not. Those files are stored in the /wp-content/themes folder; connect to your web server via SFTP, and download that folder to your computer.

Now that you have your database and WordPress files stored safely on your computer, moving them to a new host involves reversing the process, as follows:

1. **Create a new database on your new hosting account.**

   You can find the steps for creating a database in Chapter 3.

2. **Import your database backup into the new database you just created:**

   (a) *Log in to the cPanel for your hosting account.*

   (b) *Click the phpMyAdmin icon, and choose your new database on the left menu.*

   (c) *Click the Import tab at the top.*

   (d) *Click the Browse button, and select the database backup on your computer.*

   (e) *Click the Go button to import the old database into the new one.*

3. **Install WordPress in your new hosting account.**

   See Chapter 3 for the steps to install WordPress.

4. **Edit the `wp-config.php` file to include your new database name, user-name, password, and host.**

5. **Upload all that you downloaded from the `/wp-content/` folder to your new hosting account.**

6. **In your web browser, browse to your domain.**

   Your website should work, and you should be able to log in to the WordPress Dashboard using the same username and password as before, because that information is stored in the database you imported.

## Using a plugin to back up and move to a new host

A plugin that I use on a regular basis to move a WordPress website from one hosting environment to another is aptly named BackupBuddy. This plugin isn't free or available in the WordPress Plugins page; you need to pay for it. But this plugin is worth every single penny because it takes the entire backup and migration process and makes mincemeat out of it. In other words, the plugin is easy to use, and you can be done in minutes instead of hours.

You can purchase the BackupBuddy plugin from iThemes at https://ithemes. com/backupbuddy; at this writing, pricing starts at $80 per year. After purchase, you can download the plugin and install it. (See plugin installation instructions in Chapter 7.) Then follow the instructions on the WordPress Dashboard to make a backup copy of your website and move it to another server.

# 5

# The Part of Tens

# Chapter **15**

# Ten Popular WordPress Plugins

In this chapter, I list ten of the most popular plugins available for your WordPress site. This list isn't exhaustive by any means; hundreds of excellent WordPress plugins can, and do, provide multiple ways to extend the functionality of your blog. If these ten plugins aren't enough for you, you can find many more at the official WordPress Plugins page (https://wordpress.org/plugins).

**REMEMBER**

The greatest plugin of all is Akismet, which I describe in Chapter 7. Akismet is the answer to comment and trackback spam; it kills spam dead. It's installed with WordPress. Chapter 7 contains information on how to locate, download, unpack, install, activate, and manage plugins on your WordPress site.

## Custom Post Type UI

**Developer:** WebDevStudios

https://wordpress.org/plugins/custom-post-type-ui

In Chapter 12 of this book, I introduced you to the Custom Post Types feature in WordPress. Custom post types enable you to create new content types separate from posts and pages, such as movie reviews or recipes. Custom taxonomies allow you to create new types of content groups separate from categories and tags, such as genres for movie reviews or seasons for recipes.

The "Working with Custom Post Types" section in Chapter 12 shows how to create custom post types and custom taxonomies by adding several lines of code to the `functions.php` file in your WordPress theme. But not everyone is interested in, or comfortable with, digging into template files to add a bunch of code, which is where the Custom Post Type UI comes in. This plugin helps you create custom post types and taxonomies without dipping into the code at all.

The plugin provides a settings screen on the WordPress Dashboard that enables you to configure your custom post type and taxonomy. When you're finished, click the Save button to save the configuration.

# Jetpack

**Developer:** Automattic

https://wordpress.org/plugins/jetpack

Jetpack isn't one plugin; it's a suite of plugins that connects your self-hosted website running WordPress.org with the hosted WordPress.com service, bringing you many of the features that WordPress.com users enjoy. Jetpack bundles features such as the following:

>> **WordPress.com Stats:** Get information on your site visitors, such as how many there are, where they're coming from, and what content they're viewing on your site.

>> **Jetpack Comments:** Manage comments that integrate with social media login options such as Facebook, Twitter, and Google.

>> **Post by Email:** Publish posts to your blog directly from your email account.

>> **Carousel:** Transform your standard image galleries into slideshows and carousels.

>> **Spelling and Grammar:** Catch your grammar, spelling, and punctuation errors with this integrated proofreading service.

- » **VaultPress:** Manage real-time backups and security scanning for your site.

- » **Contact Form:** Insert an email contact form anywhere on your WordPress site with one click.

- » **WP.me Shortlinks:** Create a short URL for easier social sharing with the WP.me service.

- » **Tiled Galleries:** Create magazine-style image tile layouts for your photos.

- » **Custom CSS:** Customize the appearance of your site with CSS without modifying your theme files.

- » **Extra Sidebar Widgets:** Add widgets to your WordPress site, such as Easy Image and RSS.

- » **Jetpack Single Sign On:** Allow users to log in to your site by using their credentials from WordPress.com.

- » **Enhanced Distribution:** Share your content with search engines and other services in real-time.

- » **VideoPress:** Upload and insert videos on your site.

Because Jetpack runs and is hosted on the WordPress.com cloud server, updates to this suite of plugins occur automatically.

To use Jetpack, you must have a WordPress.com account (see Chapter 1).

# Limit Login Attempts Reloaded

**Developer:** WPChef

```
https://wordpress.org/plugins/limit-login-attempts-reloaded
```

This plugin limits the number of times a user can attempt to log in to your Word-Press site before they are locked out for a specified period of time. By default, without a plugin such as this one, WordPress allows unlimited login attempts, which means allowing unlimited password-cracking attempts as well. The Limit Login Attempts Reloaded plugin logs all attempts so that you can keep track of IP addresses, usernames, and email addresses that attempt to log in to your site over and over. You can add any IP, email address, or username to a blacklist that disallows any attempts from those sources on the first try. This plugin is compatible with the Multisite feature in WordPress.

# Cookie Notice for GDPR & CCPA

**Developers:** dFactory

`https://wordpress.org/plugins/cookie-notice`

General Data Protection Regulation (GDPR) is a European personal-data privacy law that was fully implemented in 2018. The law addresses the organization and processing of user data for all users in the European Union. It states that all websites must inform visitors that it is collecting and storing user information through various means. That notification must be visible to all users of the website. A similar law, the California Consumer Privacy Act (CCPA), gives consumers control of the personal information that websites collect about them. All website owners should abide by both these laws.

The Cookie Notice for GDPR & CCPA plugins enable you to set a custom message that notifies your users that you're using cookies when they visit your website. When the user clicks the I Agree button to acknowledge their consent to the storage of their personal data, your website will set a cookie so that the browser will remember their consent.

You may be wondering why your website would need to abide by these two laws. Almost every website collects data on its users. If your website does any of the following things, you should have a GDPR and CCPA notice on your website:

>> **Comments:** Websites that allow comments on articles are collecting, at minimum, users' IP addresses and email addresses, as well as the content of the comments.

>> **User Registration:** Websites that allow user registration are collecting users' IP and email addresses.

>> **Contact Forms:** Websites that have contact forms are collecting users' IP and email addresses, names, and phone numbers, as well as the content of the messages they send to the sites.

>> **Analytics:** Websites that use analytics tools are collecting user data such as IP addresses and geographical locations.

Under these laws, user consent on data collection and storage must be informed; it cannot be assumed.

# Yoast SEO

**Developer:** Team Yoast

`https://wordpress.org/plugins/wordpress-seo`

Almost everyone is concerned about search engine optimization (SEO) for their websites. Good SEO practices help the major search engines (such as Google, DuckDuckGo, and Bing) easily find and cache your blog content in their search databases so that when people search for keywords, they can find your blog in the search results. Yoast SEO helps you fine-tune your website for SEO, automatically creating optimized titles and generating HTML keywords for your individual posts. If you're a beginner, this plugin works for you out of the box, with no advanced configuration necessary. Woo-hoo! If you're an advanced user, you can fine-tune the Yoast SEO settings to your liking.

# BackupBuddy

**Developer:** iThemes

`https://ithemes.com/backupbuddy`

Starting at $80 for personal use and $199 for the unlimited version, BackupBuddy lets you back up your entire WordPress website in minutes. With this plugin, you can also determine a schedule of automated backups of your site on a daily, weekly, or monthly basis. You can store Backups from this plugin on your web hosting account; email backup files to a designated email address; transfer the backup files via SFTP to a designated SFTP server; or store the backups on Amazon's Simple Storage Service, Dropbox, or the Rackspace Cloud, if you have those accounts.

BackupBuddy backs up not only your WordPress data (posts, pages, comments, and so on), but also any theme and customized plugins you've installed (including the settings for those plugins), and it saves and backs up all WordPress settings and any widgets that you're currently using.

BackupBuddy includes an import and migration script (`importbuddy.php`) that allows you to transfer an existing site to a new domain or host within minutes. You simply download the backup file created by BackupBuddy from your Dashboard (choose BackupBuddy⇨ Backups), install the script on a new domain, and follow the steps displayed on the screen.

**TECHNICAL STUFF**

This plugin is invaluable for designers and developers who work with clients to design WordPress websites. Using BackupBuddy, you can download a backup of the site and then use the import/migration script to transfer the completed site to your client's site within minutes, saving all the customizations you did to the theme and the plugins you installed, including the settings and data you worked so hard on.

# WP Super Cache

**Developer:** Automattic

https://wordpress.org/plugins/wp-super-cache

WP Super Cache creates static HTML files from your dynamic WordPress content. Why is this plugin useful? On a high-traffic site, having cached versions of your posts and pages can decrease the load time of your website considerably. A *cached* version simply means that the content is converted to static HTML pages (as opposed to dynamically created content pulled from your database through a series of PHP commands) that are then stored on the server. This process eases the efforts the web server must take to display the content in your visitors' browsers.

You can read a helpful article written by one of the plugin's developers, Donncha O'Caoimh, at https://odd.blog/wp-super-cache.

# WooCommerce

**Developer:** Automattic

https://wordpress.org/plugins/woocommerce

E-commerce is the practice of selling products or services on your website. The WooCommerce plugin for WordPress comes in handy for that work. Whether you're selling products such as T-shirts, posters, or art, or selling services such as consulting, WooCommerce allows you to set up products and accept payment transactions on your website.

Here are some of the tasks that you can perform with the WooCommerce plugin:

>> **Accept payments.** You can accept payments from your customers from PayPal or any major credit card.

>> **Configure shipping.** If you sell physical goods that require shipping, the WooCommerce shipping feature gives you several options, including free or flat-rate shipping.

>> **Manage inventory.** Easily manage physical or digital goods (such as music, for example). You can also assign store managers to handle day-to-day inventory for large online shops.

>> **Run reports.** Keep track of your sales, reviews, stock levels, and overall store performance with WooCommerce reporting tools.

>> **Run marketing campaigns.** Run marketing campaigns with a range of discounts, coupons, use limits, and product and/or user restrictions, as well as free shipping.

>> **Configure taxes.** Configure tax settings with classes and local tax rates.

WooCommerce has a variety of add-ons, called extensions, that allow you to extend your e-commerce platform to your tastes. Its Extension library contains items such as Event Bookings, Membership Subscriptions, Product Reviews, and Product Labels.

# Google XML Sitemaps

**Developer:** Auctollo

`https://wordpress.org/plugins/google-sitemap-generator`

This plugin lets you create a Google-compliant XML site map of your entire blog. Every time you create a new post or page, the site map is updated and submitted to several major search engines, including Google, Yahoo!, and Bing. This plugin helps the search engines find and catalog new content from your site, so your new content appears in the search engines faster than it would if you didn't have a site map.

# Sucuri Security

**Developer:** Sucuri, Inc.

`https://wordpress.org/plugins/sucuri-scanner`

With the rise in popularity of the WordPress software, a nefarious group of anonymous hackers tried to take advantage of the vast number of users in the WordPress community by attempting to inject malicious code and malware into themes, plugins, and insecure and outdated files within the WordPress core code.

The Sucuri SiteCheck Malware Scanner plugin checks for malware, spam, blacklisting, and other security issues hidden inside code files. It's the best defense you have against malicious hackers and very easy to implement — and for the peace of mind that it provides you, using this plugin is worth it.

# Chapter **16**

# Ten Free WordPress Themes

T he list I present here isn't exhaustive by any means. Chapters 8, 10, and 11 give you a few more resources to find a theme that suits your needs.

All the themes in this chapter meet the following criteria:

» **They're user-friendly.** You don't have to tinker with anything to get things to look the way you want them to.

» **They're compatible with widgets.** In a word, widgets are wonderful. I cover them in Chapter 9.

» **They're free.** Some very nice premium themes are out there, but why pay if you don't have to?

» **They use valid code.** Although you may not notice it, valid code that meets W3C (https://www.w3.org) standards won't cause errors in browsers.

# Hybrid Core

**Theme designer:** Justin Tadlock

`https://themehybrid.com/hybrid-core`

Hybrid Core is more of a theme *framework*, or parent theme that can be modified endlessly to create the perfect child theme, than a theme to use straight out of the box. But don't let that description intimidate you! The theme is crazy-easy to use and very user-friendly.

By default, the Hybrid Core theme is plain and simple, but it encompasses all the WordPress features and functions you might want:

» **It's SEO-ready.** Hybrid is completely ready for search engine optimization (SEO).

» **It's highly customizable.** Hybrid has 15 custom page templates for you to choose among. Each custom page is set up slightly differently, giving you an array of options.

» **It's widget-ready.** Hybrid has multiple widgetized areas for you to drop content into, making your WordPress theme experience easy and efficient.

You can read about the Hybrid Core theme at developer Justin Tadlock's website at `https://themehybrid.com/hybrid-core`. You can also download and install the theme directly on your WordPress website by using the automatic theme installer built into your WordPress Dashboard.

TIP

Check out options for installing and tweaking WordPress themes in Chapters 9, 10, and 11. Those chapters give you information on CSS, HTML, and theme-tweaking, as well as guide you through working with parent and child themes.

# Hestia

**Theme designer:** Themeisle

`https://themeisle.com/demo/?theme=Hestia`

Hestia (see Figure 16-1) is a one-page theme built for use on a small-business website. The theme works on all devices: computers, tablets, and smartphones.

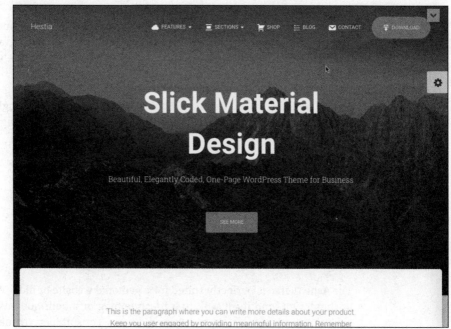

The features of Hestia include

>> A drag-and-drop interface for building content

>> SEO optimization

>> A customizer that allows you to make changes on the fly and view those changes live

>> Support for WordPress menus and custom navigation menus

>> Support for custom background colors and images

>> Built-in social networking links and sharing

# Responsive

**Theme designer:** CyberChimps

```
https://wordpress.org/themes/responsive
```

Responsive design is all the rage right now because of the emergence of mobile and tablet browsing. Responsive design ensures that a website looks perfect no matter which device a reader uses to view it. The Responsive theme by Emil Uzelac features nine page templates, including Blog, Blog Summary, and other static page templates built on a fluid grid system that adapts to the user's browsing environment.

Theme options include webmaster tools, logo management, Google font typography support, social media icons, navigation menus, and multilingual support.

# Ashe

**Theme designer:** WP Royale

https://wordpress.org/themes/ashe

The Ashe theme is perfectly suited for a personal website or blog. The look and feel of the design lend itself well to a lifestyle blog or a boutique business such as a bakery, travel agency, or health and fitness consultancy, to name a few. The design is minimal and beautiful, with soft tones and colors. This theme supports WooCommerce, for those who want to use this theme to run an online shop. The features of this theme include

» Modern, responsive design to display your content on any device

» Coded with solid technical SEO practices to assist you in reaching good SEO returns

» Capability to use the WordPress customizer to upload an image file of your logo

» Header image, full-screen slider, and Instagram slider to showcase some of your content

# Prefer Blog

**Theme designer:** Template Sell

https://wordpress.org/themes/prefer-blog

Prefer Blog is a simple blog theme that is perfect for someone who wants to create and manage a blog in WordPress. The theme works well with the Gutenberg editor. Two of the most interesting features of this theme are the built-in Author and Contact Us block patterns, which make it easy to add author information to any page or post, and a contact form (if you've installed the Contact Form 7 plugin [https://wordpress.org/plugins/contact-form-7]).

Other theme options include the following:

» Multiple sidebar options

» Featured slider

» Promotional boxes

» Custom widgets for featured posts and social icons

» Multiple color options

# BlackBird

**Theme designer:** InkThemes

https://wordpress.org/themes/blackbird

BlackBird, shown in Figure 16-2, is a responsive theme (mobile-ready) with extensive customization options. The theme allows you to

» Use your own logo

» Include your analytics code

» Customize featured text with an easy-to-use widget

» Customize background colors or images

» Incorporate post thumbnails by using the WordPress featured image feature

» Customize the header image

» Use the navigation-menu features of WordPress

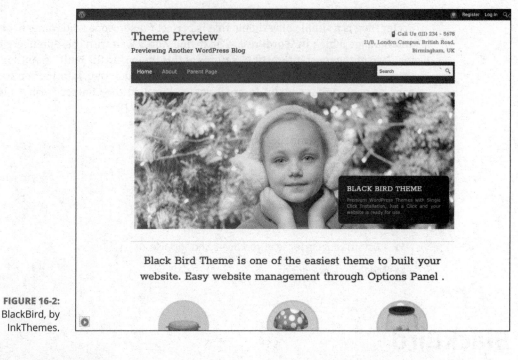

# Storefront

**Theme designer:** Automattic

`https://wordpress.org/themes/storefront`

The Storefront theme is exactly that: a theme for your store. This theme is compatible with WooCommerce and is a great free theme to get you started with your e-commerce venture. It offers several layout options, color options, responsive design, and custom widgets focused on WooCommerce features. (Read more about WooCommerce in Chapter 15.) Built-in features include

>> Custom background options

>> Custom color options

>> Custom header options

>> Custom menu options

>> Featured images

- » Footer widgets

- » Ability to set a left or right sidebar

- » Ability to remove the sidebar

Figure 16-3 shows the Storefront theme.

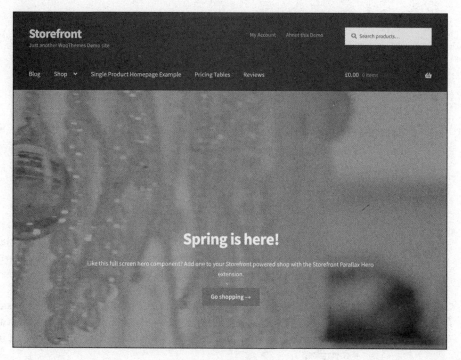

**FIGURE 16-3:**
Storefront theme.

# Sinatra

**Theme designer:** Sinatra Team

`https://wordpress.org/themes/sinatra`

You can use the Sinatra theme to create any type of website or blog. This theme is perfect for a new blogger or for someone who is running a business website, such as for creative industries, restaurants, bakeries, tech startups, and more.

The Sinatra theme works with the Gutenberg editor, as well as popular plugins such as WooCommerce, Jetpack, and Yoast SEO. Its great features include

>> Multiple layouts

>> Unlimited color options

>> Microdata integration

>> Custom background options

>> Featured images

>> Right or left sidebar

# Nisarg

**Theme designer:** Nisarg

https://wordpress.org/themes/nisarg

Nisarg is a beautiful theme for a blog. It features a nice, clean layout with easy-to-use navigation menus, a large area for you to insert your own custom header image, and support for custom backgrounds and colors.

The Nisarg theme provides custom page templates (such as no-sidebar templates, portfolio, and blog) and allows you to use default WordPress features such as custom headers, custom backgrounds, navigation menus, featured images, and post formats.

# Optics

**Theme designer:** Graph Paper Press

https://graphpaperpress.com/themes/optics/

Optics is a minimalist WordPress theme featuring a grid layout. (See Figure 16-4.) The theme has clean, simple, light elements that let your design focus on content rather than appearance. The theme uses black, white, and gray tones and a two-column layout, with content on the right and sidebar on the left.

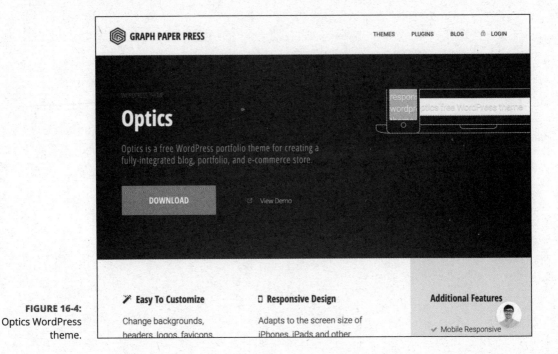

**FIGURE 16-4:**
Optics WordPress
theme.

This free theme is offered by a commercial theme company, Graph Paper Press, and to download it, you need to register for a free account on the company's website.

# Index

## N

navigation menus
building, 103, 309–315
  adding links to, 309
  CSS, 312–313
  HTML, 311
  widgets, 314–315
Dashboard, 78–105
static pages, 342
Twenty Twenty theme, 248–249
NetAdmin, 39
network pings, 23–24
Network Solutions, 33
next_posts_link() function, 283
NextGen Gallery Plugin, 178
Nginx servers, 379
Nisarg theme, 444

## O

<ol> </ol> tags, 322–323
Optics theme, 444–445
Organic Themes, 238–239

## P

<p> </p> tags, 319
Page Break blocks, 127–128
Page template (page.php), 277
Pagely, 36
Paragraph blocks, 124
background color, 138–139
bold text, 135
drop caps, 138–139
font size, 137
inline code text formatting, 135
italic text, 135
link text formatting, 135
settings management, 134–140
strikethrough text, 135
text color, 135, 138–139
passwords, 66, 150
changing, 99–100
MySQL database, 54–55, 59
SFTP, 40–42, 45
WordPress, 60

permalinks, 113–119
customizing permalink structure, 115–116
defined, 94, 113
ensuring functionality of, 117–119
pretty permalinks, 114–115, 152
RSS and, 119
search engine optimization, 365
setting management, 94–95
Permalinks Settings screen, 94–95
photo galleries, 25
photos. *See* images
PHP (Hypertext Preprocessor), 14–15, 19, 54, 260
child themes, 325–338
Multisite hosting feature, 381
safe mode, 57
start/stop commands, 266
template tags, 265–266
phpMyAdmin, 406
pictures. *See* images
pingbacks, 152–153
Plesk, 39
plugins, 195–224
activating, 210, 216–217
adding new, 104
commercial (paid) plugins, 220–223
core plugins, 202–209
  Akismet, 202–206
  Hello Dolly, 202, 206–207
defined, 9, 195–196
development community, 224
editing files, 104
email notifications, 196
free vs. paid, 222–223
image and gallery, 178
importer plugins, 412
inactive and outdated, 70
installing, 210–211
  automatic installation, 211
  error messages, 211
  manually, 211–214
  readme.txt files, 214
  reviewing descriptions of plugins, 210, 212–213
  version compatibility, 213
management of, 199–201
  activating plugins, 201
  administration page, 217

Sucuri Security plugin, 229, 436

super admin, 399

syntax highlighting, 98

# T

Table blocks, 126

Tag Cloud blocks, 130

`.tag` tag, 352

taglines, 270–271

tags, 101. *See also* templates

  converting categories to, 112

  defined, 113

  in permalinks, 115–116

  setting for posts, 152

Talking Points Memo, 18

taxonomies, 360–361

templates, 259–296

  advantages of, 259

  child themes, 325–338

    adding new template files, 326

    modifying Functions file, 337–338

    overriding parent template files, 325

    removing template files, 327

  Comments template (`comments.php`), 276, 278

  connecting template files, 263, 277–287

  converting for migration, 413–414

  defined, 260

  Footer template (`footer.php`), 262, 264, 276, 278, 281

  404 template (`404.php`), 277

  Functions file (`functions.php`), 260, 262, 280, 295

  Header template (`header.php`), 262, 264, 266–271, 278–279

  if-then-else statements, 273–275

  Main Index template (`index.php`), 260, 262, 264, 271–275, 277, 284–286

  Page template (`page.php`), 277

  Search Results template (`search.php`), 277–278

  sidebar

    defined, 264

    tag parameters for, 287–296

  Sidebar template (`sidebar.php`), 264, 271, 275, 278, 280–281, 295

  Single Post template (`single.php`), 277

  stylesheets (`style.css`), 260, 262, 264–265, 281–283, 287

tags and functions, 260, 262–263, 277–278, 283–296

  anatomy of, 265–266

  customizing posts with, 287

  defined, 262

  parameters, 268, 287–296

  for posts, 283

  viewing all, 262

  viewing and editing, 261–263

  widgets, 295–296

Text widget, 252–253

`the_author_posts_url()` function, 283

`the_category()` function, 283

`the_content()` function, 283

`the_excerpt()` function, 283

`the_title()` function, 283

themes, 9–10, 103, 225–249. *See also names of specific themes; templates; widgets*

  avoiding unsafe, 227–230

  changing, 69

  child themes, 325–340

    activating, 327–328

    adding directory for, 327

    adding new template files, 326

    customizing parent theme style, 330

    defined, 326

    images, 330–334

    loading parent theme style, 328–329

    modifying Functions file, 337–338

    overriding parent template files, 325

    preparing parent theme, 339–340

    removing template files, 337

  commercial (paid) themes, 227, 237–241

    advantages of, 237–238

    investigating, 238

    sources of, 238–241

  customizing, 69, 103

  defined, 260

  editing, 103

  free, 225–227, 437–445

    activating, 235–236

    avoiding unsafe, 227–230

    browsing, 230, 233–234

    downloading, 230–231

    favorites, 234

    Feature Filter, 230, 234

# W

W3C standards, 437

WebDevStudios, 18, 358

web-hosting services and web servers, 14–15, 34–38

  Apache `mod_rewrite` module, 117–118

  backups, 19

  bandwidth, 36–37

  cost of, 14

  CPU (central processing unit) usage, 37–38

  defined, 31, 34

  disk space, 36–37

  domain name registration, 36

  ensuring functionality of customized permalinks, 117–119

  minimum software recommendations, 15, 35

  Multisite hosting feature, 374–375

  podcasts and, 193

  pre-installed WordPress, 51–53

  services provided by, 35

  SFTP, 39–42

  technical support, 34–35

  uploading plugins, 215–216

  uploading themes, 231–232

  uploading files, 55–57

  WordPress-specific features, 35–36

websites. *See* blogs and websites

Welcome to WordPress! module, 69

widgets, 103, 249–255

  adding, 250–251

  defined, 249

  editing, 252

  navigation menus, 314–315

  rearranging, 252

  removing, 251

  RSS widget, 253–255

  Text widget, 252–253

  theme templates, 295–296

  viewing available widgets, 249

  widget areas, 249

WooCommerce plugin, 221, 223, 434–435

Woocommerce Theme Store, 240

WordCamps, 74

*WordPress All-in-One For Dummies* (Sabin-Wilson), 207

WordPress Codex, 11, 323

WordPress Events and News module, 73–74

WordPress Meetups, 74

WordPress Planet, 73

WordPress.com, 8, 12–14, 373

WordPress.org. *See also* Dashboard; plugins

  advantages of, 7–8

  defined, 8

  features of, 12–13

  minimum software recommendations, 14–15

  number of downloads, 11

  popularity of, 1

  technologies behind, 14–15, 19

  versions of, 8, 12–16, 72, 213, 404

  WordPress.com vs., 13

WP Astra, 238–239

WP Engine, 36, 51–53, 414

WP Mobile X plugin, 199

WP Recipe Maker plugin, 197

WP Super Cache plugin, 434

WP Ultimate Recipe plugin, 197

`wp_enqueue_style()` function, 329

`wp_get_archives()` function, 292–293

`wp_list_categories()` function, 293–295

`wp_list_pages()` function, 289–292

WP_Query class, 347–349

WPBeginner, 221

Writing Settings screen, 83–85

# X

XML-RPC, 84

# Y

Yoast SEO plugin, 212–218, 433

YouTube, 179, 186

# About the Author

**Lisa Sabin-Wilson** has worked with the WordPress software since its inception in 2003 and has built her career around providing technical support, hosting, and design solutions for people and organizations that use WordPress. She reaches thousands of people worldwide with her WordPress services, skills, and knowledge regarding the product. Lisa is also the author of the best-selling *WordPress All-In-One For Dummies* and *WordPress Web Design For Dummies*.

Lisa operates a few blogs online, all of which are powered by WordPress. Her personal blog (`http://lisasabin-wilson.com`) has been online since February of 2002. She and her business partner, Brad Williams, provide custom development and design services at their WordPress agency, WebDevStudios (`http://webdevstudios.com`).

When she can be persuaded away from her computer, where she is usually hard at work providing design solutions for her WordPress clients, she sometimes emerges for public speaking appearances on the topics of design, development, and WordPress. She has appeared at conferences such as the annual South By Southwest Interactive Conference, Blog World Expo, CMSExpo, Prestige Conference, and multiple WordCamp events across the country.

Lisa consults with owners of websites, both large and small. Web publishers come in thousands of different flavors, from business to personal, from creative to technical, and all points in between. Lisa is connected to thousands of them worldwide and appreciates the opportunity to share her knowledge with *WordPress For Dummies*. She hopes you find great value in it, as well!

When not designing or consulting with her clients, you can usually find her at her favorite coffee shop sipping espresso, on a mountaintop somewhere hitting the slopes with her family, or 100 feet beneath the ocean waters, scuba diving with her husband and swimming with the fishes.

You can find Lisa online at Twitter: @LisaSabinWilson.

# Dedication

For my father, Donald Sabin — thank you for your love and undying support and encouragement of my crazy choices in life. I miss you, Dad . . . rest in peace.

# Author's Acknowledgments

Many, many thanks and kudos to Matt Mullenweg and Mike Little, the WordPress core development team, and every single person involved in making WordPress the best content management system available on the Internet today. To the volunteers and testers who destroy all those pesky pre-release bugs for every new version release, the WordPress community thanks you! And to each and every WordPress plugin developer and theme designer who donates his or her time, skills, and knowledge to provide the entire WordPress community of users with invaluable tools that help us create dynamic websites, thank you a thousand times! Every person mentioned here is an invaluable asset to the overall WordPress experience; I wish I could name you all, individually, but there are literally thousands of you out there!

Extra special thanks to my friends and colleagues within the WordPress community who inspire and teach me every day.

Huge thanks to Steve Hayes and Charlotte Kughen for their support, assistance, and guidance during the course of this project. The mere fact they had to read every page of this book means that they deserve the Medal of Honor! Also, many thanks to the technical editor, Greg Rickaby, who worked hard to make sure that I look somewhat literate here.

## Publisher's Acknowledgments

**Executive Editor:** Steve Hayes

**Project Editor:** Charlotte Kughen

**Copy Editor:** Keir Simpson

**Technical Editor:** Greg Rickaby

**Production Editor:** Tamilmani Varadharaj

**Sr. Editorial Assistant:** Cherie Case

**Cover Image:** © seraficus/Getty Images

# Leverage the power

*Dummies* is the global leader in the reference category and one of the most trusted and highly regarded brands in the world. No longer just focused on books, customers now have access to the dummies content they need in the format they want. Together we'll craft a solution that engages your customers, stands out from the competition, and helps you meet your goals.

## Advertising & Sponsorships

Connect with an engaged audience on a powerful multimedia site, and position your message alongside expert how-to content. Dummies.com is a one-stop shop for free, online information and know-how curated by a team of experts.

- Targeted ads
- Video
- Email Marketing

- Microsites
- Sweepstakes sponsorship

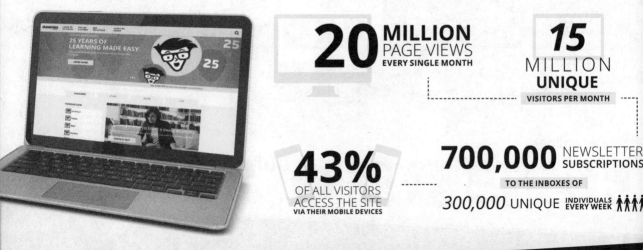

**20** MILLION
PAGE VIEWS
**EVERY SINGLE MONTH**

**15**
MILLION
**UNIQUE**
VISITORS PER MONTH

**43%**
OF ALL VISITORS
ACCESS THE SITE
**VIA THEIR MOBILE DEVICES**

**700,000** NEWSLETTER
SUBSCRIPTIONS
TO THE INBOXES OF
*300,000* UNIQUE INDIVIDUALS EVERY WEEK

# of dummies

## Custom Publishing

Reach a global audience in any language by creating a solution that will differentiate you from competitors, amplify your message, and encourage customers to make a buying decision.

- Apps
- Books
- eBooks
- Video
- Audio
- Webinars

## Brand Licensing & Content

Leverage the strength of the world's most popular reference brand to reach new audiences and channels of distribution.

## For more information, visit dummies.com/biz

# PERSONAL ENRICHMENT

| | | | | | |
|---|---|---|---|---|---|
| **Staying Sharp** | **Facebook** | **Guitar** | **Investing** | **Beekeeping** | **Digital Photography** |
| 9781119187790 | 9781119179030 | 9781119293354 | 9781119293347 | 9781119310068 | 9781119235606 |
| USA $26.00 | USA $21.99 | USA $24.99 | USA $22.99 | USA $22.99 | USA $24.99 |
| CAN $31.99 | CAN $25.99 | CAN $29.99 | CAN $27.99 | CAN $27.99 | CAN $29.99 |
| UK £19.99 | UK £16.99 | UK £17.99 | UK £16.99 | UK £16.99 | UK £17.99 |

| | | | | | |
|---|---|---|---|---|---|
| **Meditation** | **Pregnancy** | **Samsung Galaxy S7** | **iPhone** | **Crocheting** | **Nutrition** |
| 9781119251163 | 9781119235491 | 9781119279952 | 9781119283133 | 9781119287117 | 9781119130246 |
| USA $24.99 | USA $26.99 | USA $24.99 | USA $24.99 | USA $24.99 | USA $22.99 |
| CAN $29.99 | CAN $31.99 | CAN $29.99 | CAN $29.99 | CAN $29.99 | CAN $27.99 |
| UK £17.99 | UK £19.99 | UK £17.99 | UK £17.99 | UK £16.99 | UK £16.99 |

# PROFESSIONAL DEVELOPMENT

| | | | | | | |
|---|---|---|---|---|---|---|
| **Windows 10** | **AutoCAD** | **Excel 2016** | **QuickBooks 2017** | **macOS Sierra** | **LinkedIn** | **Windows 10** |
| 9781119311041 | 9781119255796 | 9781119293439 | 9781119281467 | 9781119280651 | 9781119251132 | 9781119310563 |
| USA $24.99 | USA $39.99 | USA $26.99 | USA $26.99 | USA $29.99 | USA $24.99 | USA $34.00 |
| CAN $29.99 | CAN $47.99 | CAN $31.99 | CAN $31.99 | CAN $35.99 | CAN $29.99 | CAN $41.99 |
| UK £17.99 | UK £27.99 | UK £19.99 | UK £19.99 | UK £21.99 | UK £17.99 | UK £24.99 |

| | | | | | | |
|---|---|---|---|---|---|---|
| **SharePoint 2016** | **Fundamental Analysis** | **Networking** | **Office 2016** | **Office 365** | **Salesforce.com** | **Coding** |
| 9781119181705 | 9781119263593 | 9781119257769 | 9781119293477 | 9781119265313 | 9781119239314 | 9781119293323 |
| USA $29.99 | USA $26.99 | USA $29.99 | USA $26.99 | USA $24.99 | USA $29.99 | USA $29.99 |
| CAN $35.99 | CAN $31.99 | CAN $35.99 | CAN $31.99 | CAN $29.99 | CAN $35.99 | CAN $35.99 |
| UK £21.99 | UK £19.99 | UK £21.99 | UK £19.99 | UK £17.99 | UK £21.99 | UK £21.99 |